Polypropylene-Based Biocomposites and Bionanocomposites

Scrivener Publishing
100 Cummings Center, Suite 541J
Beverly, MA 01915-6106

Publishers at Scrivener
Martin Scrivener (martin@scrivenerpublishing.com)
Phillip Carmical (pcarmical@scrivenerpublishing.com)

Polypropylene-Based Biocomposites and Bionanocomposites

Edited by
Visakh. P. M. and Matheus Poletto

Scrivener
Publishing

WILEY

This edition first published 2018 by John Wiley & Sons, Inc., 111 River Street, Hoboken, NJ 07030, USA and Scrivener Publishing LLC, 100 Cummings Center, Suite 541J, Beverly, MA 01915, USA
© 2018 Scrivener Publishing LLC
For more information about Scrivener publications please visit www.scrivenerpublishing.com.

Wiley Global Headquarters
111 River Street, Hoboken, NJ 07030, USA

For details of our global editorial offices, customer services, and more information about Wiley products visit us at www.wiley.com.

Limit of Liability/Disclaimer of Warranty
While the publisher and authors have used their best efforts in preparing this work, they make no representations or warranties with respect to the accuracy or completeness of the contents of this work and specifically disclaim all warranties, including without limitation any implied warranties of merchantability or fitness for a particular purpose. No warranty may be created or extended by sales representatives, written sales materials, or promotional statements for this work. The fact that an organization, website, or product is referred to in this work as a citation and/or potential source of further information does not mean that the publisher and authors endorse the information or services the organization, website, or product may provide or recommendations it may make. This work is sold with the understanding that the publisher is not engaged in rendering professional services. The advice and strategies contained herein may not be suitable for your situation. You should consult with a specialist where appropriate. Neither the publisher nor authors shall be liable for any loss of profit or any other commercial damages, including but not limited to special, incidental, consequential, or other damages. Further, readers should be aware that websites listed in this work may have changed or disappeared between when this work was written and when it is read.

Library of Congress Cataloging-in-Publication Data
ISBN 978-1-119-28356-0

Cover image: Pixabay.Com
Cover design by Russell Richardson

Set in size of 11pt and Minion Pro by Exeter Premedia Services Private Ltd., Chennai, India

Printed in Singapore

10 9 8 7 6 5 4 3 2 1

Contents

Preface

This book, *Polypropylene-Based Biocomposites and Bionanocomposites*, summarizes many of the recent research accomplishments in the area reflected in the title. In addition to the state-of-the-art and new challenges and opportunities in this area, an effort has been made to discuss many other topics related to polypropylene-based biocomposites and bionanocomposites. Included are subjects such as PP/cellulose-, PP/starch-, and PP/polylactic acid-based biocomposites and bionanocomposites; PP-based hybrid biocomposites and bionanocomposites; the biodegradation and flame retardancy of PP-based composites and nanocomposites; PP single-polymer composites; PP/plant-based fiber biocomposites and bionanocomposites; the development method, properties and application of PP/oil palm fiber composite; and the interfacial modification of PP-based biocomposites and bionanocomposites.

This book will be a very valuable reference source for university and college faculties, professionals, post-doctoral research fellows, senior graduate students, and researchers from R&D laboratories working in the area of polypropylene biocomposites and bionanocomposites. The various chapters in this book were contributed by prominent researchers from industry, academia and government/private research laboratories across the globe, resulting in an up-to-date record of the major findings and observations in the field.

The scope of the introductory first chapter ranges from the state-of-the-art and preparation methods of polypropylene biocomposites and bionanocomposites to environmental concerns about them. A thorough review of PP/cellulose-based biocomposites and bionanocomposites is presented in Chapter 2. The authors have divided this chapter into three parts: preparation, characterization and applications. Thermal, morphological, X-ray diffraction, rheological, viscoelastic, electrical and mechanical characterization methods are among those discussed in the characterization part of the chapter. The last part of the chapter is divided into automotive, packaging, structural, fire retardant, electrical and electronic areas of application.

In Chapter 3, various preparation and characterization methods of PP/ starch-based biocomposites and bionanocomposites are discussed by the authors. The application of these biocomposites and bionanocomposites in biomedical, packaging, automotive, military, coating, fire retardant, aerospace, and optical fields are also covered. Recycling and lifetime studies conducted on these biocomposites and bionanocomposites are equally highlighted.

Chapter 4 on polypropylene/polylactic acid (PP/PLA)-based biocomposites and bionanocomposites first introduces their preparation methods, including melt blending, melt spinning and some other newly emerging methods like autoclave preparation. Then the morphology, compatibility and crystallization of the PP/PLA-based biocomposites and bionanocomposites are discussed in order to obtain resultant materials with uniform distribution and good compatibility among different components. Also summarized are the rheological, electrical, mechanical, thermal, gas barrier, and flame retardant properties of these biocomposites and bionanocomposites, and how each of these properties is dependent upon the composition of the biocomposites and bionanocomposites.

Summarized in Chapter 5 are many of the recent research accomplishments in the area of hybrid biocomposites based on polypropylene. In this chapter, the author discusses the preparation, characterization and applications of PP-based hybrid biocomposites and bionanocomposites. Specifically discussed are their mechanical, thermal, and weathering properties and their role in improving durability, flame retardancy and thermal stability.

A good basis for understanding the biodegradation and flame retardancy of polypropylene composites and nanocomposites is provided in Chapter 6. This chapter summarizes information obtained from highly ranked journals published during the last two decades about the flame retardancy of these biocomposites and nanocomposites. The chapter begins with a short description of the specific flammability of PP, followed by an overview of fire retardants and flame testing methods and standards. Preparation principles, processing methods, properties and applications of polypropylene single-polymer composites are discussed in Chapter 7.

In Chapter 8, different subsections are devoted to topics such as types of natural fibers, the processing of plant-based fiber bionanocomposites with polypropylene, the characterization of plant fiber reinforced polypropylene, applications of plant fiber reinforced PP-based biocomposites and bionanocomposites and future perspectives in the global market. Polypropylene composite with oil palm fibers, its method of development, properties and applications, are discussed in Chapter 9. In this chapter, the authors explain pretreatment methods, cellulose extraction, composite development, characterizations, composite properties and applications. The last chapter of this book focuses on the interfacial modification of PP-based biocomposites and bionanocomposites. First, the nature of fiber-matrix interface is introduced and then the qualitative and quantitative methods for the determination of interface strength are outlined. Finally, the interface modification of PP-based biocomposites and bionanocomposites are presented in light of the existing literature.

In conclusion, the editors would like to express their sincere gratitude to all the contributors to this book, whose excellent support aided in the successful completion of this venture. We are grateful to them for the commitment and sincerity they have shown towards their contributions. Without their enthusiasm and support, the compilation of this book would not have been possible. We would also like to thank all the reviewers who have taken their valuable time to make critical comments on each chapter. We also thank the publisher John Wiley and Sons Ltd. and Scrivener Publishing for recognizing the demand for such a book, for realizing the increasing importance of the area of polypropylene-based biocomposites and bionanocomposites and for starting such a new project, which not many other publishers have handled.

Visakh. P. M.
Matheus Poletto
September 2017

Polypropylene (PP)-Based Biocomposites and Bionanocomposites: State-of-the-Art, New Challenges and Opportunities

Visakh. P. M.

Assistant Professor, Faculty of Electronic Engineering, TUSUR University, Tomsk, Russia

Abstract

This chapter presents a brief account of various topics concerning polypropylene (PP)-based biocomposites and bionanocomposites, including their preparation and state of the art. Among the topics discussed are polypropylene/cellulose-based, polypropylene/starch-based, polypropylene/polylactic acid-based and polypropylene-based hybrid biocomposites and bionanocomposites. In addition, biodegradation and flame retardancy of polypropylene-based composites and nanocomposites, easily recyclable polypropylene single-polymer composites, polypropylene/natural fiber biocomposites and bionanocomposites, and polypropylene composite with oil palm fibers are also discussed, along with method development, properties and applications and interfacial modification of polypropylene-based biocomposites and bionanocomposites.

Keywords: Polypropylene, biocomposites, bionanocomposites, cellulose, bionanomaterial, biodegradation, polypropylene, natural fiber, single polymer

1.1 Polypropylene (PP)/Cellulose-Based Biocomposites and Bionanocomposites

Composites of polypropylene with cellulose or cellulose content materials are prepared either by the treatment of cellulosic fibers or by the addition of a third component, e.g., coupling agents or compatibilizers or both,

Corresponding author: visagam143@gmail.com

Visakh. P. M. and Matheus Poletto. (eds.) Polypropylene-Based Biocomposites and Bionanocomposites, (1–22) 2018 © Scrivener Publishing LLC

to improve compatibility of the composite components PP and cellulose. Masterbatch of a composite is prepared in this way and diluted with PP matrix in melt-mixing processing equipment. A typical example of solvent casting process is the preparation of PP/kenaf fiber composite using polypropylene graft acrylic acid (PP-*g*-AA) as compatibilizer [1]. First, PP, PP-*g*-AA, and divinylbenzene (crosslinker) are dissolved in boiled xylene and then stirred for one hour after adding kenaf fibers. Finally, the xylene solvent is removed by evaporation from the mixture product to form composite. Bagheriasl *et al.* [2] prepared PP/nanocrystalline cellulose (CNC) nanocomposites via masterbatches. They produced masterbatches of CNC and the compatibilizer poly(ethylene-co-vinyl alcohol) (CO) via melt mixing and solution mixing.

Iwamoto *et al.* [3] prepared PP/microfibrillated cellulose (MFC) nanocomposite via solvent casting followed by melt-mixing process. They dissolved PP and maleated polypropylene (MAPP) in toluene at 100 °C, and then added the toluene dispersion of surfactant-coated MFC into the resulting hot solution. Haque *et al.* [4] studied the thermal degradation behavior of PP composite with candelilla bagasse fibers (CF) and reported that thermal stability was slightly improved when using CF in PP compared with neat PP. Jang and Kim [5] carried out SEM analysis of the fracture surfaces of the PP composite with hydrophobized cellulose powder by soybean oil.

They observed that the hydrophobization induced much stronger interfacial bonding between the PP matrix and cellulose powder. Nekkaa *et al.* [6] studied the morphologies of the fractured surfaces of PP composites with silane treated *Spartium junceum* fibers. They also found higher interfacial adhesion between treated fibers and PP matrix compared with untreated fibers. Krishnan *et al.* [7] carried out the morphological analysis of bionanocomposite based on PP, CNF and polystyrene (PS) by transmission electron microscopy (TEM). To find out about the dispersion and location of CNF in polymers, TEM is an effective technique. Yeo and Hwang [8] characterized the morphology of PP composites with MAPP-*g*-MFC and pristine MFC as control by SEM. They noticed that the fractured surface of the PP composites with MAPP-*g*-MFC was smoother than that obtained from the PP composites with pristine MFC.

Morales-Cepeda *et al.* [9] also studied the crystallinity of PP/cellulose composites by X-ray analysis. They also reported that the crystallinity of PP/cellulose with MAPP was slightly higher compared with uncompatibilized composite. Doumbia *et al.* [10] also studied the rheological properties of PP and PP/flax fiber composites with the same variety and volume fraction of flax fiber and also with a similar type and amount of compatibilizer.

Qiu *et al.* [11] studied the viscoelastic properties of PP/cellulose composites using MAPP as compatibilizer. The storage moduli values were increased with cellulose content. Franco-Marquès *et al.* [12] carried out mechanical characterization of lignocellulosic fiber reinforced PP composites with and without MAPP as compatibilizer. Moscoso-Sánchez *et al.* [13] also studied the mechanical properties of cellulose reinforced PP composites using MAPP as compatibilizer. They also observed similar results for compatibilized composites.

Applications of PP/cellulose composites are used in different industries such as automotive, packaging, structural, flame retardants, electrical, electronics, etc. Hao *et al.* [14] developed kenaf fiber reinforced PP composites for automotive application. They studied the open-hole and pin filled-hole effects on the tensile properties of the composites in production of automotive interior parts. Panaitescu *et al.* [15] developed PP/hemp fibers treated with silane and potassium permanganate composites for automotive parts. It is reported that potassium permanganate treatment is a cheap and effective treatment of hemp fibers which is easily applicable industrially. Hung *et al.* [16] developed composites of PP with acetylated wood particles for structural application. The composites of PP with acetylated wood particles exhibit excellent reinforcing effects on the mechanical properties and creep resistance compared with unmodified wood particle composites. Liao *et al.* [17] developed PP separators coated with cellulose aerogel based on hydroxyl ethyl cellulose via ice-segregation-induced self-assembly for improving the ionic conductivity of PP separators.

1.2   Polypropylene (PP)/Starch-Based Biocomposites and Bionanocomposites

Hamdan *et al.* [18] used this technique to prepare sago starch/polypropylene blends. It was found that blending starch with propylene using a bar blender reduced the overall mechanical properties (Young's modulus and tensile strength) with a small increase in the dynamic property (storage modulus) of the composite. The results of Tănase *et al.* [19] revealed that upon incorporation of starch into the polymer matrix, the torque of the physical blends was highly increased due to an increase in the viscosity of the starch phase. This behavior would lead to an increase in melt viscosity and power consumption during processing. The extrusion process has been widely employed in the synthesis of PP/starch biocomposites and bionanocomposites, particularly with twin-screw extruders [20, 21], although a few authors have also reported the use of single-screw extruders [22].

The barrel temperature is an important parameter as it controls the melting process. Authors have reported barrel temperature ranging from 160–180 °C for the preparation of PP/starch composites. Beckermann and his coworkers [23–25] comprehensively studied hemp fiber reinforced PP composites prepared by injection molding process. Their studies included fiber treatments and modifications, model predictions of micro-mechanics and strengths, the optimization of hemp fiber quality, and the influence of bag retting and white rot fungal treatments. Panthapulakkal and Sain [26] studied the influence of water absorption on the tensile properties of injection molded short hemp fiber/glass fiber reinforced polypropylene hybrid composites. Swelling of natural fiber as a result of prolonged exposure to water was found to lead to reduction in the stiffness of the fibers.

Liu *et al.* [27] successfully prepared starch/PP blends with improved mechanical properties by a one-step reactive compatibility technique using maleated polypropylene (MAPP) as the interfacial agent and diethanolamine (DEA) as a reactive plasticizer for starch.

In the investigation of Rosa *et al.* [28] on recycled PP blended with thermoplastic starch (TPS), addition of 30% of TPS was found to reduce tensile strength of the polymer at break. Researchers have deployed several ways in enhancing these properties in starch/PP composites. DeLeo *et al.* [29] observed that addition of clay can greatly improve the mechanical properties of PLS/polypropylene blends at high starch content.

Gupta and Alam [30] reported that addition of MAPP to potato starch/polypropylene composite improved the tensile and flexural strengths of the composite while the impact strength remained the same. The study by Obasi *et al.* [31] on the utilization of thermoplastic starch obtained from cassava starch and potato starch prepared using glycerol as a plasticizer showed that plasticized starch content exhibited an inverse relationship with the tensile strength, elongation at break and was directly related to the Young's modulus of the starch/PP blends. Liu *et al.* [32] found that addition of starch to PP improved bending strength, bending modulus, and Young's modulus but to different extents decreased yield strength, tensile strength, and elongation at break. Pure starch/PP composites should be electrical insulator. However, when fillers are added, the biocomposite can be made conductive or semiconductive.

In the work of Haydaruzzaman *et al.* [33, 34] on starch-treated coir/jute-based hybrid PP composites and 2-hydroxyethyl methacrylate + starch treated PP composites, the maximum value of dielectric constant (the measure of the ability of a material to store electrical energy) was found to be 289.39 and 335.46 respectively. The flame retardancy of starch/PP biocomposites is almost unexplored. A few reports [35] have shown that starch/PP biocomposites are better retardants than pure PP. Nie *et al.* [36] assessed the flame retardant properties of starch containing polypropylene (SCP) semi-biocomposites by limited oxygen index, UL-94 test, and cone calorimeter test.

1.3   Polypropylene (PP)/Polylactic Acid-Based Biocomposites and Bionanocomposites

Polylactic acid (PLA) is a sustainable aliphatic polyester derived from polymerization of the renewable monomer lactic acid. Besides its well-known biodegradability and renewability, the physical and mechanical properties of PLA are comparable or even better when compared with some petrochemical polymers such as polyethylene, polypropylene, polystyrene and polyethylene terephthalate. In addition to PLA/PP phases, various types of nanofillers have also been incorporated into PP/PLA biocomposites as synergists in order to develop bionanocomposites with additional functionalities like reinforcement. The main focus is on the properties of PP/PLA-based biocomposites and bionanocomposites in relation to their rheological and viscoelastic properties, mechanical and thermal properties, gas barrier properties and flame retardant behavior, etc. PP/PLA-based biocomposites and bionanocomposites can be prepared through mixing PP, PLA and/or other additives using melt blending, extrusion or spinning. Most of the PP/PLA-based biocomposites have been prepared for specimens or films through conventional melt blending, and a few studies have been reported to produce PP/PLA-based biocomposite fibers. Wojciechowska and co-authors developed a melt-spinning method for preparing PP/PLA biocomposite fiber [37]. Through optimizing the processing temperatures, PP/PLA biocomposite fibers of good quality were obtained within the whole range of component concentrations. Similar to PP/PLA-based composites, most of PP/PLA nanocomposites were prepared by melt blending, but this method is not effective enough to destroy the aggregates of nanomaterials just through mechanical shearing force.

For semicrystalline PP/PLA biocomposites, incorporation of nanofiller can also result in an altered crystallization temperature, crystallization rate and degree of crystallinity. PP/PLA-based biocomposites are electrically insulating. In order to improve their electrical conductivity, some conductive nanofillers are incorporated into PP/PLA-based biocomposites. Carbon nanotube is such a conductive nanofiller that possesses exceptional electrical conductivity as well as ultrahigh mechanical strength/modulus and outstanding thermal stability/conductivity [38, 39].

The thermal degradation behavior of PP/PLA biocomposites clearly exhibited two mass loss stages, indicating the immiscibility between the two phases. The first stage is ascribed to the thermal decomposition of PLA, and the second one is assigned to the thermal degradation of PP. The incorporation of nanoclay (Cloisite 15A) into the PP/PLA composites resulted in a notable improvement in tensile modulus, but the tensile strength and elongation at break values were reduced with the addition of the nanoclay [40]. The incorporation of PP into PLA led to an increase in the impact strength of the resultant PP/PLA composites compared to neat PLA, due to the toughening effect of PP droplets. One emerging research direction for PP/PLA-based biocomposites and bionanocomposites is focused on biomedical applications. Tanaka and coworkers used a new PP/PLA (10/90) composite mesh for groin hernia repair [41]. In comparison to PP meshes (commercial product: Prolene˚), this new composite biomaterial showed reduced inflammation and cell-mediated immune responses, indicating a better tolerance, which was probably due to the low mesh shrinkage and decreased tissue adhesion.

1.4 Polypropylene (PP)-Based Hybrid Biocomposites and Bionanocomposites

Hybridization allows designers to tailor the composite properties according to the desired structure under consideration [42]. Hybrid fillers combining two or more different types of fillers for polymer composites are sometimes very useful because they possess different properties that cannot be obtained with a single type of reinforcement [43]. In a study published by Uawongsuwan *et al.* [44] pultrusion process was used to produce long fiber pellets of jute fiber/PP and glass fiber/PP. For the making of pellets from long fiber jute and PP, four jute yarns were twisted together and passed through the impregnation die. PP-based hybrid biocomposites and bionanocomposites underwent processing at temperatures corresponding to the processing temperature of PP. The temperatures were varied in the range of 160–195 °C, depending on equipment used for shaping and composition of hybrid material. The melt temperatures did not exceed 195 °C to prevent wood degradation. This may be attributed to the improved resistance offered by the glass fibers in the composites. The mechanical properties of injection-molded short hemp fiber/glass fiber reinforced polypropylene hybrid composites were investigated by Panthapulakkal and Sain [45].

According to Uawongsuwan *et al.* [46], the effectiveness of glass fiber hybridization is highest when combined with RP-JF/PP pellet composites for tensile strength (64%), flexural strength (74%) and impact strength (948%), respectively, when comparing at 20 wt% total fiber content. Turku and Kärki [47] have also found that tensile strength and modulus of hybrid biocomposites made with softwood fiber increased as compare to biocomposite made without addition of glass fibers. However, in this case, an addition of glass fiber was found to decrease the impact strength of biocomposites.

Väntsi and Kärki [48] and Huuhilo *et al.* [49] have studied the extruded PP-based hybrid biocomposites containing 20% of mineral filler: mineral wool fibers, talc, calcium carbonate and wollastonite. In PP/bamboo fiber and PP/sisal fiber composites reinforced with glass fibers, it was found that incorporation of fibers and MAPP interrupts the linear crystallizable sequence of the PP and lowers the degree of crystallization [50, 51]. Martikka *et al.* [52] showed that the microsized mineral fillers added in concentration of 20% did not have any noticable effect on the degree of crystallinity of PP/wood fiber composites. On the other hand, nanowollastonite [53], organoclay [54] and nanosilica [55] acted as efficient nucleating agents for the crystallization of the PP matrix. The nanofillers accelerated the crystallization of the PP matrix remarkably.

The increase of heat deflection temperature upon addition of glass fibers to PP/hemp fiber or PP/sisal fiber composite was reported by Panthapulakkal and Sain [45] and Birat *et al.* [56], respectively. According to Panthapulakal and Sain [45] the HDT value of neat PP was 53 °C and was increased twice by the incorporation of hemp fibers. Gwon *et al.* [57] have found that water absorption of PP/wood fiber composites increased with addition of mineral filler; the composite made with zinc borate had highest water absorption. Increase in the water absorption of zinc borate-containing composite is attributed to the increased void fractions during the compounding process as well as high hydrophilicity of the filler.

1.5 Biodegradation and Flame Retardancy of Polypropylene-Based Composites and Nanocomposites

Biodegradable materials can be defined as materials susceptible to being assimilated by microorganisms such as fungi and bacteria. Some non-biodegradable plastics are erroneously believed to be biodegradable because they often contain biodegradable additives. Strömberg and Karlsson studied the biodegradation of polypropylene, recycled polypropylene and polylactide biocomposites exposed to a mixture of fungi and algae/bacteria in a microenvironment chamber [58]. Suharty *et al.* have shown that the addition of diammonium phosphate increased the water absorption and weight loss of a recycled PP/kenaf fiber/nano-$CaCO_3$ composite significantly [59]. Islam *et al.* have shown that the water absorption (measured in a water immersion test) of kenaf/PP and coir/PP composites increased with the addition of montmorillonite (MMT) [60]. The durability or resistance of biocomposites to various types of degradation has been the object of many studies in the field of composites. The question of the durability of the composite is inseparable from the question of degradability. Durability determines the usefulness of composite materials in a particular environment. The degree of degradation can be evaluated by estimating the change of color and appearance of the composites. In a work published by Butylina and Kärki, a comparison was made between natural and artificial weathering of PP biocomposites containing carbon black [61].

Cellulosic fibers are composed of carbon, hydrogen (fuels) and oxygen (supporter of combustion), and they are thus highly flammable and burn easily. Cellulosic fibers, being char forming material, can significantly reduce the flammability parameters of PP, including peak of heat release (pHRR), mass loss rate (MLR), total heat release (THR) and others [62–64]. However, composites have a lower decomposition temperature and shorter time to ignition [65]. Stronger interaction between the polymer and the reinforcing filler improves the thermal stability of the composite [66–68]. The crosslinking and dehydration tendency of the sulphur-containing amino acids in wool can improve char formation under combustion, making wool fibers not melting and dripping. Conzatti *et al.* report that wool fibers (20–60%) improved the thermal stability of PP especially in the presence of a compatibilizer [69]. Kim *et al.* studied the fire retardancy of a PP/wool (30%) composite containing 20% of ammonium polyphosphate (APP) [70].

Thus, the multilayered char formed during burning worked as a mass transfer barrier, slowing the escape of the volatile products generated during decomposition [52, 71]. The role of a compatibilizer in the flame retardancy of clay/PP nanocomposite is described by Gilman [72]. Tang *et al.* used a combination of organo-modified clay and a nickel catalyst (Ni-Cat) to enhance the carbonization of PP during burning [73]. The nickel catalyst is known as a good catalyst for the synthesis of CNTs. This combination was found to increase the fire retardancy of the composite significantly, which was reflected in the decreased peak of HRR and increased residue amount.

1.6   Polypropylene Single-Polymer Composites

Due to the need for environmentally friendly composite materials with good adhesion properties, single-polymer composites (SPCs) are of interest. SPCs refer to the class of composite materials in which the matrix and the reinforcement come from the same polymer. These materials have also been denoted as one polymer composites, homocomposites, all-the-same polymer composites, homogeneity composites, self-reinforced composites, or mono-materials. During the manufacture of PP traditional composites, the reinforcement is little affected in the preparation process. However, because the reinforcement in PP SPCs also comes from PP, the preparation methods for PP traditional composites are not suitable for PP SPCs. Injection molding and extrusion molding are the most popular processes in the industrial production of polymer composites, which are used to create many things. Injection molding is suitable for volume production of products with complex geometries such as packing, bottle caps, automotive parts and components, toys, etc.

Kmetty et al. [74] introduced thermoplastic elastomers (TPEs) into the injection molding of PP SPCs. A significantly wide processing window (about 90 °C) can be obtained. TPEs have been applied as matrix materials to produce discontinuous natural and man-made fiber reinforced composites. Abo El-Maaty et al. [75] were the first to prepare PP SPCs. Since then, different methods have been developed for preparing PP SPCs, including hot compaction, film tacking, coextrusion, injection molding, and a combination of them. In fact, the classification of these methods is mainly based on the main key process, since they all include many processes. Hine et al. [76] established the important parameters that control the hot compaction behavior of five woven oriented PPs. The five materials studied used different shaped oriented components (fibers and tapes), different molecular weight polymers and various weave styles, allowing the importance of these factors on hot compaction behavior to be studied. Hine et al. [77] described a route for manufacturing SPCs by combining the processes of hot compaction and film stacking. The idea is to use an interleaved film, preferably of the same polymer, placed between the layers of woven oriented elements, thereby delivering additional matrix material to the rougher interlayer region. The homoisotactic PP fibers are sandwiched between the PP-PE random copolymer films. The stacking material was introduced between the female and the male molds. Then, the fiber bundles of polymer are put between the polymer films at a given pressure and temperature.

Bárány et al. [78] prepared the real PP SPCs by film stacking method based on the principle of copolymer and used carded PP as the reinforcement and random PP copolymer as the matrix to manufacture PP SPCs with film stacking. Izer *et al.* [79] continued to investigate the film stacking of PP SPCs with β nucleated homo- and copolymer matrices. Woven PP fabric (woven from highly stretched split PP homopolymer yarns) with melting temperature of 172.4 °C was selected as reinforcement and incorporated into ca. 50% in the corresponding PP SPC. Alcock *et al.* [80] investigated the interfacial properties of the highly oriented coextruded PP tapes, the T-peel strength of PP SPCs for a given homopolymer/copolymer combination is determined by the tape draw ratio, the compaction temperature, and the drawing temperature of the tape. In Alcock's research, the temperature processing window was seen to be > 30 °C, allowing PP SPCs to be consolidated at a range of temperatures. Thus the interfacial properties of the composite can be tailored during production to suit the final application.

Khondker *et al.* [81] used injection-compression molding method and realized the production of PP SPCs. Although the injection-compression molding method has the limitations of the compression process, it also has some of the benefits of injection molding such as short cycle time. Before the injection-compression, weft-knitting technique was adopted to produce plain knitted textile fabric. In accordance with insert injection molding, the woven fabric was pre-placed like an insert on the half cavity of the moving mold plate. After mold closing the woven fabric was fixed in the middle of the whole mold cavity, and the clamping force could press the sandwiched woven fabric tightly. Injection-compression molding and insert-injection molding for SPCs are all limited by the fabric setting, and they are still not suitable for products with complex shapes. Three-dimensional parts with complex geometry cannot be produced and thus the most design-friendly and versatile processing advantages of injection molding cannot be adapted

1.7 Polypropylene/Plant-Based Fiber Biocomposites and Bionanocomposites

PBFs are natural composites in which the reinforcement constituent is distributed within a flexible matrix. Each fiber at the macroscopic level is a bundle of microfibers with the diameter in the range of 20–40 μm [82]. Injection molding is one of the most versatile processing methods for manufacturing of natural fiber PP composites. This process is usually used for producing interior parts of automobiles with the thickness of 2–5 mm like instrument panel components, consoles, door handles, and load floors. Natural fiber-PP composites are used in injection-molded products like home and office furniture, lawn and garden products, toys, housewares, power tools, sporting goods, and storage containers.

Natural fiber and PP improves interfacial adhesion, resulting in better mechanical properties. The aim of chemical treatments is to substitute some of the hydroxyl groups in the natural fiber surface with these bi-functional materials. Based on the coupling agent introduced into the fibers, there are various chemical treatments such as silane treatment, acetylation, and so on [83, 84]. Maleic anhydride is the most common coupling agent used in natural fiber-PP biocomposites. The main difference of maleic anhydride with other coupling agents is that it is mostly used to modify PP instead of natural fibers. Nowadays, different grades of maleic anhydride grafted polypropylene (MAgPP) are in the market, which are utilized to compatibilize PP with not only natural fibers, but also other reinforcements like glass fibers. However, they are highly limited by their high moisture absorption and weak interfacial adhesion between polar-hydrophilic fibers and nonpolar-hydrophobic petroleum-based plastics. The recent studies indicate an improvement in wood fiber–PP matrix bonding in the presence of compatibilizers or coupling agents [85–87].

In addition, different chemical treatments on fibers aimed at improving the adhesion between the fiber surface and matrix may not only increase the mechanical properties but also modify the surface of the fibers. Moreover, results have shown that the water absorption of natural fiber reinforced biocomposites is decreased significantly due to such chemical modifications on natural fibers as acetylation, alkali, acrylation, silane, benzoylation, permanganate, and isocyanates [88–90]. Several studies have been conducted on physical and/or chemical treatments and modification of the surface of jute fibers for improving the mechanical and physical properties of the biocomposite materials [91, 92].

For instance, a systematic study of the effect of surface treatments on the properties of PP/jute fiber has been carried out by Wang et al. [93]. Karmaker et al. have shown that swelling of an embedded jute fiber in PP, caused by water absorption, is able to fill the gaps between fibers and matrix, which were mainly formed because of the thermal shrinkage of PP melt. The fill-up of such gaps results in a higher shear strength between jute fibers and PP, and increases the mechanical properties accordingly [94]. Joseph et al. [95] comprehensively delved into the effects of processing parameters on the mechanical properties of sisal fiber reinforced PP biocomposites. Such biocomposites were prepared by both melt-mixing and solution-mixing methods. In addition, biocomposites containing longitudinally oriented fibers presented better mechanical properties than those with random and transverse orientations. The influence of chemical treatment on the tensile properties of sisal/PP biocomposites was carried out by such treatments as sodium hydroxide, urethane derivative of PPG, maleic anhydride, and permanganate [96].

The recycling of hemp fiber reinforced PP has been studied extensively for many years [97]. As expected, the length of fibers, molecular weight, and Newtonian viscosity have been decreased by reprocessing of composite. Despite the number of reprocessing cycles, the mechanical properties of recycled hemp fiber/PP biocomposites remain well preserved. Natural fibers are often considered for nonstructural applications that require low costs, low energy for processing, biodegradability, and light weight. For example, non-load-bearing indoor components in civil engineering are successfully produced with natural fiber reinforced composites because of their vulnerability to environmental attack [98].

1.8   Polypropylene Composite with Oil Palm Fibers: Method Development, Properties and Application

Chemical pretreatment involves the dissolution of lignin, hemicelluloses and crystalline part of oil palm lignocellulose with the help of mineral or organic compounds. The depolymerization is achieved by hydrolysis brought by acid or alkali. Acid hydrolysis is normally associated with the use of phosphoric, sulphuric, formic, peracetic or hydrochloric acid [99]. The effective autoclaving treatment not only disinfects, but also opens up lignocelluloses for penetration of reagents and for hydrolysis to take place.

Delignification and cellulose extraction from oil palm fibers (empty fruit bunches, mesocarps, frond and trunk) is challenging because of its inert nature and compact structure. Very strong acid and alkali may have to be used to treat the fibers and often it takes a long time to achieve the desired result. To speed up the process, multipronged approaches can be adopted such as the combination of different solvents, reactive species and assistance of electromagnetic radiations or ultrasonic and mechanical treatments. The natural plant fibers with higher cellulose content appear to potentially show better mechanical properties [100]. Cellulose whiskers (CWs) extracted from grass used as filler to fabricate polymer composite show improved composite mechanical properties and thermal stability after alkali treatment on the whiskers. Better mechanical properties are observed at 5% filler but any further loading reduces thermal stability and also the elongation due to phase separation [101]. The palm and coir fiber reinforced PP composites fabricated through injection molding show improved strength with increasing fiber loading until the optimal value of 30%, beyond which there will be a reduction in mechanical properties. This is suggested as being due to the non-uniform dispersion or weak bonding between the fibers and the matrix. The coir fiber reinforced PP composite, however, shows better mechanical properties than the palm reinforced PP composite [102].

Oil palm fibers have been utilized as biocomposites with other fibers or to reinforce other bio/synthetic polymers. The woven hybrid biocomposites using empty fruit bunch (EFB) and jute fiber have reportedly exhibited improved mechanical properties. The stability of the EFB composite is increased with the addition of jute fibers, attributable to the better stability of jute fiber than the EFB [103]. Oil palm fibers have great potential to be utilized as composites with other natural fibers and bio/synthetic polymers. The eco-friendly cellulose extraction can be improved using environmentally benign and clean solvents while purification of other biomolecules and biochemicals, such as hemicelluloses, lignin and vanillin, are put in place for more integrated bioprocesses.

1.9   Interfacial Modification of Polypropylene-Based Biocomposites and Bionanocomposites

Researchers have generally focused on improving fiber-matrix bonding by using various chemical and physical modification techniques such as silane treatment, acetylation, utilization of coupling agents and alkali treatment [104–108]. Interfacial modifications, such as alkali treatment, generally increase the spreading coefficient, indicating easy wetting. On the other hand, the treatments seem to have little effect on the work of adhesion values of the consolidated composites.

Some of the fiber modification techniques, such as alkali treatment, aim to increase the surface roughness of natural fibers in order to increase the area of contact and promote mechanical keying between fibers and polymer resins. A wide array of physical and chemical techniques can be used to modify either natural fiber surface or PP resin to increase fiber-matrix adhesion. Fiber modification techniques generally aim to change the chemical structure of fiber surface as well as the fiber surface energy, whereas matrix modification mostly involves grafting of new functional groups to increase fiber/resin compatibility. The enhancement in fiber-matrix interface is obtained as a result of increased mechanical interlocking between the fibers and the polymer resin. Gassan and Bledzki [109] investigated the alkali treatment of jute fibers and its effect on fiber structure and properties. Isometric NaOH treatment led to an increase in yarn tensile strength and modulus by 120% and 150% respectively. The change in the mechanical properties was attributed to changes in crystalline orientation, degree of polymerization and cellulose content.

Ichazo *et al.* [110] investigated the properties of polyolefin blends with acetylated sisal fibers. The influence of acetylation on the mechanical, thermal and thermo-degradational properties of sisal fiber reinforced PP, PP/HDPE and PP/HDPE with functionalized and non-functionalized EPR composites was studied. Wang *et al.* [111] investigated the effect of various chemical treatments on the mechanical properties of jute fiber reinforced recycled PP composites. These treatments include alkali, MAPP and silane treatments. The best properties were achieved with combined alkali, MAPP and silane treatment. The tensile strength and impact toughness of the composites improved considerably upon chemical treatment. The most important grafting method is the treatment of natural fibers with maleic anhydride grafted polypropylene (MAPP) copolymers. This process results in the formation of covalent bonds across fiber-matrix interface.

References

1. Suharty, N.S., Ismail, H., Diharjo, K., Handayani, D.S., and Firdaus, M., Effect of kenaf fiber as a reinforcement on the tensile, flexural strength and impact toughness properties of recycled polypropylene/halloysite composites. *Procedia Chem.* 19, 253–258, 2016.
2. Bagheriasl, D., Carreau, P.J., Dubois, C., and Riedl, B., Properties of polypropylene and polypropylene/poly(ethylene-co-vinyl alcohol) blend/CNC nanocomposites. *Compos. Sci. Technol.* 117, 357–363, 2015.
3. Iwamoto, S., Yamamoto, S., Lee, S.H., and Endo, T., Mechanical properties of polypropylene composites reinforced by surface-coated microfibrillated cellulose. *Compos. Part A: Appl. Sci. Manuf.* 59, 26–29, 2014.
4. Haque, M.M., Islam, M.S., and Islam, M.N., Preparation and characterization of polypropylene composites reinforced with chemically treated coir. *J. Polym. Res.* 19(5), 1–8, 2012.
5. Jang, S.Y., and Kim, D.S., Physical properties of polypropylene composites with hydrophobized cellulose powder by soybean oil. *J. Appl. Polym. Sci.* 133, 42929, 2016.
6. Nekkaa, S., Guessoum, M., Grillet, A.C., and Haddaoui, N., Mechanical properties of biodegradable composites reinforced with short *Spartium junceum* fibers before and after treatments. *Int. J. Polym. Mater.* 61(13), 1021–1034, 2012.

7. Krishnan, A., Jose, C., and George, K.E., Sisal nanofibril reinforced polypropylene/polystyrene blends: Morphology, mechanical, dynamic mechanical and water transmission studies. *Ind. Crops Prod.* 71, 173–184, 2015.

8. Yeo, J.S., and Hwang, S.H., Preparation and characteristics of polypropylene-graft-maleic anhydride anchored micro-fibriled cellulose: Its composites with polypropylene. *J. Adhes. Sci. Technol.* 29(3), 185–194, 2015.

9. Morales-Cepeda, A.B., Ponce-Medina, M.E., Salas-Papayanopolos, H., Lozano, T., Zamudio, M., and Lafleur, P.G., Preparation and characterization of candelilla fiber (*Euphorbia antisyphilitica*) and its reinforcing effect in polypropylene composites. *Cellulose* 22(6), 3839–3849, 2015.

10. Doumbia, A.S., Castro, M., Jouannet, D., Kervoëlen, A., Falher, T., Cauret, L., and Bourmaud, A., Flax/polypropylene composites for lightened structures: Multiscale analysis of process and fibre parameters. *Mater. Des.* 87, 331–341, 2015.

11. Qiu, W., Zhang, F., Endo, T., and Hirotsu, T., Preparation and characteristics of composites of high crystalline cellulose with polypropylene: Effects of maleated polypropylene and cellulose content. *J. Appl. Polym. Sci.* 87(2), 337–345, 2003.

12. Franco-Marquès, E., Méndez, J.A., Pèlach, M.A., Vilaseca, F., Bayer, J., and Mutjé, P., Influence of coupling agents in the preparation of polypropylene composites reinforced with recycled fibers. *Chem. Eng. J.* 166(3), 1170–1178, 2011.

13. Moscoso-Sánchez, F.J., Díaz, O.R., Flores, J., Martínez, L., Fernández, V.V.A., Barrera, A., and Canché-Escamilla, G., Effect of the cellulose of *Agave tequilana* Weber onto the mechanical properties of foamed and unfoamed polypropylene composites. *Polym. Bull.* 70(3), 837–847, 2013.

14. Hao, A., Yuan, L., and Chen, J.Y., Notch effects and crack propagation analysis on kenaf/polypropylene nonwoven composites. *Compos. Part A: Appl. Sci. Manuf.* 73, 11–19, 2015.

15. Panaitescu, D.M., Nicolae, C.A., Vuluga, Z., Vitelaru, C., Sanporean, C.G., Zaharia, C., Florea, D., and Vasilievici, G., Influence of hemp fibers with modified surface on polypropylene composites. *J. Ind. Eng. Chem.* 37, 137–146, 2016.

16. Hung, K.C., Wu, T.L., Chen, Y.L., and Wu, J.H., Assessing the effect of wood acetylation on mechanical properties and extended creep behavior of wood/recycled-polypropylene composites. *Constr. Build. Mater.* 108, 139–145, 2016.

17. Liao, H., Zhang, H., Hong, H., Li, Z., Qin, G., Zhu, H., and Lin, Y., Novel cellulose aerogel coated on polypropylene separators as gel polymer electrolyte with high ionic conductivity for lithium-ion batteries. *J. Membr. Sci.* 514, 332–339, 2016.

18. Hamdan, S., Hashim, D.M.A., Ahmad, M., and Embong, S., Compatibility studies of polypropylene (PP)–Sago starch (SS) blends using DMTA. *J. Polym. Res.* 7(4), 237–244, 2000.

19. Tănase, E.E., Popa, M.E., Râpă, M., and Popa, O., Preparation and characterization of biopolymer blends based on polyvinyl alcohol and starch. *Rom. Biotechnol. Lett.* 20(2), 10306–10315, 2015.
20. Tessier, R., Lafranche, E., and Krawczak, P., Development of novel melt-compounded starch-grafted polypropylene/polypropylene-grafted maleicanhydride/organoclay ternary hybrids. *Express Polym. Lett.* 6(11), 937–952, 2012.
21. Ferreira, W.H., Khalili, R.R., Figueira Jr., M.J.M., and Andrade, C.T., Effect of organoclay on blends of individually plasticized thermoplastic starch and polypropylene. *Ind. Crops Prod.* 52, 38–45, 2014.
22. Rosa, D.S., Guedes, C.G.F., and Carvalho, C.L., Processing and thermal, mechanical and morphological characterization of post-consumer polyolefins/thermoplastic starch blends. *J. Mater. Sci.*, 42, 551–557, 2007.
23. Beckermann, G.W., and Pickering, K.L., Engineering and evaluation of hemp fibre reinforced polypropylene composites: Fibre treatment and matrix modification. *Compos. Part A: Appl. Sci. Manuf.* 39, 979–988, 2008.
24. Beckermann, G.W., and Pickering, K.L., Engineering and evaluation of hemp fibre reinforced polypropylene composites: Micro-mechanics and strength prediction modeling. *Compos. Part A: Appl. Sci. Manuf.* 40, 210–217, 2009.
25. Beckermann, G.W., Pickering, K.L., Alam, S.N., and Foreman, N.J., Optimising industrial hemp fibre for composites. *Compos. Part A: Appl. Sci. Manuf.* 38, 461–468, 2007.
26. Panthapulakkal, S., and Sain, M., Injection-molded short hemp fiber/glass fiber-reinforced polypropylene hybrid composites-mechanical, water absorption and thermal properties. *J. Appl. Polym. Sci.* 103, 2432–2441, 2007.
27. Liu, L., Yu, Y., and Song, P., Improved mechanical and thermal properties of polypropylene blends based on diethanolamine-plasticized corn starch via *in situ* reactive compatibilization. *Ind. Eng. Chem. Res.* 52, 16232–16238, 2013.
28. Rosa, D.S., Guedes, C.G.F., and Carvalho, C.L., Processing and thermal, mechanical and morphological characterization of post-consumer polyolefins/thermoplastic starch blends. *J. Mater. Sci.* 42, 551–557, 2007.
29. DeLeo, C., Pinotti, C.A., Goncalves, M.C., and Velankar, S., Preparation and characterization of clay nanocomposites of plasticized starch and polypropylene polymer blends. *J. Polym. Environ.* 19, 689–697, 2011.
30. Gupta, A.P., and Alam, A., Study of flexural, tensile, impact properties and morphology of potato starch/polypropylene blends. *Int. J. Adv. Res.* 2(11), 599–604, 2014.
31. Obasi, H.C., Egeolu, F.C., and Oparaji, O.D., Comparative analysis of the tensile and biodegradable performances of some selected modified starch filled polypropylene blends. *Am. J. Chem. Mater. Sci.* 2(2): 6–13, 2015.
32. Liu, L., Yu, Y., and Song, P., Improved mechanical and thermal properties of polypropylene blends based on diethanolamine-plasticized corn starch via *in situ* reactive compatibilization. *Ind. Eng. Chem. Res.* 52, 16232–16238, 2013.
33. Haydaruzzaman, A.H.K., Hossain, M.A., Khan, M.A., Khan, R.A., and Hakim, M.A., Fabrication and characterization of jute reinforced

polypropylene composite: Effectiveness of coupling agents. *J. Compos. Mater.* 44(16), 1945–1962, 2010.

34. Haydaruzzaman, A.H.K., Hossain, M.A., Khan, M.A., and Khan, R.A., Mechanical properties of the coir fiber-reinforced polypropylene composites: Effect of the incorporation of jute fiber. *J. Compos. Mater.* 44(4), 401–416, 2010.

35. Matkó, Sz., Toldy, A., Keszei, S., Anna, P., Bertalan, Gy., and Marosi, Gy., Flame retardancy of biodegradable polymers and biocomposites. *Polym. Degrad. Stab.* 88, 138–145, 2005.

36. Nie, S., Song, L., Guo, Y., Wu, K., Xing, W., Lu, H., and Hu, Y., Intumescent flame retardation of starch containing polypropylene semibiocomposites: Flame retardancy and thermal degradation. *Ind. Eng. Chem. Res.* 48(24), 10751–10758, 2009.

37. Wojciechowska, E., Fabia, J., Slusarczyk, C., Gawlowski, A., Wysocki, M., Graczyk, T., Processing and supermolecular structure of new iPP/PLA fibres. *Fibres Text. East. Eur.* 13, 126–128, 2005.

38. Yoon, J.T., Lee, S.C., and Jeong, Y.G., Effects of grafted chain length on mechanical and electrical properties of nanocomposites containing polylactide-grafted carbon nanotubes. *Compos. Sci. Technol.* 70, 776–782, 2010.

39. Moniruzzaman, M., and Winey, K.I., Polymer nanocomposites containing carbon nanotubes. *Macromolecules 39*, 5194–5205, 2006.

40. Ebadi-Dehaghani, H., Khonakdar, H.A., Barikani, M., and Jafari, S.H., Experimental and theoretical analyses of mechanical properties of PP/PLA/clay nanocomposites. *Compos. Part B-Eng.* 69, 133–144, 2015.

41. Tanaka, K., Mutter, D., Inoue, H., Lindner, V., Bouras, G., Forgione, A., Leroy, J., Aprahamian, M., and Marescaux, J., *In vivo* evaluation of a new composite mesh (10% polypropylene/90% poly-L-lactic acid) for hernia repair. *J. Mater. Sci.-Mater. Med.* 18, 991–999, 2007.

42. Samal, S.K., Mohanty, S., Nayak, S.K., Polypropylene-bamboo/glass fiber hybrid composites: Fabrication and analysis of mechanical, morphological, thermal, and dynamic mechanical behavior. *J. Plast. Compos.* 28, 2729–2747, 2009.

43. Nurdina, A.K., Mariatti, M., and Samayamutthirian, P. Effect of single-mineral filler and hybrid-mineral filler additives on the properties of polypropylene composites. *J. Vinyl. Addit. Technol.* 15, 20–28, 2009.

44. Uawongsuwan, P., Yang, Y., and Hamada, H., Long jute fiber-reinforced polypropylene composite: Effects of jute fiber bundle and glass fiber hybridization. *J. Appl. Polym. Sci.* 132(15), 41819, 2015.

45. Panthapulakkal, S., and Sain, M., Injection-molded short hemp fiber/glass fiber-reinforced polypropylene hybrid composites: Mechanical water absorption and thermal properties. *J. Appl. Polym. Sci.* 103, 2432–2441, 2007.

46. Uawongsuwan, P., Yang, Y., and Hamada, H., Long jute fiber-reinforced polypropylene composite: Effects of jute fiber bundle and glass fiber hybridization. *J. Appl. Polym. Sci.* 132(15), 41819, 2015.

47. Turku, I., and Kärki, T., The effect of carbon fibers, glass fibers and nanoclay of wood flour-polypropylene composite properties. *Eur. J. Wood Prod.* 72, 73–79, 2014.

48. Väntsi, O, and Kärki, T., Utilization of recycled mineral wool as filler in wood-polypropylene composites. *Constr. Build. Mater.* 55, 220–226, 2014.

49. Huuhilo, T., Martikka, O., Butylina, S., and Kärki, T., Mineral fillers of wood-plastic composites. *Wood Mater. Sci. Eng.* 5, 34–40, 2010.

50. Samal, S.K., Mohanty, S., Nayak, S.K., Polypropylene-bamboo/glass fiber hybrid composites: Fabrication and analysis of mechanical, morphological, thermal, and dynamic mechanical behavior. *J. Plast. Compos.* 28, 2729–2747, 2009.

51. Nayak, S.K., and Mohanty S., Sisal glass fiber reinforced PP hybrid composites: Effect of MAPP on the dynamic mechanical and thermal properties. *J. Reinf. Plast. Comp.* 29(10), 1551–1568, 2010.

52. Martikka, O., Huuhilo, T., Butylina, S., and Kärki, T., The effect of mineral fillers on the thermal properties of wood-plastic composites. *Wood Mater. Sci. Eng.* 7, 107–114, 2012.

53. Farhadinejad, Z., Ehsani, M., Khosravian, B., and Ebrahimi, G., Study of thermal properties of wood plastic composite reinforced with cellulose micro fibril and nano inorganic fiber filler. *Eur. J. Wood Prod.* 70, 823–828, 2012.

54. Lee, H., and Kim, D.S., Preparation and physical properties of wood/polypropylene/clay nanocomposites. *J. Appl. Polym. Sci.* 111(6), 2769–2776, 2009.

55. Safwan, M.M., Lin, O.H., and Akil, H.M., Preparation and characterization of palm kernel shell/polypropylene biocomposites and their hybrid composites with nanosilica. *Bioresources* 8(2), 1539–1550, 2013.

56. Birat, K.C., Panthapulakkal, S., Kronka, A., Agnelli, J.A.M., Tjong, J., and Sain, M., Hybrid biocomposites with enhanced thermal and mechanical properties for structural applications. *J. Appl. Polym. Sci.* 132(34), 42452, 2015.

57. Gwon, J.G., Lee, S.Y., Chun, S.J., Doh, G.H., and Kim, J.H., Physical and mechanical properties of wood-plastic composites hybridized with inorganic fillers. *J. Compos. Mater.* 46(3), 301–309, 2011.

58. Strömberg, E., and Karlsson, S., The effect of biodegradation on surface and bulk property changes of polypropylene, recycled polypropylene and polylactic biocomposites. *Int. Biodeter. Biodegr.* 63, 1045–1053, 2009.

59. Suharty, N.S., Almanar, I.P., Sudirman, Dihardjo, K., and Astasari, N., Flammability, biodegradability and mechanical properties of biocomposites waste polypropylene/kenaf fiber containing nano $CaCO_3$ with diammonium phosphate. *Procedia Chem.* 4, 282–287, 2012.

60. Islam, M.S., Hasbullah, N.A.B., Hasan, M., Talib, Z.A., Jawaid, M., and Haafiz, M.K.M., Physical, mechanical and biodegradable properties of kenaf/coir hydrid fiber reinforced polymer nanocomposites. *Mater. Today Commun.* 4, 69–76, 2015.

61. Butylina, S., and Kärki, T., Resistance to weathering of wood-polypropylene and wood-wollastonite-polypropylene composites made with and without carbon black. *Pigm. Resin Technol.* 43, 185–193, 2014.

62. Helwig, M.D., and Paukszta, D., Flammability of composites based on polypropylene and flax fibers. *Mol. Cryst. Liq. Cryst.* 254(1), 373–380, 2000.
63. Borysiak, S., Paukszta, D., and Helwig, M., Flammability of wood–polypropylene composites. *Polym. Degrad. Stab.* 91, 3339–3343, 2006.
64. Kozłowski, R., and Władyka-Przybylak, M., Flammability and fire resistance of composites reinforced by natural fibers. *Polymer. Adv. Tech.* 19, 446–453, 2008.
65. Arao, Y., Nakamura, S., Tomita, Y., Takakuwa, K., Umemura, T., and Tanaka T., Improvement on fire retardancy of wood flour/polypropylene composites using various fire retardants. *Polym. Degrad. Stab.* 100, 79–85, 2014.
66. Morgan, A.B., and Wilkie, C.A., *Flame Retardant Polymer Nanocomposites*, Wiley-Interscience, 2007.
67. Albano, C., Gonzáez, J., Ichazo, M., and Kaiser, D., Thermal stability of blends of polyolefins and sisal fiber. *Polym. Degrad. Stab.* 66, 179–190, 1999.
68. Fu, S., Song, P., Yang, H., Jin, Y., Lu, F., Ye, J., *et al.*, Effect of carbon nanotubes and its functionalization on the thermal and flammability properties of polypropylene/wood flour composites. *J. Mater. Sci.* 45, 3520–3528, 2010.
69. Conzatti, L., Giunco, F., Stagnaro, P., Patrucco, A., Marano, C., Rink, M., and Marsano, E., Composites based on polypropylene and short wool fibres. *Compos. Part A* 47, 165–171, 2013.
70. Kim, N.K., Lin, R.J.T., and Bhattacharyya, D., Effects of wool fibres, ammonium polyphosphate and polymer viscosity on the flammability and mechanical performance of PP/wool composites. *Polym. Degrad. Stab.* 119, 167–177, 2015.
71. Gilman, J.W., Jackson, C.L., Morgan, A.B., and Harris, R.H., Flammability properties of polymer-layered-silicate nanocomposites, polypropylene and polystyrene nanocomposites. *Chem. Mater.* 12, 1866–1873, 2000.
72. Gilman, J.W., Flammability and thermal stability studies of polymer layered-silicate (clay) nanocomposites. *Appl. Clay Sci.* 15, 31–49, 1999.
73. Tang, T., Chen, X., Chen, H., Meng, X., Jiang, Z., and Bi, W., Catalyzing carbonization of polypropylene itself by supported nickel catalyst during combustion of polypropylene/clay nanocomposite for improving fire retardancy. *Chem. Mater.* 17, 2799–2802, 2005.
74. Kmetty, Á., Bárány, T., and Karger-Kocsis, J., Injection moulded all-polypropylene composites composed of polypropylene fibre and polypropylene based thermoplastic elastomer. *Compos. Sci. Technol.* 73(73), 72–80, 2012.
75. Abo El-Maaty, M.I., Bassett, D.C., Olley, R.H., Hine, P.J., and Ward, I.M., The hot compaction of polypropylene fibres. *J. Mater. Sci.* 31(5), 1157–1163, 1996.
76. Hine, P.J., Ward, I.M., Jordan, N.D., Olley, R., and Bassett, D.C., The hot compaction behavior of woven oriented polypropylene fibres and tapes. I. Mechanical properties. *Polymer* 44(4), 1117–1131, 2003.
77. Hine, P.J., Olley, R.H., and Ward, I.M., The use of interleaved films for optimising the production and properties of hot compacted, self reinforced polymer composites. *Compos. Sci. Technol.* 68(6), 1413–1421, 2008.

78. Bárány, T., Izer, A., and Czigány, T., On consolidation of self-reinforced polypropylene composites. *Plast. Rubber Compos.* 35(9), 375–379, 2006.
79. Izer, A., Bárány, T., and Varga, J., Development of woven fabric reinforced all-polypropylene composites with beta nucleated homo- and copolymer matrices. *Compos. Sci. Technol.* 69(13), 2185–2192, 2009.
80. Alcock, B., Cabrera, N.O., Barkoula, N-M., Loos, J., and Peijs, T., Interfacial properties of highly oriented coextruded polypropylene tapes for the creation of recyclable all-polypropylene composites. *J. Appl. Polym. Sci.* 104(1), 118–129, 2007.
81. Khondker, O.A., Yang, X., Usui, N., and Hamada, H., Mechanical properties of textile-inserted PP/PP knitted composites using injection–compression molding. *Compos. Part A: Appl. Sci. Manuf.* 37(12), 2285–2299, 2006.
82. Gibson, L.J., The hierarchical structure and mechanics of plant materials. *J. R. Soc. Interface* 9(76), 2749–2766, 2012.
83. Faruk, O., Bledzki, A.K., Fink, H.-P., and Sain, M., Biocomposites reinforced with natural fibers: 2000–2010. *Prog. Polym. Sci.* 37(11), 1552–96, 2012.
84. Mohanty, A.K., Misra, M., and Drzal, L.T. (Eds.), *Natural Fibers, Biopolymers, and Biocomposites*, CRC Press, 2005.
85. Schloesser, T., and Knothe, J., Vehicle parts reinforced with natural fibres. *Kunstst-Plast Eur.* 87, 25–26, 1997.
86. Colberg, M., and Sauerbier, M., Injection moulding of natural fibre-reinforced plastics. *Kunstst-Plast Eur.* 87(12), 1780–1782, 1997.
87. Schneider, J., Myers, G., Clemons, C., and English, B., Biofibres as reinforcing fillers in thermoplastic composites. *Eng. Plast. (UK)* 8(3), 207–222, 1995.
88. Lu, J.Z., Qinglin, W., and McNabb Jr., H.S., Chemical coupling in wood fiber and polymer composites: A review of coupling agents and treatments. *Wood Fiber Sci.* 32(1), 88–104, 2000.
89. Kokta, B., Maldas, D., Daneault, C., and Beland, P., Composites of poly(vinyl chloride) and wood fibers. Part II: Effect of chemical treatment. *Polym. Compos.* 11(2), 84–89, 1990.
90. Quiroga, A., Marzocchi, V., and Rintoul, I., Influence of wood treatments on mechanical properties of wood–cement composites and of Populus Euroamericana wood fibers. *Compos. Part B: Eng.* 84, 25–32, 2016.
91. Seki, Y., Innovative multifunctional siloxane treatment of jute fiber surface and its effect on the mechanical properties of jute/thermoset composites. *Mater. Sci. Eng. A* 508(1), 247–52, 2009.
92. Mwaikambo, L.Y., and Ansell, M.P., Chemical modification of hemp, sisal, jute, and kapok fibers by alkalization. *J. Appl. Polym. Sci.* 84(12), 2222–2234, 2002.
93. Wang, X., Cui, Y., Xu, Q., Xie, B., and Li, W., Effects of alkali and silane treatment on the mechanical properties of jute-fiber-reinforced recycled polypropylene composites. *J. Vinyl Addit. Technol.* 16(3), 183–8, 2010.
94. Karmaker, A., Hoffmann, A., and Hinrichsen, G., Influence of water uptake on the mechanical properties of jute fiber-reinforced polypropylene. *J. Appl. Polym. Sci.* 54(12), 1803–1807, 1994.

95. Joseph, P., Joseph, K., and Thomas, S., Effect of processing variables on the mechanical properties of sisal-fiber-reinforced polypropylene composites. *Compos. Sci. Technol.* 59(11), 1625–1640, 1999.

96. Dwivedi, U., and Chand, N., Influence of MA-g-PP on abrasive wear behaviour of chopped sisal fibre reinforced polypropylene composites. *J. Mater. Process. Technol.* 209(12), 5371–5375, 2009.

97. Bourmaud, A., and Baley, C., Rigidity analysis of polypropylene/vegetal fibre composites after recycling. *Polym. Degrad. Stab.* 94(3), 297–305, 2009.

98. Kalia, S., Thakur, K., Celli, A., Kiechel, M.A., and Schauer, C.L., Surface modification of plant fibers using environment friendly methods for their application in polymer composites, textile industry and antimicrobial activities: A review. *J. Environ. Chem. Eng.* 1(3), 97–112, 2013.

99. Coral Medina, J.D., Woiciechowski, A.L., Zandona Filho, A., Bissoqui, L., Noseda, M.D., de Souza Vandenberghe, L.P., Zawadzki, S.F., and Soccol, C.R., Biological activities and thermal behavior of lignin from oil palm empty fruit bunches as potential source of chemicals of added value. *Ind. Crops Prod.* 94, 630–637, 2016.

100. Akhtar, M.N., Sulong, A.B., Nazir, M.S., Majeed, K., Khairul Fadzly Radzi, M., Ismail, N.F., and Raza, M.R., Kenaf-biocomposites: Manufacturing, characterization, and applications, in: *Green Biocomposites: Manufacturing and Properties*, Jawaid, M., Sapuan, S.M., and Alothman, O.Y. (Eds.), pp. 225–254, Springer International Publishing: Cham, 2017.

101. Pandey, J.K., Chu, W.S., Kim, C.S., Lee, C.S., and Ahn, S.H., Bio-nano reinforcement of environmentally degradable polymer matrix by cellulose whiskers from grass. *Compos. Part B: Eng.* 40, 676–680, 2009.

102. Haque, M.M., Hasan, M., Islam, M.S., and Ali, M.E., Physico-mechanical properties of chemically treated palm and coir fiber reinforced polypropylene composites. *Bioresour. Technol.* 100, 4903–4906, 2009.

103. Jawaid, M., Abdul Khalil, H.P.S., and Alattas, O.S., Woven hybrid biocomposites: Dynamic mechanical and thermal properties. *Compos. Part A: Appl. Sci. Manuf.* 43, 288–293, 2012.

104. Xie, Y., Hill, C.A.S., Xiao, Z., Militz, H., and Mai, C., Silane coupling agents used for natural fiber/polymer composites: A review. *Compos. Part A: Appl. Sci. Manuf.* 41, 806–819, 2010.

105. Rowell, R.M., Acetylation of natural fibers to improve performance. *Mol. Cryst. Liq. Cryst.* 418, 153–164, 2004.

106. Mohanty, S., Nayak, S.K., Verma, S.K., and Tripathy, S.S., Effect of MAPP as a coupling agent on the performance of jute–PP composites. *J. Reinf. Plast. Compos.* 23, 625–637, 2004.

107. Onal, L., and Karaduman, Y., Mechanical characterization of carpet waste natural fiber-reinforced polymer composites. *J. Compos. Mater.* 43, 1751–1768, 2009.

108. Gassan, J., and Bledzki, A.K., Possibilities for improving the mechanical properties of jute/epoxy composites by alkali treatment of fibres. *Compos. Sci. Technol.* 59, 1303–1309, 1999.

109. Gassan, J., and Bledzki, A.K., Alkali treatment of jute fibers: Relationship between structure and mechanical properties. *J. Appl. Polym. Sci.* 71, 623–629, 1999.
110. Ichazo, M.N., Albano, C., and Gonzalez, J., Behavior of polyolefin blends with acetylated sisal fibers. *Polym. Int.* 49, 1409–1416, 2000.
111. Wang, X., Cui, Y., Xu, Q., Xie, B., and Li, W., Effects of alkali and silane treatment on the mechanical properties of jute-fiber-reinforced recycled polypropylene composites. *J. Vinyl Addit. Technol.* 16, 183–188, 2010.

2

Polypropylene (PP)/Cellulose-Based Biocomposites and Bionanocomposites

Md. Minhaz-Ul Haque

Department of Applied Chemistry and Chemical Engineering, Islamic University, Kushtia, Bangladesh

Abstract

Composites of polypropylene with cellulose and lignocellulosic fibers obtained from different resources (e.g., jute, cotton, sisal, kenaf, bamboo, etc.) are exclusively part of this chapter, in which different aspects of polypropylene (PP) and cellulosic fiber-based biocomposites and bionanocomposites are described. Various preparation methods, such as solvent casting or melt-mixing of biocomposites and nanocomposites, are described along with examples. This chapter also describes different characterizations such as thermal (differential scanning calorimetry and thermogravimetric analysis), morphological, X-ray diffraction, rheological, viscoelastic, electrical and mechanical properties of PP/cellulose-based biocomposites and bionanocomposites. Different resources of cellulose fibers, including their compatibilizing systems with PP matrix, are also presented. The last part of the chapter presents various applications such as automotive, packaging, structural, fire retardant, electrical and electronic applications of PP/cellulose composites. Readers of this chapter will become acquainted with various aspects of PP/cellulose composites and will also understand why biocomposites and bionanocomposites of PP/cellulose occupy a large portion of polymer composite materials.

Keywords: Biocomposites, bionanocomposites, cellulose, morphology, mechanical properties, polymer matrix composite (PMC), polypropylene, thermal properties

2.1 Introduction

Cellulose or cellulosic materials abundant in nature are promising reinforcement for many polymers because of their high mechanical properties.

Corresponding author: minhaz@acct.iu.ac.bd; minhaz1978@gmail.com

---

Visakh. P. M. and Matheus Poletto. (eds.) Polypropylene-Based Biocomposites and Bionanocomposites, (23–54) 2018 © Scrivener Publishing LLC

Cellulose reinforced polymer composites are generally applied in load-bearing applications [1–6]. Compared to inorganic fillers (e.g., silica, clay, glass fibers, etc.), cellulose or lignocellulosic materials are selectively used in polymer composites because of their numerous advantages (e.g., availability, renewability, sustainability, biodegradability, light weight, low cost, etc.). Hence, composite of polypropylene (PP) with cellulosic materials is a part of polymer-based cellulose composites. Because of its low density, low production cost, design flexibility, and recyclability, polypropylene is one of the most important and widely used polyolefins for matrix material in polymer composites [7]. Incorporation of cellulose into PP matrix not only imparts biodegradability and improved mechanical properties to the PP/cellulose composites, it can also reduce the cost of PP/cellulose composite materials. However, owing to its nonpolar hydrophobic character, PP is incompatible with hydrophilic cellulose materials. As a result, the mixed material displays poor mechanical properties because of inhomogeneous dispersion of the cellulose fibers into PP matrix as well as poor interfacial adhesion between PP and cellulose fibers. To improve the dispersion and interfacial adhesion a compatibilizer, namely maleic anhydride grafted PP (MAPP), is commonly used for PP/cellulose-based composites. Table 2.1 shows a list of different lignocellulosic materials, including various compatibilizing systems used in the preparation of PP/cellulose-based composites.

Discussed in this chapter are different aspects of PP/cellulose-based composites such as composite preparation methods, characterization and applications. In the beginning of this chapter, different preparation methods of PP/cellulose-based biocomposites and bionanocomposites are described. Thereafter, different characterizations such as morphological, thermal, rheological, and mechanical properties of PP/cellulose-based biocomposites and bionanocomposites are discussed. The last part of the chapter presents various applications of PP/cellulose composites and a conclusion successively.

2.2 PP/Cellulose-Based Biocomposites and Bionanocomposites

2.2.1 Preparation

Owing to their hydrophilic nature, cellulosic materials are generally incompatible with hydrophobic polypropylene. Consequently, composites of PP with cellulose fibers, owing to their incompatibility, lead to poor mechanical properties. Thus, composites of polypropylene with cellulose

Table 2.1 Various resources of cellulosic fibers used to prepare PP composites.

Type of fibers	Compatibilizers or modifiers	Refs.
jute fibers	maleated polypropylene (MAPP)	[8]
tunicate whiskers	MAPP	[9]
spent coffee ground	silanization with (3-glycidyloxypropyl) tri-methoxysilane, MAPP	[10]
cellulose fibers	polyethylene oxide (PEO), oxidatively degraded polypropylene (DgPP), MAPP and 1,6-diisocyanatohexane (DIC)	[11–14]
coconut fibers	Alkali treatment	[15, 16]
corn fibers	MAPP	[17]
hemp fibers	poly[styrene-b-(ethylene-co-butylene)-b-styrene] (SEBS)	[18]
CNC	poly(ethylene-co-vinyl alcohol)	[19]
rayon cellulose fibers	polymer waxes	[20]
palm fibers	maleated polypropylene (MAPP)	[21]
pulque fibers	alkali treatment	[22]
oak and pine wood	maleated polypropylene (MAPP)	[23]
microfibrillated cellulose	silane coupling agent	[24]
ramie fibers	silicone oil	[25]
lyocell fibers	maleated polypropylene (MAPP)	[26]

(Continued)

Table 2.1 Cont.

Type of fibers	Compatibilizers or modifiers	Refs.
rice husk	styrene ethylene butadiene styrene-grafted maleic anhydride (SEBS-g-MA), MAPP	[27]
bleached sulfite and bleached kraft fibers	maleic anhydride grafted coupling agent (MAPP)	[28]
bamboo cellulose fibers	MAPP	[29]
eucalyptus Kraft pulp	polyethylenimine	[30]
rice-husk fibers, bagasse fibers	maleated polypropylene (MAPP)	[31]
sisal fibers	maleated polypropylene (MAPP)	[32]
kenaf fibers	polypropylene grafted acrylic acid (PP-g-AA)	[33]
cotton cellulose	maleated polypropylene (MAPP)	[34]
microcrystalline cellulose	silane aminopropyltriethoxysilane, MAPP	[35]

or cellulose content materials are prepared either by the treatment of cellulosic fibers or by the addition of a third component, e.g., coupling agents or compatibilizers or both, to improve compatibility of the composite components PP and cellulose. Different chemical treatments of cellulosic fibers are carried out to impart hydrophobic character on the surface of cellulose or lignocellulosic fibers to facilitate homogeneous dispersion of cellulosic fibers into PP matrix and also assist to improve interfacial adhesion between PP matrix and cellulose fibers. Important chemical treatments of cellulosic fibers include esterification with anhydrides (maleic, propionic, phthalic, crotonic and succinic) [36]; surface hydrophobization of cellulose by soybean oil through transesterification reaction procedure [37]; acetylation [38]; benzoyl chloride treatment [39]; alkalization and heat treatment [31]; TEMPO-mediated oxidation [29]; oxidized with $NaIO_4$ to produce cellulose aldehyde [15]; treatment of ramie fibers with epoxy-silicone oil [25]; silane treatment [18]; and silanization of hemp fibers (HF) with g-Aminopropyltriethoxysilane (APS), g-Glycidoxypropyltrimethoxysilane (GPS) and g-Methacryloxypropyltrimethoxysilane (MPS), etc. [40]. A typical example of silane treatment of lignocellulosic fibers, namely hemp fibers, is described as follows: A solution of 2 wt% silane (APS, MPS or GPS) is prepared in 80/20 ethanol/water mixture and the pH of the solution is adjusted close to 5 by acetic acid. An amount of 100 g hemp fibers is added to 250 ml silane solution containing 2 wt% silane, kept at room temperature for 24 h and mixed from time to time and then decanted. A curing step at 120 °C for 1 h is applied to all silane-treated fibers. The silane-treated hemp fibers are dried in air for 24 h [40].

2.2.1.1 Biocomposites Preparation

Composites of polypropylene are prepared by either solvent-casting and/or melt-mixing processes [33, 41]. In solvent-casting process all starting materials are dissolved in a suitable solvent and then the cellulosic materials are dispersed in PP solution with the help of a stirrer. Finally, the mixture product is evaporated to release the solvent to form composite. Sometimes, a masterbatch of a composite is also prepared in this way and diluted with PP matrix in melt-mixing processing equipment. A typical example of a solvent-casting process is the preparation of PP/kenaf fiber composite using polypropylene graft acrylic acid (PP-g-AA) as compatibilizer [33]. First, PP, PP-g-AA, and divinyl benzene (crosslinker) are dissolved in boiled xylene and then stirred for one hour after adding kenaf fibers. Finally, the xylene solvent is removed by evaporation from the mixture product to form composite

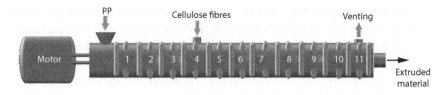

Figure 2.1 A sketch of an extruder showing PP/cellulose composite processing.

For preparing PP/cellulose composites by melt-mixing process different plastic processing equipment is used such as LaboPlastomill [26], Brabender [42], internal mixer [43], extruder [28, 35], etc. The properties of prepared composites also depend on the various parameters of the processing equipment such as processing temperature, processing time, rotor or screw speed, etc. Among the different processing equipment, extruder, owing to its wide industrial application, is very commonly used for PP/cellulose composites. For example, an extruder preparation of PP/cellulose-based composite prepared by Bengtsson *et al.* [28] can be described as follows: Dried cellulose fibers (moisture content below 1 wt%) were compounded with PP matrix using a co-rotating twin-screw extruder. Then the PP, MAPP (0–3 wt%), and lubricant (1 wt%) were dry-mixed and added into the extruder through feeder 1 (Figure 2.1). The cellulose fibers (40–60 wt%) were added into the extruder through a side feeder 2. To feed the materials a K-Tron gravimetric feeder (Niederlenz, Switzerland) was used. The PP, MAPP and lubricants in feeder 1 were fed at 175 °C temperature (zone 1). The cellulose fibers were fed at 180 °C temperature (zone 4) through a twin-screw side feeder operating at 100 rpm. The temperatures at different zones were between 175 and 190 °C. The screw speed was 100 rpm, the melt pressure at the die varied between 4 and 27 bar depending on material blend, and the material output was 3.75 kg/h. Vacuum venting at 180 °C temperature (zone 11) was used to vent volatile compounds. The samples were extruded through a rectangular die with the dimensions of 5–30 mm and cooled in a sizing die.

2.2.1.2   Bionanocomposites Preparation

In the cell wall of lignocellulosic materials (wood and agricultural waste), cellulose exists as aggregates of nanosize fibrils. These nanofibrils are embedded in a matrix of non-cellulosic material [44]. Stable aqueous suspensions of these cellulose nanofibrils (CNFs or MFCs) (Figure 2.2) or nanocrystals (CNCs) can be obtained by high shear disintegration of cellulose pulp by a number of passes through a homogenizer (mechanical method) [45, 46] and acid hydrolysis of cellulose pulp (chemical method)

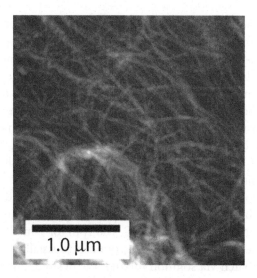

Figure 2.2 Atomic force microscopic image of MFC.

[47] respectively. Among them, cellulose nanofibers, owing to their longer length, show a much higher aspect ratio than that of nanocrystals.

In spite of the interesting properties of cellulose nanomaterials, such as their abundance in nature, biodegradability, high strength, high surface area, etc., their application as a reinforcing agent in hydrophobic polymer nanocomposites is a challenge. Strong interparticle interactions due to hydrogen bonds lead cellulose nanomaterials to form agglomerates, sometimes as large as a few micrometers, when incorporated into nonpolar PP matrix. Consequently, due to a lack of strong interface between the matrix and substrate, achieving enhanced mechanical properties with more widespread polymers, such as hydrophobic polyolefins, is challenging [48]. Thus, dispersion is a crucial matter to improve the properties of PP/cellulose bionanocomposites. So, cellulose nanofibers or nanocrystals are usually modified to impart hydrophobic character on the surfaces or a compatibilizer is used to avoid agglomerations in PP/cellulose-based bionanocomposite.

Bionanocomposites of polypropylene are also prepared by either solvent-casting and/or melt-mixing processes. In solvent-casting process, modified nanocellulose is generally dispersed in a suitable solvent (toluene, xylene, etc.) and PP including compatibilizer is added into the solvent. The solution mixture is strongly stirred. Finally, bionanocomposite is obtained by the evaporation of solvent from the mixture. In melt-mixing process, a masterbatch of PP/nanocellulose is usually prepared like solvent-casting process or in an internal mixer and then diluted with more PP in an extruder. Several examples of PP/cellulose-based bionanocomposite preparation methods are as follows: Ljungberg et al. [9] prepared PP/cellulose nanocomposites by mixing solubilized PP in hot toluene (110 °C) with CNC dispersed in toluene. Toluene was evaporated completely at 80 °C in a ventilated oven to obtain nanocomposite. The resulting nanocomposite materials were hot pressed at 200 °C for 1 min at a pressure of 6.9×10^5 Pa to ensure a constant film thickness. The samples were removed from the press plates and quenched in cold water.

Bagheriasl et al. [19] prepared PP/CNC nanocomposites via masterbatches. They produced masterbatches of CNC and the compatibilizer poly(ethylene-co-vinyl alcohol) (CO) via melt mixing and solution mixing. The masterbatch of CNC and CO was compounded in an Brabender Plasti-Corder internal mixer at 210 °C, 100 rpm for 7 min under nitrogen atmosphere. Then, the product was ground into small granules. To prepare the masterbatch via solution mixing, the desired amount of CO was solubilized in DMF in which CNC was previously dispersed. The mixture was then poured onto a tray to release solvent by evaporation and ground into powder. The nanocomposites of PP/CO/CNC were prepared in an internal mixer at 210 °C, 100 rpm for 10 min under nitrogen atmosphere by melt-compounding PP matrix with the masterbatches. Thereafter, test specimens from the melt-compounded sample were prepared by compression mold.

Iwamoto et al. [7] prepared PP/MFC nanocomposite via solvent casting followed by melt-mixing process. They dissolved PP and MAPP in toluene at 100 °C, and then added the toluene dispersion of surfactant-coated MFC into the resulting hot solution. The mixture was dried to release solvent. The dry mass was melt-compounded using a twin-screw extruder at 170 °C and 30 rpm. Finally, the nanocomposites were obtained by hot compression or injection molding the extruded mass.

Recently, a new solid-state shear pulverization (SSSP) strategy for producing PP/cellulose nanocomposite has also been developed by Iyer et al. [49]. They found that the developed nanocomposite of PP with nanocellulose exhibited excellent dispersion of cellulose nanocrystals and remarkable property enhancements. They fed PP pellets into a pulverizer through a feeder at a feeding rate of 90 g/h. Then they fed cellulose nanocrystals (5–10 wt%) into the pulverizer through another feeder at different feed rates depending on the desired filler content in the final composite. The PP pellets and CNC were mixed in the pulverizer with a screw speed of 200 rpm. During pulverization a recirculating ethylene glycol/water mix at −7 °C cooling system was used to cool the pulverizer barrels. The pulverizer has spiral conveying and mixing elements that cause intimate mixing of PP pellets and CNC powder. The screws employed in the pulverizer were designed to impart high specific energy to the material with the mixing zone containing one reverse, two neutral, and three forward kneading elements and the pulverization zone containing three forward, two neutral, and two reverse shearing elements.

2.2.2   Characterization

2.2.2.1   Thermal Characterizations

2.2.2.1.1   Differential Scanning Calorimetry (DSC)

Different thermal parameters, such as melting (T_m) and crystallization temperatures (T_c), heat of fusion (ΔH_f), and percentage crystallinity (Xc) characteristics of PP matrix and composites, are studied by DSC technique. The relative percentage crystallinity in PP and composite is usually calculated according to the following equation:

$$Xc = (\Delta H_f / \Delta H°_f W) \times 100 \qquad (2.1)$$

where ΔH_f is the heat of fusion of the neat PP or PP/cellulose composite, $\Delta H°_f$ is the heat of fusion for 100% crystalline PP ($\Delta H°_f = 207$ J/g), and W is the mass fraction of neat PP in the composite.

From DSC data, it has been reported that the melting and crystallization temperatures of PP are at about 160 °C and 110–120 °C, respectively, using the heating and cooling rate 10 °C/min [50]. It is reported that melting peak temperature and crystallization temperature of PP matrix are affected by cellulose fibers. Although cellulose fibers have very little effect on the melting peak temperature, the crystallization temperature of PP in composites increased with the increase of cellulose content. This is attributed to the fact that cellulose fibers can act as a nucleating agent for PP matrix during the nucleation stage, resulting in an increase in the crystallization temperature T_c with increasing cellulose content.

It is also found that the nucleating effect of cellulose is increased by the presence of maleated polypropylene (MAPP). This increasing nucleating effect is likely due to the stronger interaction of cellulose fibers with PP matrix compatibilized by MAPP, which improved the nucleating activity of cellulose fibers for PP [11]. Although cellulose can act as nucleating agent for PP matrix, the presence of cellulose decreased the crystallinity of the PP matrix in composites [51]. The cause of decreased crystallinity is ascribed to the interference of cellulose fibers with the crystallization during the crystal growth, which results in a decrease in the overall crystallinity level of the PP matrix [50]. It has also been reported that the crystallinity of

PP matrix in composite is only increased a little by the presence of highly crystalline cellulose.

Ljungberg *et al.* [9] investigated the thermal properties of PP nanocomposites with untreated and surface-treated CNC with MAPP and nonpolar surfactant (phosphoric ester of polyoxyethylene-9-nonylphenyl ether) by means of DSC. The PP matrix, including its nanocomposites with and without treated CNC, displayed similar thermograms with small differences in their melting temperatures. The untreated and surfactant-treated nanocomposites displayed double crystallization peaks compared with neat PP and MAPP-treated CNC nanocomposite. The results showed that the untreated CNC and surfactant-treated CNC acted as nucleating agents for the PP matrix. It was also observed that the untreated CNC has the largest nucleating effect. The MAPP-treated CNC did not act as a nucleating agent for PP in nanocomposite. This is an opposite behavior of MAPP-treated nanocomposite compared with microcomposite in which MAPP improved the nucleating activity of cellulose fibers for PP, as mentioned above. From this analysis, it is concluded that the nucleation of PP is influenced by the surface characteristics of the CNC. In addition, the nanocomposite films of untreated and surfactant-treated CNC (6 wt%) displayed equivalent enthalpy values to the neat PP. Whereas, the nanocomposite consisting of MAPP-treated CNC (6 wt%) showed a significant decrease in these values.

2.2.2.1.2   Thermogravimetric Analysis (TGA)

Thermogravimetric analysis is an important technique to study the thermal resistance of polymer and composite materials. Qiu *et al.* [11] studied the thermal resistance of PP/cellulose composites with and without MAPP as compatibilizer. It has been reported that the thermal stability of PP/cellulose without compatibilizer is similar with neat PP. Thermal stability of PP/cellulose composite is improved in the presence of MAPP. The cause of improvement of thermal stability is attributed to the strong interfacial interaction between cellulose fiber surfaces with PP matrix compatibilized by MAPP. Morales-Cepeda *et al.* [51] studied the thermal degradation behavior of PP composite with candelilla bagasse fibers and reported that thermal stability was slightly improved when using CF in PP compared with neat PP. Figure 2.3 presents the thermogravimetric curves (TGA and DTGA) of cellulose, PP, and PP/cellulose with and without MAPP. In Figure 2.3 it is seen that composite with and without MAPP shows a three-step degradation behavior. The first step, in the range of 190–355 °C, is ascribed to thermal depolymerization of hemicellulose and cleavage of

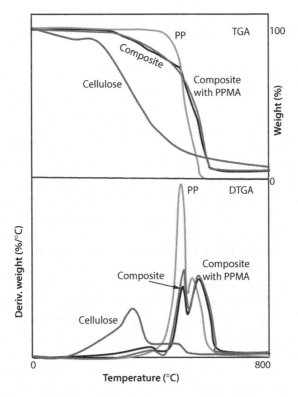

Figure 2.3 TGA and DTGA curves for cellulose, PP, and PP/cellulose with and without PPMA.

glycosidic linkages of cellulose at 344 °C [52, 53]. The second step, showing peaks at 493 °C and 491 °C with and without MAPP, respectively, in the region of 450–520 °C is ascribed to the thermal decomposition of PP. The third step, showing peak at 474 °C in the region of 520–600 °C is due to further breakage of the decomposition products of stage 2 [52].

Yang *et al.* [54] studied the isothermal and non-isothermal decompositions of cellulose nanofiber-filled polypropylene composites by means of thermogravimetric analysis. They reported that the maximum thermal degradation temperature is increased and the thermal degradation onset temperature is decreased as the cellulose content increased in the composite. This is because of the lower thermal stability of nanocellulose fillers than that of neat PP. They also reported that the thermal degradation of the nanocomposite was hindered at higher temperature conditions because of the increased residual mass content of the cellulose nanofibril fillers compared with the PP matrix.

2.2.2.2   Morphological Characterizations

The properties of fiber reinforced polymer composites depend on the quality of the stress transfer in the interphase. And the stress transfer depends on the compatibility between fibers and polymer. The interphase is a transition region over which the physical and mechanical properties change from the bulk properties of fibers to the bulk properties of polymer [26, 57]. Therefore, interfacial adhesion of fibers and matrix is very essential. The interfacial adhesion of fibers and polymer can be verified by scanning electron microscopy (SEM) analysis of cryogenically fracture surfaces of the composites. Jang and Kim [37] carried out SEM analysis of the fracture surfaces of the PP composite with hydrophobized cellulose powder by soybean oil. They observed that the hydrophobization induced much stronger interfacial bonding between the PP matrix and cellulose powder.

Li *et al.* [25] studied the morphologies of the fractured surfaces of PP composites with epoxy-silicone oil treated ramie fibers by SEM. They found higher interfacial adhesion between treated ramie fibers and PP matrix compared with untreated ramie fibers. Nekkaa *et al.* [58] studied the morphologies of the fractured surfaces of PP composites with silane treated *Spartium junceum* fibers. They also found higher interfacial adhesion between treated fibers and PP matrix compared with untreated fibers.

Many other researchers [28, 32] carried out scanning electron microscopy analysis of PP composites with cellulose fibres using MAPP as compatibilizer. Figure 2.4 and Figure 2.5 represent SEM images of the fracture

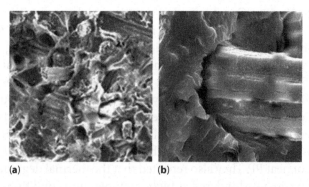

(a)   (b)

Figure: 2.4 SEM images of the fracture surface of composites with cellulose fibers without coupling agent at (a) low magnification and (b) higher magnification.

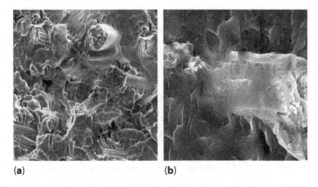

(a)   (b)

Figure: 2.5 SEM images of the fracture surface of composites with cellulose fibers without coupling agent at (a) low magnification and (b) higher magnification.

surface of the composites with cellulose fibres with and without coupling agent, respectively. The compatibilizing effect of MAPP can be easily noticed in the images. In the image of composite without MAPP the cellulose fibers are easily visible and many of them were pulled out from the polypropylene matrix during fracture (Figure 2.4a). Distinct gaps between the cellulose fibers and PP matrix are also seen at higher magnified image (Figure 2.4b). This indicates poor interfacial adhesion. Whereas in the images of composite with MAPP, the cellulose fibers are largely wetted with PP matrix and most of the fibers are fractured (Figure 2.5a), indicating better interfacial adhesion between the cellulose fibers and PP matrix. The close association of cellulose fibers and PP matrix is also evidenced at higher magnified image (Figure 2.5b).

Krishnan *et al.* [59] carried out the morphological analysis of bionanocomposite based on PP, CNF and polystyrene (PS) by transmission electron microscopy (TEM). To find out the dispersion and location of CNF in polymers TEM is an effective technique. In the TEM images they observed that CNF are located at the interface between the blend of PP and PS polymers. In the images they also observed that at low concentration (0.5 wt%) cellulose nanofibers were dispersed in the blend polymers without any agglomeration.

Krishnan *et al.* [59] also used SEM to study the morphology of the fractured cross sections of the tensile samples of cellulose nanofiber reinforced PP/PS blend composite having 0.5 and 5 wt% fiber concentration. In the SEM images they observed that spherical domains of PS are surrounded by the PP continuous phase and 0.5 wt% of cellulose nanofibers improved the adhesion with homogeneous structure. However, at a higher concentration of CNF (5 wt%) some agglomeration of CNF was observed.

Yeo and Hwang [43] used SEM to characterize the morphology of PP composites with MAPP-g-MFC and pristine MFC as control. They noticed that the fractured surface of the PP composites with MAPP-g-MFC was smoother than that obtained from the PP composites with pristine MFC. They also observed that the uniform distribution of MAPP-g-MFC in PP matrix and MAPP-g-MFC were tightly connected with the PP matrix in what seems to be the continuous phase. This is an indication that MAPP anchored on the MFC surface acts as a compatibilizer between pristine MFC and PP matrix.

Bagheriasl et al. [19] studied the morphology of PP/CNC with poly(ethylene-co-vinyl alcohol) (EVA) as compatibilizer by SEM and TEM analysis. In SEM analysis they observed that the composite of PP and CNC with comaptibilizer EVA is free from agglomerates as compared with PP/CNC composite without compatibilizer. In TEM analysis they observed that CNCs were well dispersed into PP matrix and form a network.

Ljungberg et al. [60] used SEM to characterize the dispersion quality of untreated as well as surface-treated CNC with MAPP and nonpolar surfactant (phosphoric ester of polyoxyethylene-9-nonylphenyl ether) in atactic polypropylene nanocomposites. They found that surfactant-treated CNCs were well dispersed in PP matrix. They observed that the nanocomposite of PP with surfactant-treated CNC was transparent whereas the nanocomposites of PP/CNC and PP with MAPP-treated CNC were not transparent. Although in the SEM micrographs of the fracture surfaces of PP/CNC and PP with MAPP-treated CNC nanocomposites some agglomerates of CNC were observed, no agglomerate of CNC appeared in the SEM image of the fracture surfaces of PP with surfactant-treated CNC nanocomposite.

2.2.2.3   X-Ray Diffraction

The study of crystal morphology and crystallinity of polymers and composites by X-ray diffractometry is a widely used technique. Moscoso-Sánchez et al. [55] studied crystallinity of the PP/MAPP composite as a function of cellulose wt% incorporated into the polymer matrix either unfoamed or submitted to a treatment with foaming agent. A typical XRD patterns of the composites are shown in Figure 2.6. In the XRD patterns, the diffraction peaks observed at about $2\theta = 14.1°$, $16.8°$, $18.5°$ and $22°$ are due to the presence of isotactic polypropylene [56]. The XRD pattern of the PP/MAPP composite containing cellulose is almost similar with PP and the diffraction peak at $2\theta = 21.7°$ related to plane (200) of cellulose is not observed because it overlaps the diffraction peak at $2\theta = 22°$ of the isotactic polypropylene.

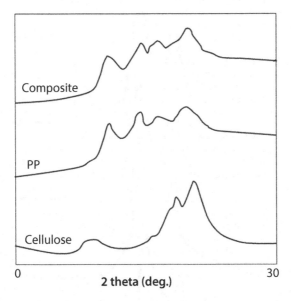

0   30
2 theta (deg.)

Figure 2.6 XRD patterns of cellulose, PP and composite PP/MAPP with cellulose [55].

The crystallinity of the composite samples was estimated based on the more intense diffraction peaks observed in the XRD patterns at $2\theta = 14.1°$ and $22°$ with respect to the less intense diffraction peak observed at $2\theta = 18.5°$. The calculation was done using the following formula:

$$X_C = 100 \; (I_{CR} - I_{AM})/I_{CR} \qquad (2.2)$$

where I_{CR} is the intensity of the maximum diffraction peak measured as the height of the crystalline diffraction peak at $2\theta = 14.1°$ or $2\theta = 22°$ (represents both the crystalline and amorphous materials) and I_{AM} is the height of the smaller diffraction peak measured at $2\theta = 18.6°$ (represents the amorphous material only). Quantitative estimations of the total crystallinity of the PP/cellulose with MAPP composite showed that unlike DSC analysis the crystallinity of the composites increased with the wt% of cellulose incorporated into the PP matrix. The increments in the crystallinity of the composites with and without foaming agent are about 9% and 5%, respectively for the 30 wt.% of cellulose content in the composites.

Morales-Cepeda *et al.* [51] also studied the crystallinity of PP/cellulose composites by X-ray analysis. They also reported that the crystallinity of PP/cellulose with MAPP was slightly higher compared with uncompatibilized composite. Ljungberg *et al.* [9] characterized PP nanocomposites with

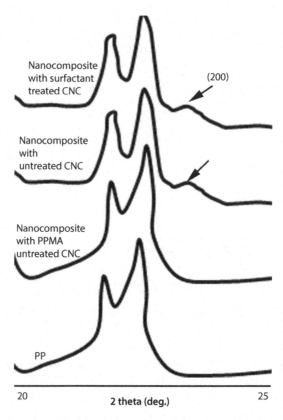

Nanocomposite
with surfactant
treated CNC

(200)

Nanocomposite
with
untreated CNC

Nanocomposite
with PPMA
untreated CNC

PP

20

2 theta (deg.)

25

Figure 2.7 A part of X-ray diffractograms of PP and nanocomposites reinforced with treated and untreated CNC [9].

untreated and surface-treated CNC with MAPP and nonpolar surfactant (phosphoric ester of polyoxyethylene-9-nonylphenyl ether) by X-ray diffraction in order to identify the crystal morphology. A part of X-ray diffractograms with the intensity as a function of the scattering angle 2θ are shown in Figure 2.7.

In the X-ray diffractogram of neat PP, five representative peaks are observed by the planes (110), (040), (130), (111) and (041) of α-phase of PP at the scattering angles 14°, 17°, 18.5°, 21° and 22°, respectively. These five peaks are also observed in the diffractograms of the nanocomposites. However, the two nanocomposites reinforced with surfactant-treated CNC and untreated CNC displayed two additional peaks at angles 16° and 23° in the diffractograms. These two peaks correspond with the most representative peaks by the plane (300) of the β-phase of PP and the main cellulose peak by the plane (200) corresponding to the

Iβ-phase respectively. Neither the neat PP film nor the nanocomposite with MAPP-treated CNC displayed the presence of the β-phase. It is also noticed that the intensity of the β-phase is larger for the untreated CNC nanocomposite compared with surfactant-treated CNC nanocomposite. It is concluded that the hydrophilic character of the CNC surfaces favors the appearance of the β-phase.

2.2.2.4 Rheological Characterizations

The study of the flow of the polymer composite is part of rheology. The fibers move and rotate with the flow of the polymer matrix while processing the composites. The flow properties of the composite materials depend on the fiber length, stiffness, strength, volume fraction and nature of the fiber matrix adhesion [61]. Figure 2.8 and Figure 2.9 show typical rheological behaviour curves. Chen *et al.* [62] studied the rheological behavior of PP and its composites with steam-treated and untreated rice straw fibers. From this analysis they observed that the rice straw fibers have a great effect on the storage modulus (G′), loss modulus (G″) and complex viscosity (η*) of PP as a function of angular frequency (ω). The storage modulus, loss modulus and viscosity of PP composites filled with untreated and treated rice straw fibers are higher than that of neat PP. However, the values of G′, G″ and η* of the PP composites with treated rice straw fibers decrease compared with untreated rice straw fibers. This behavior is more obvious at low frequency. Apparently neat PP and composites with treated and untreated rice straw fibers display typical shear thinning behavior, with a plateau region being observed at low frequency. This result indicates that the dispersion of rice straw fibers in the PP matrix and interface adhesion

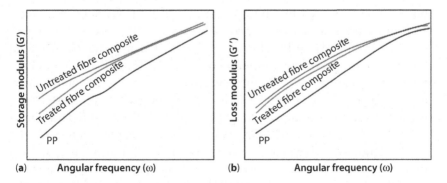

Figure 2.8 Typical rheological curves of PP and composites with treated and untreated cellulose (a) storage modulus (G′), and (b) loss modulus (G″) [62].

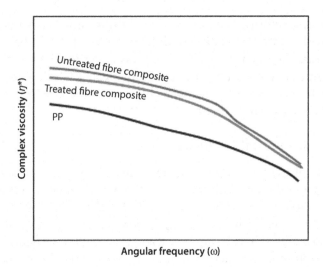

Figure 2.9 Complex viscosity (η^*) of PP and composites with treated and untreated cellulose [62].

between PP matrix and treated rice straw fibers are improved by the steam treatment of rice straw fibers. Thus, the pretreatment can increase the compatibility of PP and rice straw fibers.

Doumbia *et al.* [63] also studied the rheological properties of PP and PP/flax fiber composites with the same variety and volume fraction of flax fiber and also a similar type and amount of compatibilizer. They also found that similar behavior of composite materials, i.e., the storage modulus, loss modulus and complex viscosity of composites, are much more than that of neat PP matrix.

Bagheriasl *et al.* [19] studied the rheological properties of PP/CNC nanocomposite with poly(ethylene-co-vinyl alcohol) (EVA) as compatibilizer. They observed that both η^* and G' of PP/CNC nanocomposites with and without compatibilizer containing 5 wt% CNC are slightly higher compared with the neat PP. The use of EVA in PP matrix slightly increased the value of η^* and the incorporation of 5 wt% CNC in PP/15 EVA did not increase the value of η^* when the masterbatch was prepared via melt mixing. Whereas, the nanocomposite for which the EVA/CNC masterbatch was prepared via solution mixing (PP/15EVA/5CNC) exhibits a shear-thinning behavior for η^* without any plateau region at low frequencies and G' that tends towards a plateau at low frequencies. This result is an indication of the formation of a network and a transition from liquid to solid-like behavior. This is attributed to the better dispersion of the CNC in the copolymer EVA in solution.

2.2.2.5   Viscoelastic Characterizations

Dynamic mechanical thermal analysis (DMTA) is commonly used to study the viscoelastic behavior of polymeric materials. The dynamic mechanical properties, such as storage or viscoelastic modulus (E′), loss modulus (E″) and the mechanical loss factor (tan δ) of polymer, are evaluated by this analysis to study relevant stiffness and damping characteristics of polymers for different applications. Joseph *et al.* [64] studied the viscoelastic properties of PP/sisal fiber composite in terms of fiber content (10–30 wt%), fiber length and fiber treatment in the temperature ranges of −20 to 100 °C. It is reported that both the E′ value and E″ value of PP/sisal fiber composites increased with increasing fiber content in the composites. The composites showed the highest values of E′ and E″ when the sisal fiber length was 2 mm. It is also reported that PP composites with treated sisal fibers exhibited the highest values of storage and loss moduli compared with untreated sisal fiber composites. Qiu *et al.* [65] studied the viscoelastic properties of PP/cellulose composites using MAPP as compatibilizer. The storage moduli values were increased with cellulose content. The composites consisting of MAPP displayed slightly higher values of storage moduli compared with the composites without MAPP. Botev *et al.* [66] studied the viscoelastic behavior of PP/basalt fiber composites with MAPP. They also reported that storage moduli values of the composites consisting of MAPP were higher than that of uncompatibilized composites.

2.2.2.6   Electrical Characterizations

Polyolefins, mainly polyethylene or polypropylene, are used to fabricate most conventional lithium-ion battery separators because of their high mechanical strength and high electrochemical stability [67]. However, their ionic conductivity is poor. Electrolyte infiltration of PP separators is difficult due to its low polarity and low surface energies. This leads to poor compatibility among the PP separators, the electrolyte and electrodes in batteries. To improve the ionic conductivity of PP separators, Liao *et al.* [68] coated PP separators with cellulose aerogel based on hydroxyl ethyl cellulose via ice-segregation-induced self-assembly. They evaluated the electrochemical performances of the separator by using a cell consisting of the coated separator, lithium foil as the counter and reference electrodes, and LiFePO$_4$ as the cathode. They observed that the porous cellulose aerogel-coated separator exhibited superior dimensional stability, electrolyte uptake, and hence a higher ionic conductivity and battery cling performance than that of its non-coated counterpart. In the analysis of Zhang *et al.* [67], the electrolyte wettability and ionic conductivity of cellulose-based composite was explored as lithium-ion battery separator obtained via an electrospinning technique followed by a dip-coating process. The results demonstrate that the composite possessed good electrolyte wettability, excellent heat tolerance, and high ionic conductivity. It was also observed that the cells consisting of the composite separator displayed better rate capability and enhanced capacity retention compared with the cells consisting of the commercialized PP separator under the same conditions.

2.2.2.7 Mechanical Characterizations

The mechanical properties of polymer composites largely depend on the dispersion of fibers in the polymer matrix and their interfacial interaction. The PP matrix has no functional group to interact with the cellulosic fibers, which have –OH (hydroxyl) groups on the surface. When no coupling agent is added the compatibilization between PP matrix and cellulose fibers follows the mechanisms based on mechanical anchoring due to the porous surface of the fibers [69]. Whereas, when PP is modified with coupling agents such as MAPP then the compatibilization between PP and cellulose fibers is achieved by chemical mechanism forming of ester bonds between the –OH groups of cellulose fibers and anhydride group of MAPP [70]. The latter one, i.e., chemical mechanism, ensures better stress transfer from PP to cellulose fibers. Franco-Marquès *et al.* [71] carried out mechanical characterization of lignocellulosic fiber reinforced PP composites with and without MAPP as compatibilizer. The PP composite consisting of 50 wt% of fibers and MAPP (6 wt% of fibers) displayed the value of ultimate tensile strength of about 50 MPa, which was about 81% and 59% higher than the plain PP and noncoupled 50 wt% reinforced composite, respectively. Although the stiffness and Young's modulus values of PP matrix were increased with increasing fiber content, no significant differences in the values of Young's modulus of compatibilized and uncompatibilized composites were observed. The elongation capacity (ε_t) of composite was lower with increasing fiber content in the composite, as the rigid fibers restrict the chain mobility of PP matrix. The addition of MAPP induced a higher deformation capacity of the composite, resulting in more intensive chemical anchoring of the polymer matrix on the surface of the fibers and a better transfer of stresses from the matrix to the fiber. Moscoso-Sánchez *et al.* [55] also studied the mechanical properties of cellulose-reinforced PP composites using MAPP as compatibilizer. They also observed similar results for compatibilized composites.

Qiu *et al.* [11] studied the effects of molecular weight of PP and the content of the compatibilizer (MAPP) on the tensile properties of PP/

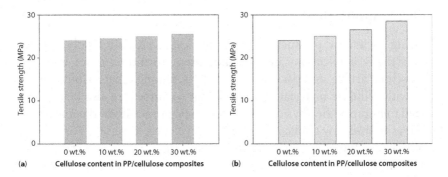

Figure 2.10 Tensile strength of PP and its composites with (a) untreated fibers and (b) treated fibers [72].

cellulose composites. They observed that the molecular weight of PP has very little effect on tensile strength both in the non-compatibilized and compatibilized composites at lower MAPP content. However, at high MAPP content, the composites with higher molecular weight of PP revealed larger tensile strength. Compared with the non-compatibilized composites, the tensile strength of the compatibilized composite increased noticeably with increasing MAPP content. They also observed that the tensile strength of the composite was greatly enhanced by increasing MAPP content, showing a maximum at around 10 wt% MAPP content.

El-Abbassi *et al.* [72] studied the influence of the alkali treatment on the mechanical behavior of alfa fiber reinforced PP composites. Tensile strength of PP and its composites with untreated and treated cellulose fibers are shown in Figure 2.10. It is observed that the use of fibers improves the tensile strength of the polypropylene matrix. The results also showed that the tensile strength remains approximately constant for untreated fiber composites, while it significantly increases for treated fiber composites (Fig. 2.10b). This significant increase is ascribed to the elimination of wax, lignin and impurities by alkali treatment, which increases the surface roughness of the fibers.

Yeo and Hwang [43] studied tensile strength and tensile modulus of the nanocomposites of MAPP-*g*-MFC and pristine MFC. They reported that the tensile strength of the nanocomposites with pristine MFC did not show a distinguishing change but those of nanocomposites with MAPP-*g*-MFC was significantly increased with increasing cellulosic filler content due to the enhanced interfacial adhesion of the composites by MAPP, which was anchored on the MFC surface.

2.2.3   Applications

Owing to its low production cost and design flexibility, polypropylene finds many applications in the automotive, furniture, office appliance, household article, and packaging industries, among others. Composites of PP with cellulose materials from different resources are also being developed to improve the physical and mechanical properties of the materials as well as to find new applications for the developed materials. For example, in Thailand PP/vetiver grass powder composites are being used in many applications such as household articles, furniture and office appliances, etc. [73]. The applications of PP/cellulose composites in different sectors, such as automotive, packaging, structural, fire retardants, electrical, electronics, etc., are mentioned in the following sections successively.

2.2.3.1   Automotive Applications

Some automobile manufacturers in Europe (such as Audi, Ford and BMW) have already started to use cellulosic materials, such as bast fibers, wood flours, etc., in PP matrix as a replacement for glass fibers [74]. Research is also going on to develop PP/cellulose-based composites using different fiber resources for automotive applications. Hao *et al.* [75] developed kenaf fiber reinforced PP composites for automotive application. They studied the open-hole and pin filled-hole effects on the tensile properties of the composites in the production of automotive interior parts. Based on the mechanical properties of the composite measured in terms of uniaxial tensile, open-hole tensile, and pin filled-hole tensile strength, it is reported that the developed composite is relatively ductile and insensitive to the notch. Panaitescu *et al.* [40] developed PP/hemp fibers treated with silane and potassium permanganate composites for automotive parts. It is reported that potassium permanganate treatment, which is a cheap and effective treatment of hemp fibers, is easily applicable industrially. It is also reported that potassium permanganate treated hemp fibers have a good effect on the mechanical properties of composites. Other composite materials, such as rice husk reinforced PP foam with MAPP and SEB-*g*-MA as compatibilizers [76], cellulose fiber reinforced PP composite foams [77], and PP/flax fiber composite were also developed for automotive applications [63].

2.2.3.2   Packaging Applications

Every year a huge amount of plastic waste is produced all over the world, most of which is from packaging materials. Hence, the recent trend is to develop packaging materials based on bioresources and renewable systems.

Because of the high porosity, cost effectiveness, high chemical/thermal resistance, ease of processing and availability of cellulose, Ummartyotin and Pechyen [34] developed PP/cellulose composites using MAPP as compatibilizer for packaging application. The composite of PP matrix with 20 wt% of cellulose derived from cotton was designed as active packaging with the additional feature of microwavable properties. It is observed that thermal and mechanical properties of PP/cellulose composite were better than that of PP matrix. Hence, the crystal formation and crystallinity (50%) of composite indicated that the composite can be used as packaging for microwave application.

2.2.3.3   Structural Applications

One of the important applications of natural fiber reinforced polymer composites is in the building industries, e.g., fencing, industrial flooring, railings, etc. [78]. Hung et al. [79] developed composites of PP with acetylated wood particles for structural application. The composites of PP with acetylated wood particles exhibit excellent reinforcing effects on the mechanical properties and creep resistance compared with unmodified wood particle composites. It is also reported that the moisture content of the composites decreased with increasing the content of acetylated wood particle, while the flexural and tensile properties of the composites were improved with increasing the content of acetylated wood particles up to 13%. Based on the properties of the acetylated wood/PP composites, it is suggested that the composite offer a high-performance alternative to conventional wood plastic composites for building and construction applications.

2.2.3.4   Fire Retardant Applications

PP/cellulose composites are commonly used for outdoor applications, and hence their overall performance is affected by outdoor weather conditions such as sunlight, wetting, temperature changes, etc. Another significant issue for PP/cellulose composites are their thermal stability and fire retardancy properties. The influence of fire retardant on the photooxidation of polypropylene is reported [80, 81]. Recently, Turku and Kärki [82] studied the influence of fire retardants, such as aluminum tri-hydrate, zinc borate, melamine, graphite, and titanium dioxide, on the durability and flammability of polypropylene/cellulose-based composites. They found that fire retardant-loaded composite had better weather resistance compared with the unfilled composites.

2.2.3.5   Electrical and Electronic Applications

Polypropylene has found its application as separator in lithium-ion battery cell. However, its poor ionic conductivity owing to its incompatibility with the electrolyte and electrode reduces its performance. Liao *et al.* [68] developed PP separators coated with cellulose aerogel based on hydroxyl ethyl cellulose via ice-segregation-induced self-assembly for improving the ionic conductivity of PP separators. They reported that the developed porous cellulose aerogel-coated PP separator has superior dimensional stability, electrolyte uptake, and a higher ionic conductivity and better cycling performance compared with its non-coated PP separators. The PP/cellulose-based separators via an electrospinning technique followed by a dip-coating process were also developed by Zhang *et al.* [67]. They reported that the composite possessed good electrolyte wettability, excellent heat tolerance, and high ionic conductivity. They also reported that the cells using the developed composite separator showed better rate capability and enhanced capacity retention compared with the cells using the commercialized PP separator under the same conditions.

2.3   Conclusion

The study of polypropylene/cellulose-based composites has drawn great attention because of their manifold applications in different sectors and new application possibility in other areas. As polypropylene and cellulose fibers are incompatible owing to their opposite characters, chemical modification of cellulose fibers or a coupling agent is necessary in the composite fabrication for obtaining good physical and mechanical properties of the composite materials. To overcome the incompatibility problem of PP matrix and cellulose fibers, the most promising approach is to create covalent bonds between PP matrix and cellulose fibers. Although different chemical treatments of lignocellulosic fibers (e.g., alkali treatment, silane treatment, etc.) are efficient methods of chemical modification of fibers to achieve compatibility between PP matrix and fibers, it is obvious from the above studies that maleated coupling agents extensively used with PP matrix as chemical bond can be created between the anhydride groups of maleated coupling agents and the hydroxyl groups of cellulose fibers. It can be concluded that the application area of composites can be broadened by focusing attention on the development of cheap and easily applicable new coupling agents from natural products, such as lignin, by examining the precise reaction mechanism happening at the fiber-matrix interface. It can also be concluded that although PP/cellulose-based biocomposites find many applications, such as automotive, furniture, etc., the industrial application of PP/cellulose-based bionanocomposites has not yet been started. Hence, it is necessary to develop an economic processing route for PP/cellulose-based bionanocomposites for industrialization of these materials.

References

1. Nunez, A.J., Sturm, P.C., Kenny, J.M., Aranguren, M.I., Marcovich, N.E., and Reboredo, M.M., Mechanical characterization of polypropylene–wood flour composites. *J. Appl. Polym. Sci.* 88(6), 1420–1428, 2003.

2. Cantero, G., Arbelaiz, A., Mugika, F., Valea, A., and Mondragon, I., Mechanical behavior of wood/polypropylene composites: Effects of fibre treatments and ageing processes. *J. Reinf. Plast. Compos.* 22(1), 37–50, 2003.

3. Bledzki, A.K., and Gassan, J., Composites reinforced with cellulose based fibres. *Prog. Polym. Sci.* 24(2), 221–274, 1999.

4. Bledzki, A.K., Faruk, O., and Huque, M., Physico-mechanical studies of wood fiber reinforced composites. *Polym. Plast. Technol. Eng.* 41(3), 435–451, 2002.

5. Bledzki, A.K., Letman, M., Viksne, A., and Rence, L., A comparison of compounding processes and wood type for wood fibre-PP composites. *Compos. Part A: Appl. Sci. Manuf.* 36(6), 789–797, 2005.

6. Ichazo, M.N., Albano, C., Gonzalez, J., Perera, R., and Candal, M.V., Polypropylene/wood flour composites: Treatments and properties. *Compos. Struct.* 54(2), 207–214, 2001.

7. Iwamoto, S., Yamamoto, S., Lee, S.H., and Endo, T., Mechanical properties of polypropylene composites reinforced by surface-coated microfibrillated cellulose. *Compos. Part A: Appl. Sci. Manuf.* 59, 26–29, 2014.

8. Singh, A.A., and Palsule, S., Jute fiber reinforced chemically functionalized polypropylene self-compatibilizing composites by Palsule process. *J. Compos. Mater.* 50(9), 1199–1212, 2016.

9. Ljungberg, N., Cavaillé, J.Y., and Heux, L., Nanocomposites of isotactic polypropylene reinforced with rod-like cellulose whiskers. *Polymer* 47(18), 6285–6292, 2006.

10. García-García, D., Carbonell, A., Samper, M.D., García-Sanoguera, D., and Balart, R., Green composites based on polypropylene matrix and hydrophobized spend coffee ground (SCG) powder. *Compos. Part B: Eng.* 78, 256–265, 2015.

11. Qiu, W., Endo, T., and Hirotsu, T., Structure and properties of composites of highly crystalline cellulose with polypropylene: Effects of polypropylene molecular weight. *Eur. Polym. J.* 42(5), 1059–1068, 2006.

12. Miyazaki, K., and Nakatani, H., Additive effect of poly(ethylene oxide) on polypropylene/fibrous cellulose composite: Effects of additive amount of poly(ethylene oxide) on Young's modulus and morphology. *J. Appl. Polym. Sci.* 114(3), 1656–1663, 2009.

13. Miyazaki, K., Moriya, K., Okazaki, N., Terano, M., and Nakatani, H., Cellulose/polypropylene composites: Influence of the molecular weight and concentration of oxidatively degraded and maleated polypropylene compatibilizers on tensile behavior. *J. Appl. Polym. Sci.* 111(4), 1835–1841, 2009.

14. Qiu, W., Zhang, F., Endo, T., and Hirotsu, T., Isocyanate as a compatibilizing agent on the properties of highly crystalline cellulose/polypropylene composites. *J. Mater. Sci.* 40(14), 3607–3614, 2005.

15. Haque, M.M., Islam, M.S., and Islam, M.N., Preparation and characterization of polypropylene composites reinforced with chemically treated coir. *J. Polym. Res.* 19(5), 1–8, 2012.

16. Leao, R.M., Luz, S.M., Araujo, J.A., and Novack, K., Surface treatment of coconut fiber and its application in composite materials for reinforcement of polypropylene. *J. Nat. Fibers* 12(6), 574–586, 2015.

17. Kumar, N.R., Rao, C.R., Srikant, P., and Rao, B.R., Mechanical properties of corn fiber reinforced polypropylene composites using Taguchi method. *Mater. Today: Proc.* 2(4), 3084–3092, 2015.

18. Panaitescu, D.M., Vuluga, Z., Ghiurea, M., Iorga, M., Nicolae, C., and Gabor, R., Influence of compatibilizing system on morphology, thermal and mechanical properties of high flow polypropylene reinforced with short hemp fibers. *Compos. Part B: Eng.* 69, 286–295, 2015.

19. Bagheriasl, D., Carreau, P.J., Dubois, C., and Riedl, B., Properties of polypropylene and polypropylene/poly(ethylene-co-vinyl alcohol) blend/CNC nanocomposites. *Compos. Sci. Technol.* 117, 357–363, 2015.

20. Franciszczak, P., and Bledzki, A.K., Tailoring of dual-interface in high tenacity PP composites—Toughening with positive hybrid effect. *Compos. Part A: Appl. Sci. Manuf.* 83, 185–192, 2016.

21. Kahraman, R., Abbasi, S., and Abu-Sharkh, B., Influence of epolene G-3003 as a coupling agent on the mechanical behavior of palm fiber-polypropylene composites. *Int. J. Polym. Mater.* 54(6), 483–503, 2005.

22. Sultan, M.T., Haque, M.M.U., Maniruzzaman, M., and Alam, M.A., Composites of polypropylene with pulque fibres: Morphology, thermal and mechanical properties. *J. Thermoplast. Compos. Mater.* 28(12), 1615–1626, 2015.

23. Ashori, A., and Nourbakhsh, A., Reinforced polypropylene composites: Effects of chemical compositions and particle size. *Bioresour. Technol.* 101(7), 2515–2519, 2010.

24. Ifuku, S., and Yano, H., Effect of a silane coupling agent on the mechanical properties of a microfibrillated cellulose composite. *Int. J. Biol. Macromolec.* 74, 428–432, 2015.

25. Li, X., He, L., Zhou, H., Li, W., and Zha, W., Influence of silicone oil modification on properties of ramie fiber reinforced polypropylene composites. *Carbohydr. Polym.* 87(3), 2000–2004, 2012.

26. Lee, S.H., Wang, S., Pharr, G.M., and Xu, H., Evaluation of interphase properties in a cellulose fiber-reinforced polypropylene composite by nanoindentation and finite element analysis. *Compos. Part A: Appl. Sci. Manuf.* 38(6), 1517–1524, 2007.

27. Yeh, S.K., Hsieh, C.C., Chang, H.C., Yen, C.C., and Chang, Y.C., Synergistic effect of coupling agents and fiber treatments on mechanical properties and moisture absorption of polypropylene–rice husk composites and their foam. *Compos. Part A: Appl. Sci. Manuf.* 68, 313–322, 2015.

28. Bengtsson, M., Le Baillif, M., and Oksman, K., Extrusion and mechanical properties of highly filled cellulose fibre–polypropylene composites. *Compos. Part A: Appl. Sci. Manuf.* 38(8), 1922–1931, 2007.

29. Wang, S., Lin, Y., Zhang, X., and Lu, C., Towards mechanically robust cellulose fiber-reinforced polypropylene composites with strong interfacial interaction through dual modification. *RSC Adv.* 5(63), 50660–50667, 2015.

30. De la Orden, M.U., Sánchez, C.G., Quesada, M.G., and Urreaga, J.M., Novel polypropylene–cellulose composites using polyethylenimine as coupling agent. *Compos. Part A: Appl. Sci. Manuf.* 38(9), 2005–2012, 2007.

31. Nourbakhsh, A., Ashori, A., and Tabrizi, A.K., Characterization and biodegradability of polypropylene composites using agricultural residues and waste fish. *Compos. Part B: Eng.* 56, 279–283, 2014.

32. Kaewkuk, S., Sutapun, W., and Jarukumjorn, K., Effects of interfacial modification and fiber content on physical properties of sisal fiber/polypropylene composites. *Compos. Part B: Eng.* 45(1), 544–549, 2013.

33. Suharty, N.S., Ismail, H., Diharjo, K., Handayani, D.S., and Firdaus, M., Effect of kenaf fiber as a reinforcement on the tensile, flexural strength and impact toughness properties of recycled polypropylene/halloysite composites. *Procedia Chem.* 19, 253–258, 2016.

34. Ummartyotin, S., and Pechyen, C., Microcrystalline-cellulose and polypropylene based composite: A simple, selective and effective material for microwavable packaging. *Carbohydr. Polym.* 142, 133–140, 2016.

35. Samat, N., Marini, C.D., Maritho, M.A., and Sabaruddin, F.A., Tensile and impact properties of polypropylene/microcrystalline cellulose treated with different coupling agents. *Compos. Interface.* 20(7), 497–506, 2013.

36. Borysiak, S., Supermolecular structure of wood/polypropylene composites: I. The influence of processing parameters and chemical treatment of the filler. *Polym. Bull.* 64(3), 275–290, 2010.

37. Jang, S.Y., and Kim, D.S., Physical properties of polypropylene composites with hydrophobized cellulose powder by soybean oil. *J. Appl. Polym. Sci.* 133(6), 42929, 2016.

38. Hung, K.C., Wu, T.L., Chen, Y.L., and Wu, J.H., Assessing the effect of wood acetylation on mechanical properties and extended creep behavior of wood/recycled-polypropylene composites. *Constr. Build. Mater.* 108, 139–145, 2016.

39. Joseph, P.V., Joseph, K., and Thomas, S., Short sisal fiber reinforced polypropylene composites: The role of interface modification on ultimate properties. *Compos. Interface.* 9(2), 171–205, 2002.

40. Panaitescu, D.M., Nicolae, C.A., Vuluga, Z., Vitelaru, C., Sanporean, C.G., Zaharia, C., Florea, D., and Vasilievici, G., Influence of hemp fibers with modified surface on polypropylene composites. *J. Ind. Eng. Chem.* 37, 137–146, 2016.

41. Renner, K., Móczó, J., Suba, P., and Pukánszky, B., Micromechanical deformations in PP/lignocellulosic filler composites: Effect of matrix properties. *Compos. Sci. Technol.* 70(7), 1141–1147, 2010.

42. Yang, H.S., Gardner, D.J., and Nader, J.W., Characteristic impact resistance model analysis of cellulose nanofibril-filled polypropylene composites. *Compos. Part A: Appl. Sci. Manuf.* 42(12), 2028–2035, 2011.

43. Yeo, J.S., and Hwang, S.H., Preparation and characteristics of polypropylene-graft-maleic anhydride anchored micro-fibriled cellulose: Its composites with polypropylene. *J. Adhes. Sci. Technol.* 29(3), 185–194, 2015.

44. Klemm, D., Heublein, B., Fink, H.P., and Bohn, A., Cellulose: Fascinating biopolymer and sustainable raw material. *Angew. Chem. Int. Ed.* 44(22), 3358–3393, 2005.

45. Herrick, F.W., Casebier, R.L., Hamilton, J.K., and Sandberg, K.R., Microfibrillated cellulose: Morphology and accessibility, *J. Appl. Polym. Sci.: Appl. Polym. Symp. (United States)*, Vol. 37, No. CONF-8205234-Vol. 2, 1983.

46. Nakagaito, A.N., and Yano, H., The effect of morphological changes from pulp fiber towards nano-scale fibrillated cellulose on the mechanical properties of high-strength plant fiber based composites. *Appl. Phys. A* 78(4), 547–552, 2004.

47. Bondeson, D., Mathew, A., and Oksman, K., Optimization of the isolation of nanocrystals from microcrystalline cellulose by acid hydrolysis. *Cellulose* 13(2), 171–180, 2006.

48. Suzuki, K., Sato, A., Okumura, H., Hashimoto, T., Nakagaito, A.N., and Yano, H., Novel high-strength, micro fibrillated cellulose-reinforced polypropylene composites using a cationic polymer as compatibilizer. *Cellulose* 21(1), 507–518, 2014.

49. Iyer, K.A., Schueneman, G.T., and Torkelson, J.M., Cellulose nanocrystal/polyolefin biocomposites prepared by solid-state shear pulverization: Superior dispersion leading to synergistic property enhancements. *Polymer* 56, 464–475, 2015.

50. Kuboki, T., Foaming behavior of cellulose fiber-reinforced polypropylene composites in extrusion. *J. Cell. Plast.* 50(2), 113–128, 2013.

51. Morales-Cepeda, A.B., Ponce-Medina, M.E., Salas-Papayanopolos, H., Lozano, T., Zamudio, M., and Lafleur, P.G., Preparation and characterization of candelilla fiber (*Euphorbia antisyphilitica*) and its reinforcing effect in polypropylene composites. *Cellulose* 22(6), 3839–3849, 2015.

52. Joseph, P.V., Joseph, K., Thomas, S., Pillai, C.K.S., Prasad, V.S., Groeninckx, G., and Sarkissova, M., The thermal and crystallisation studies of short sisal fibre reinforced polypropylene composites. *Compos. Part A: Appl. Sci. Manuf.* 34(3), 253–266, 2003.

53. Poletto, M., Ornaghi, H.L., and Zattera, A.J., Native cellulose: Structure, characterization and thermal properties. *Materials* 7(9), 6105–6119, 2014.

54. Yang, H.S., Kiziltas, A., and Gardner, D.J., Thermal analysis and crystallinity study of cellulose nanofibril-filled polypropylene composites. *J. Therm. Anal. Calorim.* 113(2), 673–682, 2013.

55. Moscoso-Sánchez, F.J., Díaz, O.R., Flores, J., Martínez, L., Fernández, V.V.A., Barrera, A., and Canché-Escamilla, G., Effect of the cellulose of *Agave tequilana* Weber onto the mechanical properties of foamed and unfoamed polypropylene composites. *Polym. Bull.* 70(3), 837–847, 2013.

56. Lima, M.F.S., Vasconcellos, M.A.Z., and Samios, D., Crystallinity changes in plastically deformed isotactic polypropylene evaluated by X-ray diffraction and differential scanning calorimetry methods. *J. Polym. Sci. Part B: Polym. Phys.* 40(9), 896–903, 2002.

57. Kim, J.K., and Mai, Y.W. (Eds.), *Engineered Interfaces in Fiber Reinforced Composites*, Elsevier, 1998.

58. Nekkaa, S., Guessoum, M., Grillet, A.C., and Haddaoui, N., Mechanical properties of biodegradable composites reinforced with short *Spartium junceum* fibers before and after treatments. *Int. J. Polym. Mater.* 61(13), 1021–1034, 2012.

59. Krishnan, A., Jose, C., and George, K.E., Sisal nanofibril reinforced polypropylene/polystyrene blends: Morphology, mechanical, dynamic mechanical and water transmission studies. *Ind. Crops Prod.* 71, 173–184, 2015.

60. Ljungberg, N., Bonini, C., Bortolussi, F., Boisson, C., Heux, L., and Cavaillé, J.Y., New nanocomposite materials reinforced with cellulose whiskers in atactic polypropylene: Effect of surface and dispersion characteristics. *Biomacromolecules* 6(5), 2732–2739, 2005.

61. Kalaprasad, G., and Thomas, S., Melt rheological behavior of intimately mixed short sisal–glass hybrid fiber reinforced low density polyethylene composites. II. Chemical modification. *J. Appl. Polym. Sci.* 89(2), 443–450, 2003.

62. Chen, M., Ma, Y., Xu, Y., Chen, X., Zhang, X., and Lu, C., Isolation and characterization of cellulose fibers from rice straw and its application in modified polypropylene composites. *Polym. Plast. Technol. Eng.* 52(15), 1566–1573, 2013.

63. Doumbia, A.S., Castro, M., Jouannet, D., Kervoëlen, A., Falher, T., Cauret, L., and Bourmaud, A., Flax/polypropylene composites for lightened structures: Multiscale analysis of process and fibre parameters. *Mater. Des.* 87, 331–341, 2015.

64. Joseph, P.V., Mathew, G., Joseph, K., Groeninckx, G., and Thomas, S., Dynamic mechanical properties of short sisal fibre reinforced polypropylene composites. *Compos. Part A: Appl. Sci. Manuf.* 34(3), 275–290, 2003.

65. Qiu, W., Zhang, F., Endo, T., and Hirotsu, T., Preparation and characteristics of composites of high crystalline cellulose with polypropylene: Effects of maleated polypropylene and cellulose content. *J. Appl. Polym. Sci.* 87(2), 337–345, 2003.

66. Botev, M., Betchev, H., Bikiaris, D., and Panayiotou, C., Mechanical proper-
    ties and viscoelastic behavior of basalt fiber-reinforced polypropylene. *J. Appl.
    Polym. Sci.* 74(3), 523–531, 1999.
67. Zhang, J., Liu, Z., Kong, Q., Zhang, C., Pang, S., Yue, L., Wang, X., Yao, J., and
    Cui, G., Renewable and superior thermal-resistant cellulose-based composite
    nonwoven as lithium-ion battery separator. *ACS Appl. Mater. Interfaces* 5(1),
    128–134, 2012.
68. Liao, H., Zhang, H., Hong, H., Li, Z., Qin, G., Zhu, H., and Lin, Y., Novel
    cellulose aerogel coated on polypropylene separators as gel polymer electro-
    lyte with high ionic conductivity for lithium-ion batteries. *J. Membr. Sci.* 514,
    332–339, 2016.
69. Herrera-Franco, P., and Valadez-Gonzalez, A., A study of the mechanical
    properties of short natural-fiber reinforced composites. *Compos. Part B: Eng.*
    36(8), 597–608, 2005.
70. Mutjé, P., Vallejos, M.E., Girones, J., Vilaseca, F., López, A., López, J.P., and
    Méndez, J.A., Effect of maleated polypropylene as coupling agent for polypro-
    pylene composites reinforced with hemp strands. *J. Appl. Polym. Sci.* 102(1),
    833–840, 2006.
71. Franco-Marquès, E., Méndez, J.A., Pèlach, M.A., Vilaseca, F., Bayer, J., and
    Mutjé, P., Influence of coupling agents in the preparation of polypropylene
    composites reinforced with recycled fibers. *Chem. Eng. J.* 166(3), 1170–1178,
    2011.
72. El-Abbassi, F.E., Assarar, M., Ayad, R., and Lamdouar, N., Effect of alkali
    treatment on Alfa fibre as reinforcement for polypropylene based eco-
    composites: Mechanical behaviour and water ageing. *Compos. Struct.* 133,
    451–457, 2015.
73. Hirunpraditkoon, S., and García, A.N., Kinetic study of vetiver grass powder
    filled polypropylene composites. *Thermochim. Acta*, 482(1), 30–38, 2009.
74. Bledzki, A.K., Faruk, O., and Sperber, V.E., Cars from bio-fibres. *Macromol.
    Mater. Eng.* 291(5), 449–457, 2006.
75. Hao, A., Yuan, L., and Chen, J.Y., Notch effects and crack propagation analy-
    sis on kenaf/polypropylene nonwoven composites. *Compos. Part A: Appl. Sci.
    Manuf.* 73, 11–19, 2015.
76. Yeh, S.K., Hsieh, C.C., Chang, H.C., Yen, C.C., and Chang, Y.C., Synergistic
    effect of coupling agents and fiber treatments on mechanical properties and
    moisture absorption of polypropylene–rice husk composites and their foam.
    Compos. Part A: Appl. Sci. Manuf. 68, 313–322, 2015.
77. Kuboki, T., Mechanical properties and foaming behavior of injection molded
    cellulose fiber reinforced polypropylene composite foams. *J. Cell. Plast.* 50(2),
    129–143, 2013.
78. Clemons, C., Wood-plastic composites in the United States: The interfacing of
    two industries. *Forest Prod. J.* 52(6), 10, 2002.
79. Hung, K.C., Wu, T.L., Chen, Y.L., and Wu, J.H., Assessing the effect of wood
    acetylation on mechanical properties and extended creep behavior of wood/

recycled-polypropylene composites. *Constr. Build. Mater.* 108, 139–145, 2016.

80. Gardette, J.L., Sinturel, C., and Lemaire, J., Photooxidation of fire retarded polypropylene. *Polym. Degrad. Stab.* 64(3), 411–417, 1999.

81. Sinturel, C., Philippart, J.L., Lemaire, J., and Gardette, J.L., Photooxidation of fire retarded polypropylene. I. Photoageing in accelerated conditions. *Eur. Polym. J.* 35(10), 1773–1781, 1999.

82. Turku, I., and Kärki, T., Accelerated weathering of fire-retarded wood-polypropylene composites. *Compos. Part A: Appl. Sci. Manuf.* 81, 305–312, 2016.

3

Polypropylene (PP)/Starch-Based Biocomposites and Bionanocomposites

Saviour A. Umoren[1,*] and Moses M. Solomon[2]

[1]Centre of Research Excellence in Corrosion, Research Institute, King Fahd University of Petroleum and Minerals, Dhahran, Saudi Arabia
[2]Corrosion Research Laboratory, Department of Mechanical Engineering, Faculty of Engineering, Duzce University, Duzce, Turkey

Abstract

The vision of replacing conventional plastics which are derived from non-renewable sources with bioplastics of natural origin is essential to eco-system preservation and a secure future. This century has witnessed an increase in the production and utilization of biocomposites and bionanocomposites. Various methods of preparation and characterization of polypropylene/starch-based biocomposites and bionanocomposites are discussed in this chapter. Application of PP/starch-based biocomposites and bionanocomposites in biomedical, packaging, automotive, military, coating, fire retardant, aerospace, and optical sectors are also covered. The recycling and lifetime studies conducted on PP/starch-based biocomposites and bionanocomposites are equally highlighted.

Keywords: Polypropylene, starch, biocomposites, bionanocomposites, extrusion, plastics, biopropylene, polymer

3.1 Introduction

It is rather difficult to imagine life without plastics; on our daily lives, we utilize plastic materials (cups, eyeglasses, cell phones, televisions, cars, etc.) for several purposes which make life better. Ever since 1907, when Leo H. Baekeland synthesized the first synthetic plastic, Bakelite, through the chemical reaction of phenol and formaldehyde, many other plastics have been produced and

*Corresponding author: umoren@kfupm.edu.sa

Visakh. P. M. and Matheus Poletto. (eds.) Polypropylene-Based Biocomposites and Bionanocomposites, (55–84) 2018 © Scrivener Publishing LLC

several modifications done to enhance its properties. Before 1954, polymer scientists made numerous attempts to produce other plastics from polyolefins of which only the polyethylene family was known for commercial purposes. In 1955, F. J. Natta, an Italian scientist, announced the discovery of stereospecific polypropylene. However, Natta made a significant breakthrough when the commercial production of polypropylene was made through the use of the so-called Ziegler-Natta catalyst ($Al(C_2H_5)$ + $TiCl_4$). The polymer can be molded or extruded into plastic materials with good flexibility, toughness, light weight, and heat resistance. Because of these properties polypropylene materials enjoyed endless applications in industrial and household textiles, home furnishings and indoor-outdoor carpets, fabrics for diapers, laboratory and medical equipment, packaging and labeling materials, as well as stationeries. It is ranked the second most essential plastic with revenue expected to exceed US\$ 145 billion by 2019 [1]. However, non-renewable petroleum resources, which will increase in cost as they are constantly depleted, still remain the primary source of polypropylene and the polymer is non-biodegradable. Hence, the fears of exhausting non-renewable sources and environmental concerns associated with their disposal by end users have increased interest in producing biodegradable plastics of renewable origin.

Starch is a polymer of natural origin composed of two homopolymers of D-glucose, amylase, mostly linear α-D(1,4')-glucan and branched amylopectin having the same backbone as amylose but with α-1,6'-linked branched points [2]. The starch chain is characterized with a lot of hydroxyl functional groups; two secondary hydroxyl functional groups at C-2 and C-3 and one primary hydroxyl functional group at C-6 when it is not linked (Figure 3.1). It is semicrystalline and completely biodegradable in nature. The properties of starch are similar to those of petroleum-based polymers [3, 4] but differ due to the fact that the granular form cannot be processed in the same manner as petroleum-based polymers. Meanwhile, starch is highly hydrophilic, brittle in the absence of suitable plasticizer, and possesses poor mechanical properties [5, 6]. Significant effort has been made to address these issues, one such effort being compositing with other polymers known to possess desirable properties. A composite is a material consisting of two or more chemically distinct constituents on a minute scale, having a distinct interface separating them, and with properties different from the properties of any of the components working individually [7]. A highly biodegradable composite with excellent mechanical properties can be obtained by blending starch with polycaprolactone or polylactic acid [8], but would not be cost effective. A low-cost composite with excellent mechanical properties can be made from starch and polyolefins such as polyethylene and polypropylene [9].

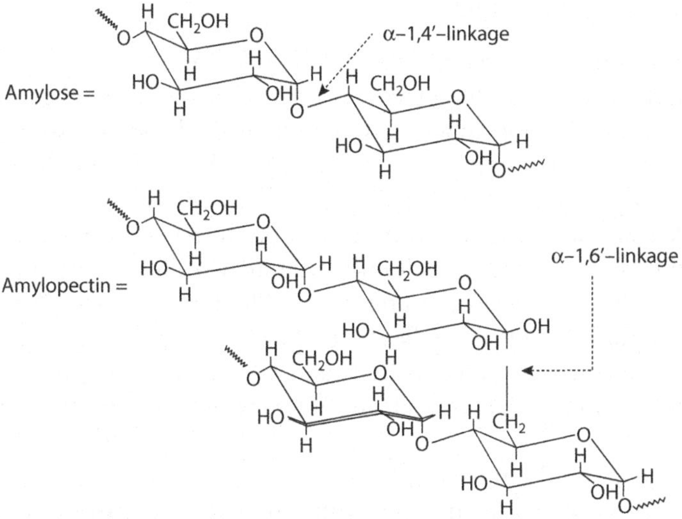

Figure 3.1 Molecular structure of starch [10].

Bionanocomposite can be defined as a system that can be degraded biologically, which consists of two or more constituents with at least one of the constituents having a dimension of less than 100 nm. The main difference between bionanocomposites and the conventional biocomposites is the high surface-to-volume ratio of the nanoparticles. This property makes bionanocomposites uniquely different from conventional biocomposites. However, synthesis of bionanocomposites in which the nanoparticles are uniformly dispersed in the base matrix is challenging due to the fact that nanoparticles tend to aggregate during synthesis. Many synthetic techniques have been employed by scientists, including milling, mixing, *in-situ* polymerization, etc. This chapter discusses the synthesis, characterization, and application of PP/starch-based biocomposites and bionanocomposites.

3.2   PP/Starch Biocomposites and Bionanocomposites

3.2.1   Preparation

Generally, composites processing begins by either mixing or compounding followed by shaping and finishing. The commonly used techniques for the preparation of starch-based composites are presented below.

3.2.1.1   Mill Processing

Milling is a dry processing technique of grinding materials into powder. The mechanical ball milling technique dates back to 1966 when Benjamin [11] used it to develop an alloy combining oxide dispersing strengthening γ' precipitation hardening in a nickel-based super alloy intended for gas turbine application. Generally, the primary objective of milling is to reduce particle size and ensure uniform dispersion. In recent times, high energy ball mills, such as tumbler ball mills, vibratory mills, planetary mills, and attritor mills, have been the preferred milling machines in synthesizing nanocomposites and are believed to produce composites with unique properties which are difficult to obtain by other conventional methods [12]. This technique has several advantages which include:

    i. It is simple and versatile,
    ii. It produces composite with improved compatibilization between constituents,
    iii. It increases the degree of dispersion,
    iv. Radical chain scissions within polymer particles are induced due to the combination of mechanical effects such as impact, compressive and shear forces.

The major disadvantages of this technique are the contamination of powder by the balls and the problem of consolidating the powder product without coarsening the nanocrystalline microstructure. The procedure involves loading both the materials to be milled and the grinding medium (usually hardened steel ball) into a milling chamber and milling for the desired length of time. Figure 3.2 illustrates the principle of mechanical milling. However, the contamination problem is often given as a reason to disqualify the use of this method for the preparation of starch/PP biocomposites and bionanocomposites. Meanwhile, Kazayawoko *et al.* [13] reported the use of this technique in the synthesis of wood-fiber/PP composites.

3.2.1.2   Intermix and Brabender

Composites and nanocomposites can be prepared by using Brabender mixer [14, 15]. It is believed that a high degree of blending and mixing for optimum properties can be achieved with a Brabender. A Brabender mixer consists of a mixer back stand with a gear unit and a mixer bowl. An electronic safety system which guides the drive unit from damage due to overload is always provided. The Brabender blades are made of special steel

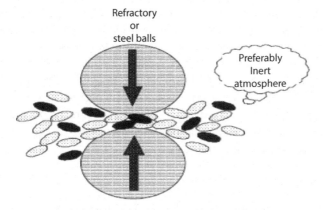

Figure 3.2 Schematic representation of the principle of mechanical milling.

Figure 3.3 Brabender model 50 EHT.

and counter-rotating screws in which counter-rotation towards each other at different speeds provide excellent compounding and mixing characteristics. Heating as well as cooling is done either through a cold circulation thermostat or electrically with air cooling. The photo in Figure 3.3 shows a Brabender model 50 EHT.

Hamdam *et al.* [15] used this technique to prepare sago starch/polypropylene blends. It was found that blending starch with propylene using a Brabender reduced the overall mechanical properties (Young's modulus and tensile strength) with a small increase in the dynamic property (storage modulus) of the composite. Hamdam *et al.* observed from SEM experiments that the starch granules were partially melted and remained as a separate phase forming heterogeneous two-phase mixture in the polymer matrix. It was also found that the starch/polypropylene blend became relatively viscous in nature with rising temperature. This is in line with the recent report of Tănase *et al.* [16] on the processability characteristics and melt rheology of starch/polyvinyl alcohol blends using a Brabender. The results of Tănase *et al.* revealed that upon incorporation of starch into the polymer matrix, the torque of the physical blends highly increased due to an increase in the viscosity of the starch phase. This behavior would lead to an increase in melt viscosity and power consumption during processing. Melt viscosity is very essential during processing operations involving melt flow, such as extrusion, compression molding and injection molding, and it is important that polymer blends be processed smoothly with minimum power consumption.

3.2.1.3   Melt Blending

This technique is fast becoming the preferred and standard method for starch/PP biocomposites and bionanocomposites synthesis [17–20]. Compared with milling and Brabending methods, melt blending is compatible with industrial polymer processing techniques like extrusion and injection molding. It creates a more uniform distribution of starch particles in the polymer matrix and as a result produces composites with improved mechanical properties. The uniformly distributed starch nanoparticles in polypropylene matrix are shown in the SEM image in Figure 3.4.

Melt blending involves the processing of mixture of the polymers (starch and polypropylene) and reinforcing materials above the melting point of the polymers in a processing chamber, typically an extruder. During processing, the polymer chains diffuse in between aggregated nanofillers and form a nanostructured system that is controlled by processing conditions such

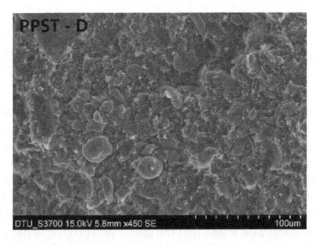

Figure 3.4 SEM image of the fracture surface of PP/starch blends. (Reprinted with permission from [9])

as temperature, shear, mixing conditions, and residence time. Temperature influences melting and flow rate of melt, shear induces platelets delamination from the reinforcing agents, while extended residence time is essential to allow the diffusion of polymer chains into the interlayer spaces and to obtain an exfoliated morphology [21]. However, strong shear and long residence time will give rise to polymer chain degradation and hence have to be balanced during processing.

3.2.1.4 Extrusion

The word extrusion is derived from the Latin word *extrudere* (*ex* means out while *trudere* means to push) and is a processing technique whereby heat is applied on dry powder, granular, or reinforced plastics and forces the melt through an orifice in a die. The basic configuration of a single-screw extruder is shown in Figure 3.5. The twin-screw extruder works according to the same principles as the single-screw extruder but differs in that it has two rotating screws in its barrel. Twin-screw is costlier than the single-screw extruder but offers better mixing, greater shear and higher die pressure than the single-screw extruder.

The twin-screw extruder has three different kinds of setups: (a) intermeshing counter-rotating, (b) intermeshing co-rotating, and (c) non-intermeshing counter-rotating. These setups are shown in Figure 3.6. In the co-rotating design, the materials are forced through a nip (a gap between two screws) from which mixing takes place. However, the disadvantage is that some materials may move forward down the barrel without passing through the nip. In the counter-rotating design, the materials are transferred from one screw to another. This design gives a more thorough mixing than the co-rotating design.

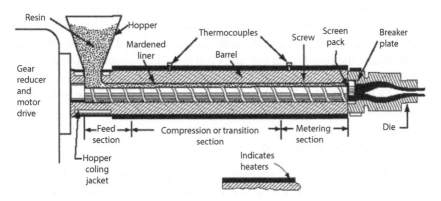

Figure 3.5 Configuration of a typical single-screw extruder.

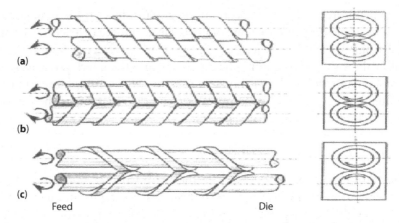

Figure 3.6 Common types of twin-screw extruders: (a) intermeshing co-rotating, (b) intermeshing counter-rotating, and (c) non-intermeshing counter-rotating.

Extrusion process has been widely employed in the synthesis of PP/starch biocomposites and bionanocomposites, particularly the twin-screw extruders [8, 17, 22], although a few authors have also reported the use of single-screw extruders [23]. This is because of the excellent mixing effect of the twin-screw extruder, which allows the starch granules to be homogenously distributed and wetted in the polypropylene melt, consequently improving composite property. The barrel temperature is an important parameter as it controls the melting process. Authors have reported barrel temperature ranging from 160–180 °C for the preparation of PP/starch composites [8, 17, 22].

When using extrusion technique for the preparation of PP/starch composite, certain precautions have to be taken. Before processing, it should be ensured that the rupture disk in the extruder is good. Most extruders have a rupture disk in a hollow threaded fastener situated near the front end of the barrel which protects the bolts holding the die in place. This disk bursts if the pressure near the die is too much and could be extremely dangerous. The disk fails at varying pressure. The pressure applied in most reported PP/starch composites synthesis using extrusion ranged between 11 to 68.9 MPa [22, 23]. Adequate ventilation should also be provided to remove fumes and smoke produced by burning polypropylene.

3.2.1.5   Injection Molding

Injection is the principal processing technique for converting plastics into molded parts. It uses a machine called the injection molding machine to melt its feeds and forces the melt into a mold. The output of this process is high, with molded parts requiring little or no finishing. Injection molding has minimal warping and shrinkage and can be used for recycled materials. Nevertheless, the technique is not very appropriate for short production runs as the machines are very costly and complicated.

Basically, the injection process consists of five steps: first, the mold closes; second, the screw moves forward and the non-return valve on the front end of the screw prevents backward movement of plasticated materials along screw flight. As a result, the screw behaves as a ram and pushes the hot melt into the mold cavity. Third, the screw maintains pressure through the nozzle until the material is cooled and set. Fourth, timers stop injection pressure and the screw returns to pick up fresh feed from the hopper and, fifth, the mold opens and ejector pins eject the molded parts from the mold. Most often, these steps are described in terms of *time cycle*. Based on this, there are four elements in the injection molding *time cycle*.

1. *Fill time*: this is the time taken for air to be displaced from the mold cavity by the melt.
2. *Pack time*: this is the time required to maintain sufficient pressure to fill the mold with melt and achieve gate freeze.
3. *Dwell time*: this is the time required for the formed part to cool or set enough to allow safe ejection.
4. *Dead time*: this is the time required for the mold to open, eject the molded part, and close again.

Great work has been done on the production of polypropylene/starch composites using the injection processing techniques. Beckermann and his coworkers [24–26] comprehensively studied hemp fiber reinforced PP composites prepared by injection molding process. Their studies included fiber treatments and modifications, model predictions of micro-mechanics and strengths, the optimization of hemp fiber quality, and the influence of bag retting, and white rot fungal treatments. The effect of surface treatment on the injection molded pineapple leaf fiber reinforced polycarbonate composites was assessed [27]. The modified pineapple leaf fiber composite produced improved mechanical properties. The results from the thermogravimetric analysis revealed that the thermal stability of the composites was lower than that of neat polycarbonate resin and that the thermal stability decreased with increasing pineapple leaf fiber content. Panthapulakkal and Sain [28] studied the influence of water absorption on the tensile properties of injection molded short hemp fiber/glass fiber reinforced polypropylene hybrid composites. Swelling of natural fiber as a result of prolonged exposure to water was found to lead to reduction in the stiffness of the fibers. The loss in strength and modulus values of the hemp fiber composites was believed to be due to the inability of the swelled natural fiber to carry the stress transferred from the matrix through the disrupted interface as a result of water absorption. Also, Bledzki *et al.* [29] studied the different separation processes (mechanical, refiner and enzymatic separation) with injection molded hemp and partially with flax and wheat straw reinforced PP composites. The authors noted that thermomechanical processed hemp fiber-PP composites possessed better mechanical properties compared to other processes and composites.

3.2.1.6   In-Situ Polymerization Technique

This method has to do with the swelling of a nanofiller within a liquid monomer or monomer solution to form a fine and thermodynamically stable reinforcing phase within a polymer matrix. *In-situ* polymerization was first used by the Toyota researchers in the early nineties to disperse aluminosilicate in polyamide matrix on a nanometer level. Since then, it has enjoyed wide application in composite and nanocomposite synthesis [30–35]. The first step in *in-situ* polymerization technique involves the dispersion of nanoparticles in liquid monomer or low molecular weight precursor or their solutions to obtain a homogenous mixture. The polymerization reaction can be initiated by exposing the mixture to an appropriate source of heat or sunlight, by the diffusion of a suitable initiator or by an organic initiator or catalyst fixed through cation exchange inside the interlayer before the swelling step. It is essential that the filler bears chemical functional groups through which the catalyst or co-catalyst can react. Also, to prevent catalyst deactivation, the filler has to be pretreated.

In-situ polymerization method has several advantages:

i.   There is thermodynamic compatibility at the polymer matrix reinforcement interface.
ii. The reinforcement surfaces are free from contamination; hence stronger matrix dispersion bond can be achieved.
iii. It can be used for the preparation of insoluble and thermally unstable polymer composites which cannot be synthesized via other synthetic techniques.
iv. It is suitable for the preparation of polymer composites with high nanotube loading, as it provides excellent miscibility with almost any type of polymer.

However, for starch composite, since starch molecular chains are synthesized during the plant growth, this method is limited to chemically modified starch bionanocomposites. Liu *et al.* [20] successfully prepared starch/PP blends with improved mechanical properties by a one-step reactive compatibility technique using maleated polypropylene (MAPP) as the interfacial agent and diethanolamine (DEA) as a reactive plasticizer for starch. It was found that the addition of MAPP greatly reduced the size of the starch domains in the PP matrix and enhanced the interfacial bonding between starch and the polymer matrix because of *in-situ* reactions among MAPP, starch, and DEA.

3.2.2   Characterizations

3.2.2.1   *Tensile Characterizations*

Tensile properties are important indicators of the strength of a material. Commonly tested tensile properties for plastics are tensile, compressive, flexural, and shear strength. These properties are usually tested for using the universal testing machine shown in Figure 3.7. Changing from one test to another requires altering the direction and speed of the crosshead and switching the fixtures which hold the samples.

The tensile strength gives information on the extent to which a material can withstand forces pulling it apart. The compressive strength indicates the quantity of force needed to rupture a material; the flexural strength

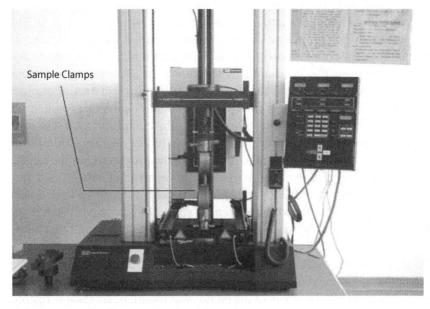

Figure 3.7 Tensile testing machine (Instron 4465).

measures how much force can be applied to a material before it breaks while the shear strength shows the maximum force required to produce a fracture completely separating the movable part of a material from the stationary.

Plasticized starch (PLS) has poor tensile characteristics. Although blending with PP can improve the tensile properties, at high PLS content, the properties remain poor. In the investigation of Rosa et al. [23] on recycled PP blended with thermoplastic starch (TPS), addition of 30% of TPS was found to reduce tensile strength of the polymer at break. Researchers have deployed several ways to enhance these properties in starch/PP composites. DeLeo et al. [8] observed that addition of clay can greatly improve the mechanical properties of PLS/polypropylene blends at high starch content. Unmodified and organically modified montmorillonite (MMT) clays and Cloisite 30B respectively, were added to blends of glycerol-plasticized starch and polypropylene, compatibilized using maleated polypropylene. At high PLS content, the addition of clay was found to increase the tensile strength and tensile modulus by an order of magnitude but reduced the ultimate elongation only slightly. In the work of Hanifi et al. [18] on starch/PP nanocomposites compatibilized with maleic anhydride grafted PP (MAPP) with various percentages (0, 3, and 5) of modified clay (Cloisite 30B), it was observed that in the presence of 5 wt% nanoclay, the values of tensile strength (15.5 MPa) and elastic modulus (10.2 MPa) increased by 9.1 and 70% respectively while that of elongation at break (4.2%) decreased by 49%. Gupta and Alam [9] reported that addition of MAPP to potato starch/polypropylene composite improved the tensile and flexural strengths of the composite while impact strength remains the same.

The study on the utilization of thermoplastic starch obtained from cassava starch and potato starch prepared using glycerol as a plasticizer by Obasi et al. [36] showed that plasticized starch content exhibited an inverse relationship with the tensile strength, elongation at break and directly related to Young's modulus of the starch/PP blends. Addition of MAPP to the blends was found to improve all the properties but the tensile strength and elongation at break were still lower than the neat PP. A comparison of potato starch/PP with cassava starch/PP blends revealed that potato starch/PP exhibited higher tensile strength than cassava starch/PP blends and is probably due to fiber content. Liu et al. [20] found that addition of starch to PP improved bending strength, bending modulus, and Young's modulus but to different extents decreased yield strength, tensile strength, and elongation at break. Upon plasticization of the starch with DEA, the elongation at break of the blends was found to remarkably restore to 220%, while both the yield strength and tensile strength merely changed slightly. Addition of 15 wt% MAPP to the DEA plasticized starch/PP blend improved the Young's modulus and tensile strength respectively by 56.3% and 76.5%. Mantia *et al.* [37], however, found that by using manganese stearate to assist the dispersion of sago starch in PP, the tensile strength and elongation at break decreased by 12.5% and 27%, respectively.

3.2.2.2   Thermal Characteristics

Thermal properties explain the response of a material to applied heat. Numerous methods abound for the measurement of thermal properties. For starch/PP composites and nanocomposites, commonly used techniques are differential scanning calorimetry (DSC), differential thermal analysis (DTA), and thermal gravimetric analysis (TGA). The DSC gives information on how a material's heat capacity is affected on subjecting it to heat. The DTA allows detection of changes in material's temperature caused by chemical modification while the TGA measures the changes in the mass of a material with temperature.

Figure 3.8 shows the TGA curves of (a) pristine starch and (b) neat PP, PP/starch (PCS), PP/DEA/starch (PDS), PP/DEA/starch with different MAPP contents (PDPS) in nitrogen condition obtained by Liu *et al.* [20]. The decomposition of pristine starch started at about 190 °C (T_{onset}, defined as the temperature where 5 wt% mass loss occurs) and underwent a sharp mass loss at 310 °C (T_{max}, defined as the temperature where the maximum weight loss takes place), leaving 16.6 wt% of residue char at 550 °C due to the high-char-forming ability of starch. By contrast, neat PP is much more thermally stable, showing T_{onset} of about 426 °C and T_{max} of about 470 °C, but at 500 °C, the polymer degrades substantially with a char residue of as low as 0.29 wt%. This agrees with Al-Mulla *et al.* [38], who found from the TGA curves of recycled PP and MAPP and the compatibilized and uncompatibilized blends   recorded at a heating rate of 20 °C/min from ambient temperature to 700 °C in a nitrogen atmosphere that the blends and recycled PP typically degraded completely between 400 and 600 °C.

Liu *et al.* observed that introduction of 30 wt% starch into PP led to two-step decomposition. T_{onset} of PCS shifted to a much lower temperature of 294 °C, and two T_{max} values at 328 and 474 °C (~ 4 °C higher than that of pure PP) occur respectively, indicating the degradation of cornstarch and PP. Incorporation of DEA into the starch/PP blends was found to further decrease T_{onset} (~ 264 °C) and the first T_{max} (T_{max1} ~ 307 °C) because of the lower thermal stability and volatilization of DEA at elevated temperature. Nevertheless, the second T_{max} (T_{max2}) belonging to degradation of the PP matrix was seen to further increase up to 476 °C (6 °C higher than that of

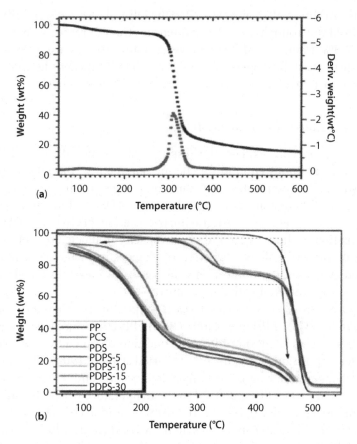

Figure 3.8 TGA curves for (a) pristine starch and (b) neat PP, PP/starch (PCS), PP/DEA/ starch (PDS), PP/DEA/starch with different MAPP contents (PDPS) under nitrogen conditions. (Reprinted with permission from [20])

the PP matrix), indicating an enhanced thermal stability due to the thermal protection action of the char residue formed by degradation of starch. Meanwhile, with the addition of MAPP and its increasing loading level, both T_{onset} and T_{max} values were found to first increase and then decrease slightly, which means that the presence of MAPP has limited effects on the thermal properties of the PP/starch.

Khanna *et al.* [39] found from DTA and TGA curves of PP grafted with starch that the composite had a single stage of decomposition. Differential thermal gravimetry (DTG) curves showed a split in the peak of the initial decomposition and a fast rising wavering curve with small peak maxima at 432 °C at a decomposition rate of 1.035 mg/min. The decomposition region traversed by the grafted PP in DTA curve showed an endothermic peak, indicating a continuous decomposition instead of a sharp peak absorbing less energy (89.5 mJ/mg). The melting endothermic peak was found to start at 60 °C and gave a sharp peak at 151 °C with 169 mJ/mg of energy absorbed. However, the % weight loss of the grafted polymer was found to be higher than that of pristine PP at corresponding temperatures.

3.2.2.3   Morphological Characterizations

Optical or electron micron microscopy has quite often been used as a method of direct observation to show the degree of miscibility in starch/PP blends. The nature of encapsulation of starch granules in PP matrix defined the type of degradation. Enzymatic degradation would be impossible if starch granule is completely encapsulated in the polymer matrix. Scanning electron microscopy observations of starch/PP blend samples [9, 17, 19, 20, 22, 36, 39, 40] revealed that starch granules are not completely encapsulated in PP matrix in starch/PP biocomposites and bionanocomposites. The starch granules are in an open and accessible structure, loosely adhering to PP matrix.

Generally, starch disperses poorly in PP matrix because of poor interfacial action between starch (polar compound) and PP (nonpolar compound), which would have a negative influence on the mechanical properties. SEM and STEM experiments have shown that addition of MAPP [20, 39] and/or organoclay [19, 22] to starch/PP blends played a vital role in reducing the interfacial energy and promoting the interfacial adhesion between starch and PP matrix.

3.2.2.4   X-Ray Diffraction

X-ray diffraction uses wave phenomena to measure the atomic spacing within a material. Crystalline materials adopt a regular pattern in the arrangement of atoms or ions in space. If X-ray radiation passes through this "ordered" arrangement, diffraction occurs and the incidence radiation is reflected from the individual planes of the atoms at a special angle, θ. X-ray thus gives information about the interatomic spacing and structural regularity of a material.

Biocomoposite of TPS/PP + MAPP has been reported to reflect XRD radiations at 2θ = 15°, 18°, 20°, 22.5°, and 23°, which corresponds to the 110, 040, 130, 111 and 131 + 041 lattice planes of monoclinic α-phase of PP respectively [22]. The work of Tessier et al. [17] on the influence of polar organoclay (C30B) and nonpolar organoclay (C20A) on the X-ray diffraction pattern of PP-g-starch/5 wt% MAPP blend revealed that the primary

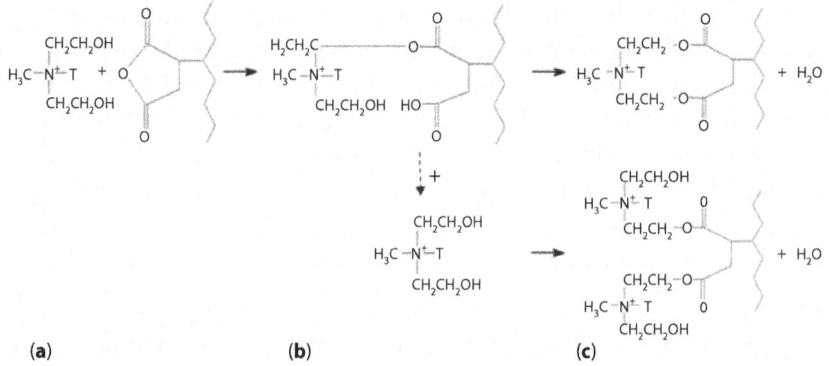

Figure 3.9 Expected chemical reactions of (a) C30B polar organo-modified clay and MAPP compatibilizer, (b) formation of an ester and acid groups, (c) final ester chemical structure. (Adapted from [17])

diffraction (d_{100}) of neat C20A and C30B appear around $2\theta = 4$ and $5.4°$ with interlayer spacing of 1.55 and 0.81 nm respectively. Upon addition of C20A to PP-g-starch/MAPP blend, the XRD peak shifts to around $2\theta = 2.56°$, showing an increase in interlayer spacing in the composite arising from polymer intercalation within clay platelets. However, in C30B/PP-g-starch/MAPP composite, the XRD spectra show no characteristic signal, indicating the existence of clay platelets exfoliation. The maleic anhydride of the MAPP compatibilizer can react with both the polar C30B and the starch phase of the matrix according to the chemical reactions presented in Figure 3.9 [17]. The hydroxyl group (–OH) in C30B can react with the MAPP (–CO–O–CO–) to give an ester (–COO–) and an acid (–COOH) (Figure 3.9a,b). The acid group can then react with another hydroxyl group of the surfactant to form a second ester and a water molecule (Figure 3.9c). More so, maleic anhydride can react with the numerous hydroxyl groups of the starch which is mainly constituted of amylase and amylopectine. These chemical reactions favor the formation of a surfactant/MAPP/starch network which makes the platelet separation, intercalation and exfoliation possible, thereby increasing the dynamic storage modulus. In the case of the nonpolar organo-modified clay (C20A), the long alkyl chains of the surfactant intercalate the platelets, forming a paraffin film that promotes exfoliation. In the case of C20A, the exfoliation mechanism is mechanically driven (by polypropylene chain insertion) rather than chemically driven. Since C30B organoclay offers a better degree of clay dispersion and exfoliation level than C20A, it is the preferred reinforcing agent in starch/PP biocomposite.

3.2.2.5 Spectroscopic Characterizations

Information on the compatibility among constituents in a composite can be obtained through spectroscopic characterization. The use of Fourier transmission infrared spectroscopy (FTIR) is common in starch/PP blends characterization. Figure 3.10 shows the FTIR spectrum of PP and its blends with different compositions of starch using MAPP as the coupling agent. Characteristic peaks typical of starch are evidenced at 3100–3600 cm^{-1} (O-H stretching from starch), 1749 cm^{-1} (C=O stretching from starch), 1653 cm^{-1} (intramolecular hydrogen bonding), 1255 cm^{-1} (-O-C(O)- and C-OH stretching), 1151 cm^{-1} (C-O-C stretching), and 1025 cm^{-1} (C-O stretch vibration). The peak at about 1740 cm^{-1} is a strong indicator of the presence of anhydride group in PP backbone [41]. The CH$_3$ bands at 2950 cm^{-1}, 2868 cm^{-1} and 1375 cm^{-1}, the CH$_2$ bands at 2917 cm^{-1}, 2837 cm^{-1} and 1457 cm^{-1}, and three isotactic peaks at 1167 cm^{-1}, 997 cm^{-1} and 973 cm^{-1} are confirmation of the presence of polypropylene component in all the blends. MAPP characteristic peaks of symmetric and asymmetric stretching vibrations would appear at about 1784 and 1861 cm^{-1} respectively in a typical MAPP FTIR spectrum [20, 38]. These peaks would disappear in a compatible starch/PP-MAPP blend due to the formation of ester groups through the reaction between hydroxyl and anhydride groups.

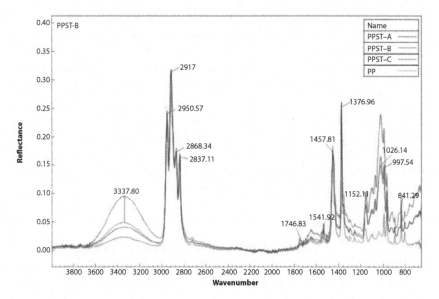

Figure 3.10 FTIR spectrum of samples PP, PPST-A, PPST-B and PPST-C. (Reprinted with permission from [9])

3.2.2.6 Rheological Characterizations

Rheological evaluation methods have been applied to the study of the phase separation of partially miscible or immiscible polymer blends. Among the commonly used techniques is the time-temperature superposition (TTS) principle. The TTS principle states that, when the viscoelastic behavior of a material (for instance complex viscosity or complex elasticity modulus) is studied, a change of temperature is completely equivalent to a shift of the logarithmic time scale (the so-called time shift). In principle, TTS principle should hold for a homogenous mixture and fail for partially miscible blends like polymer blends due to different temperature dependencies of the components. Surprisingly, TTS had been reported to hold for polymer blends like chitosan-based composites, HDPE/LDPE, PS/PP, etc. [42, 43]. The simple reason for this observation may be that the activation energies of the blend's components are not too far apart. Nevertheless, failure of the TTS principle can still offer information on the specific interactions that have taken place in a homogenous system.

Tessier *et al.* [17] used this technique to examine the dynamic rheological behavior of 95/5 and 80/20 wt% PP-g-starch/MAPP blends recorded at different temperatures, namely 170, 180 and 190 °C. The curves of complex viscosity as a function of the shifted frequency at 180 °C were plotted. A very good superposition was obtained for the blends containing MAPP2 (polypropylene-grafted maleic anhydride with graft efficiency of 1%) and MAPP3 (polypropylene-grafted maleic anhydride with graft efficiency of 1.4%). Their rheological behavior was found to depend neither on the temperature nor on the blend composition (i.e., MAPP weight content), suggesting excellent miscibility. A peculiar behavior was noticed in the case of the blend containing MAPP1 (polypropylene-grafted maleic anhydride with graft efficiency of 0.5%); a shift appeared at lower frequencies and depended on both the temperature and the MAPP weight content. Such a deviation may be attributed to a partial, lower miscibility of the blend, particularly when the compatibilizer weight content was high. The failure of TTS principle should not be used as a strict principle to judge phase separation as it may fail in the miscible state.

3.2.2.7 Viscoelastic Characterizations

Viscoelasticity is a characteristic of material that makes it behave both as a viscous and elastic material when subjected to deformation. For composites, viscoelastic properties are dependent on temperature, time, and frequency. When a material is subjected to constant stress, it experiences a decrease in the elastic modulus over a period of time due to the molecular rearrangement in an attempt to minimize the localized stresses. Dynamic mechanical analysis is used in the study of viscoelasticity property of a material. It involves the application of a small oscillatory stress on the material and measurement of the resultant strain. The relationship between the applied stress and the resultant strain is represented by the complex dynamic modulus, G

$$G = G' + iG'' \tag{4.1}$$

where i = ; G' is the storage modulus which is defined as the stress in phase with the strain in a sinusoidal deformation divided by the strain; G'' is the loss modulus and is defined as the stress out of phase with the strain divided by the strain. The ratio of G'/G'' is referred to as the loss tangent, $tan\delta$. $Tan\delta$ helps to discern transition peaks occurring in blends and to know whether peak shift is a function of temperature.

$$G' = \frac{\sigma_0}{\varepsilon_0} \cos\delta \tag{4.2}$$

$$G'' = \frac{\sigma_0}{\varepsilon_0} \sin\delta \tag{4.3}$$

where σ_0 and ε_0 are the amplitude of stress and strain and δ is the phase between them.

Hamma et al. [44] reported that the G' and G'' of two starch-grafted-polypropylene composites with trade names G906PF and G906PJ increased with increase in frequency. This increase could be due to the unraveling of the entanglements, which allowed for a large extent of relaxation to occur. The elastic and viscous response of the composites was observed to show the same order of magnitude except at frequency where G906PJ and G906PF exhibited gel behavior. Furthermore, the values of the modulus of G906PJ were found to be higher than those of G906PF in the investigated frequency range. Hamdan et al. [15] noticed from tan δ versus temperature curves for sago starch/PP blend that $tan\delta$ increased with rise in temperature. An increase in tan δ with increasing temperature is an indicator that the blend was becoming more viscous in nature as the temperature increased.

Al-Mulla et al. [38] observed that the increase in G' and G'' of recycled PP/starch blends with and without MAPP as compatibilizer was prominent at high frequencies (>1 rad s^{-1}). The slopes of G' and G'' and for the compatibilized and uncompatibilized recycled PP/starch blends were found to be slightly greater than those of recycled PP alone. This trend could be an indication of the formation of new chemical structures in both the compatibilized and uncompatibilized blends due to chemical reactions. Al-Mulla *et al.* also reported that the G' and G'' of the compatibilized blends in the low- and high-frequency regions were larger compared to those of the uncompatibilized blends and recycled PP, which was an indication that the molecular structure, molar mass and distribution of components vary between the compatibilized and uncompatibilized blends and recycled PP. The higher G' and G'' values of the compatibilized and uncompatibilized blends relative to those of recycled PP indicated that maleated PP, used as a compatibilizer, controls the viscoelastic properties of the blends. This result suggested that the incorporation of MAPP into the compatibilized blends improved the mechanical properties of the system over the frequency range investigated.

3.2.2.8   Electrical Characterizations

Polymeric materials are electrical insulators due to the predominantly covalent type of bonds that characterize them and limit their electrical conductivity. Ordinarily, pure starch/PP composites should be electrical insulator. However, when fillers are added, the biocomposite can be made conductive or semiconductive. In the work of Haydaruzzaman *et al.* [45, 46] on starch-treated coir/jute-based hybrid PP composites and 2-hydroxyethyl methacrylate + starch-treated PP composites, the maximum value of dielectric constant (the measure of the ability of a material to store electrical energy) was found to be 289.39 and 335.46 respectively. The dielectric constant somewhat decreased with increasing frequency and this may be due to interfacial polarization. For starch-treated coir/jute-based hybrid PP composite, the AC conductivity was found to increase with increase in frequency and the highest value obtained was at 6.5 Log Hz.

3.2.3   Applications

The improvement in certain properties like biodegradability of starch/PP blends compared to pure PP has made them an excellent candidate for applications in the durable goods market in various sectors where traditional PP is used. The potential areas of usage include the medical, automotive, packaging, coating, optical, aerospace, fire retardant, and structural fields.

3.2.3.1   Biomedical Applications

Polypropylene has been extensively used in the medical field in medical devices, drug delivery systems, nonwoven fabrics, and packaging for medical devices, drugs, and solutions. It has gained application in disposable and prefilled syringes, diagnostic cuvettes, sample cups, parenteral kit parts, needle shields, centrifuge tubes, connectors, surgical trays, infectious waste containers and bags, clamps, spine supports, and drapes. It has equally enjoyed usage in contact lens molding cups, needle disposal containers, medication spoons, analytical test strips and blood oxygenator membrane, as well as infant feeding tubes. PP plastics possess good resistance to solvents and autoclave heat; have high tensile strength and stiffness; and have a long flex life, high heat distortion temperature and low density and moisture vapor transmission rate. These excellent characteristics allow PP to be widely utilized in medical applications. However, the non-biodegradability of PP has always been a thing of serious concern. Starch/PP composite is biodegradable, a property that projects it as an excellent replacement for pure PP in medical applications.

3.2.3.2   Packaging Applications

The use of starch/PP plastics for packaging applications is one of the many strategies to minimize the environmental impact of pure PP. One third of pure PP is used for packaging purposes. It is used in packaging biscuits, crisps, sweets, bread, pasta, dairy products, convenience foods, snacks, and dried fruits. PP films are useful as flexible packaging for shirts, hosiery and shrink-wrap applications for toys, games, hardware, frozen foods, and cigarette wrap. On July 18, 2008, Cereplast sought permission from the Food and Drug Administration (FDA) to use starch/PP resin under the trade name Biopropylene™ (also known as CP-Bio-PP) to produce containers and utensils intended to contact all types of food. The FDA gave Cereplast a license on August 5, 2008, which allows sheets, plates, utensils, and food containers used in serving food to customers in quick service restaurants to be made from Bioproplene™ [47]. Starch/PP plastics also find usage in the manufacturing of shopping bags, checkout bags, green bin liners, overwrap packages as well as breathable and mulch films.

3.2.3.3   Automotive Applications

The production of vehicles in which parts are made from bioplastics is an exciting new area of development in the auto industry. In October 2007, Cereplast Inc., a Hawthorne, California-based maker of proprietary bio-based plastics, introduced Biopropylene that can replace traditional polypropylene in many automotive applications. Cereplast was honored with the inaugural AUTOPLAST SPEICON award by the Society of Plastics Engineers (SPE) at AUTOPLAST 2007 in Mumbi, India, for the highest contribution in plastic material development used in the auto industry [48]. The automotive industry is a major user of polypropylene in the manufacturing of vehicle components such as interior panels, air conditioning equipment, etc. At present, the automotive group PSA Peugeot Citroën is exploring the use of Biopropylene in the production of sustainable automobiles. Sandrine Raphanaud, the manager of the MAATEO project remarked, "Biopropylene is a very interesting and promising material that could assist PSA in reducing the carbon footprint of our automobiles" [48].

3.2.3.4   Military Applications

Starch/PP plastic is used in making safety helmets. In October 10, 2012, Cereplast Inc. announced that the company had begun selling safety helmets (Figure 3.11) made from Biopropylene H-101 [49]. Starch/PP plastic is also a promising candidate for making military belts, packs, and pouches.

3.2.3.5   Coating Industry

The coating industry has been under intense pressure from consumers and regulatory bodies to reduce the environmental impact of its products and processes. Chomarat, a family-owned textile group with an international distribution, recently rolled out the world premiere of a new line of biodegradable coated fabric called OFLEX™ Bio-based, made from Gaïalene®

Figure 3.11 Safety helmets made from Biopropylene.

produced by Roquette. Chomarat, specializing in coated textiles, developed OFLEX™ by using thermoplastic polypropylene (TPP) and Gaïalene®, a sustainable thermoplastic plant-based resin made from starch by Roquette. The OFLEX™ Bio-based line consists of a textile material or foam coated with biobased TPP. The coating lends itself readily to dyeing, is free of plasticizers, and is recyclable in the polyolefin stream, providing numerous design options with a high level of performance. OFLEX™ is highly flexible and soft and serves as an alternative to coated PVCs and leathers. The soft touch and ease of dyeing opens up new prospects in the traditional markets of leather products, luggage, telephony, sport and leisure activities and event furnishings.

3.2.3.6   Fire Retardant Applications

The flame retardancy of starch/PP biocomposites is almost unexplored. A few reports [50, 51] have shown that starch/PP biocomposites are better retardant than pure PP. Nie *et al.* [50] assessed the flame retardant properties of starch containing polypropylene (SCP) semi-biocomposites by limited oxygen index, UL-94 test, and cone calorimeter test. The results of cone calorimetry testing showed that the peak of heat release rate and total heat released by SCP decreased substantially compared to pure PP. Matkó *et al.* [51] noted that PP composite reinforced with 50% wood flakes and 10% ammonium polyphosphate had the potential to be used as industrial flame retardant.

3.2.3.7   Aerospace Applications

Aerospace engineers rely to a large extent on patented FDM technology for prototyping, tooling, and part manufacturing. Recently, Cereplast Inc. announced a new bioplastic resin grade, Biopropylene A150D, an injection molding grade manufactured with 51% post-industrial algae biomass for aerospace applications [52]. The Biopropylene is said to have low to no odor due to the discovery of a post-industrial process that significantly reduces the distinctive smell that is inherent to algae biomass. Biopropylene A150D can be processed on existing conventional electric and hydraulic reciprocating screw injection molding machines, and is recommended for thin-wall injection molding applications. Biopropylene A150D meets CONEG (Coalition of Northeastern Governors) and RoHS (Restriction of Hazardous Substances) requirements and can be processed into certain aerospace products like housings, lenses, chassis components, subsequent molds, panels, enclosure and containers, turbine blades, and bezels.

3.2.3.8   Optical Applications

Cereplast Biopropylene has found application in many areas of electronics. It can be tailored into user's products such as CD/DVD cases, consumer electronics, cell-phone casings, and computer housings. Biopropylene is also a potential candidate to replace the traditional polypropylene capacitors.

3.2.3.9   Recycling and Life-Cycle Assessments

Biopropylene is recyclable as per ASTM 6866 standard. Life-cycle assessment (LCA) is defined by ISO (International Standards Organization) as the compilation and evaluation of the inputs, outputs and the potential environmental impacts of a product system throughout its life cycle (ISO 14040). That is to say that LCA identifies the material and energy usage, emissions and waste flows of a product, process or service over its entire life cycle in order for its environmental impacts to be ascertained. Starch/PP plastic has minimal impact on the environment compared to conventional PP. The LCA of Cereplast Biopropylene 101 bioplastic and four different conventional plastics, including low density polyethylene (LDPE), polyethylene terephthalate (PET), high impact polystyrene (HIPS) and polypropylene (PP), revealed that Biopropylene is far better than the conventional plastics with significant percentage reduction in global warming potential (GWP) compared to conventional PP [53]. Tables 3.1 and 3.2 summarize the GWP and impact on resource depletion for starch/PP plastic and conventional plastics.

Table 3.1 The global warming potential (GWP) of each plastic with and without the inclusion of biogenic carbon [54].

Plastic product	GWP (lbs CO_2 eq/lb)	GWP including biogenic carbon (lbs CO_2 eq/ib)	Change
Starch/PP	2.33	1.73	−25.45%
PP	2.53	2.54	0.35%
PET	3.27	3.28	0.01%
HIPS	4.09	4.12	0.58%
LDPE	2.66	2.67	0.53%

Table 3.2 The impact on resource depletion for each plastic [54].

Plastic	Abiotic depletion (lbs Sb eq/lb)
Starch/PP	0.0283
PP	0.0369
PET	0.0386
HIPS	0.0443
LDPE	0.0377

3.3   Conclusion

Research and development of starch/PP biocomposite and bionanocomposite plastics have been catalyzed by the fear of exhausting crude oil, increase in the cost of petroleum-based PP, and the growing environmental concerns about the end users' products of conventional PP. Starch/PP plastics are biodegradable and eco-friendly with significant percentage reduction on the carbon footprint on the environment. Starch/PP biocomposites and bionanocomposites can be made using various synthetic approaches, however, cost and starch dispersability on the polymer matrix are the watchwords for the choice of technique. Biopropylene has good mechanical, thermal, and barrier properties in addition to biodegradability and thus offers great promise to extend its use into wider, more demanding applications.

References

1. Ceresana, Market Study: Polypropylene, 3rd edition, www.ceresana.com/en/market-studies/plastics/polypropylene/, 2015.
2. Pareta, R., and Edirisinghe, M.J., A novel method for the preparation of starch films and coatings. *Carbohydr. Polym.* 63, 425–431, 2006.
3. Potts, J.E., Environmentally degradable plastics, in, *Kirk-Othmer Encyclopedia of Chemical Technology*, 3rd ed., suppl. vol., pp. 638–668, Wiley, New York, 1981.
4. Huang, J.C., Shetty, A.S., and Wang, S.W., Biodegradable plastics: A review. *Adv. Polym. Technol.* 10, 23–30, 1990.
5. Griffin, G.J.L., Starch polymer blends. *Polym. Degrad. Stab.* 45, 241, 1994.

6. Sathya, K., and Syed, S.H.R., Overview of starch based plastic blends. *J. Plast. Flim. Sheet.* 22(1), 39–58, 2006.
7. Agarwal, B., Broutman, L., and Chandrashekhara, K., *Analysis and Performance of Fiber Composites*, 3rd ed., Wiley, New Jersey, 2006.
8. DeLeo, C., Pinotti, C.A., Goncalves, M.C., and Velankar, S., Preparation and characterization of clay nanocomposites of plasticized starch and polypropylene polymer blends. *J. Polym. Environ.* 19, 689–697, 2011.
9. Gupta, A.P., and Alam, A., Study of flexural, tensile, impact properties and morphology of potato starch/polypropylene blends. *Int. J. Adv. Res.* 2(11), 599–604, 2014.
10. Lu, D.R., Xiao, C.M., and Xu, S.J., Starch-based completely biodegradable polymer materials. *Express Polym. Lett.* 3(6), 366–375, 2009.
11. Benjamin, J.S., Mechanical alloying. *Sci. Am.* 234(5), 40–48, 1976.
12. Yadav, T.P., Yadav, R.M., and Singh, D.P., Mechanical milling: A top down approach for the synthesis of nanomaterials and nanocomposites. *Nanosci. Nanotechnol.* 2(3), 22–48, 2012.
13. Kazayawoko, M., Balatinecz, J.J., and Matuana, L.M., Surface modification and adhesion mechanisms in woodfiber-polypropylene composites. *J. Mater. Sci.* 34, 6189–6199, 1999.
14. Kim, Y.C., and Lee, C.Y., Effect of the starch content on the silicate dispersion and rheological properties of polypropylene/starch/silicate composites. *Korean Chem. Eng. Res.* 46(1), 106–111, 2008.
15. Hamdan, S., Hashim, D.M.A., Ahmad, M., and Embong, S., Compatibility studies of polypropylene (PP)–Sago starch (SS) blends using DMTA. *J. Polym. Res.* 7(4), 237–244, 2000.
16. Tănase, E.E., Popa, M.E., Râpă, M., and Popa, O., Preparation and characterization of biopolymer blends based on polyvinyl alcohol and starch. *Rom. Biotech. Lett.* 20(2), 10306–10315, 2015.
17. Tessier, R., Lafranche, E., and Krawczak, P., Development of novel melt-compounded starch-grafted polypropylene/polypropylene-grafted maleicanhydride/organoclay ternary hybrids. *Express Polym. Lett.* 6(11), 937–952, 2012.
18. Hanifi, S., Ahmadi, S., and Oromiehie, A., Mechanical properties and biodegradability of polypropylene/starch reinforced nanoclay blends. *Iranian J. Polym. Sci. Technol.* 26(2), 139–148, 2013.
19. Abreu, A.S., Oliveira, M., and Machado, A.V., Thermoplastic starch-polypropylene reinforced with clay, in: *XIV Latin American Symposium on Polymers*, October 12–16th, 2014.
20. Liu, L., Yu, Y., and Song, P., Improved mechanical and thermal properties of polypropylene blends based on diethanolamine-plasticized corn starch via *in situ* reactive compatibilization. *Ind. Eng. Chem. Res.* 52, 16232–16238, 2013.
21. Dennis, H.R., Hunter, D.L., Chang, D., Kim, S., White, J.L., Cho, J.W., and Paul, D.R., Effect of melt processing conditions on the extent of exfoliation in organoclay-based nanocomposites. *Polymer* 42, 9513–9522, 2001.

22. Ferreira, W.H., Khalili, R.R., Figueira Junior, M.J.M., and Andrade, C.T., Effect of organoclay on blends of individually plasticized thermoplastic starch and polypropylene. *Ind. Crops Prod.* 52, 38–45, 2014.

23. Rosa, D.S., Guedes, C.G.F., and Carvalho, C.L., Processing and thermal, mechanical and morphological characterization of post-consumer polyolefins/thermoplastic starch blends. *J. Mater. Sci.* 42, 551–557, 2007.

24. Beckermann, G.W., and Pickering, K.L., Engineering and evaluation of hemp fibre reinforced polypropylene composites: Fibre treatment and matrix modification. *Compos. Part A: Appl. Sci. Manuf.* 39, 979–988, 2008.

25. Beckermann, G.W., and Pickering, K.L., Engineering and evaluation of hemp fibre reinforced polypropylene composites: Micro-mechanics and strength prediction modeling. *Compos. Part A: Appl. Sci. Manuf.* 40, 210–217, 2009.

26. Beckermann, G.W., Pickering, K.L., Alam, S.N., and Foreman, N.J., Optimising industrial hemp fibre for composites. *Compos. Part A: Appl. Sci. Manuf.* 38, 461–468, 2007.

27. Threepopnatkul, P., Kaerkitcha, N., and Athipongarporn, N., Effect of surface treatment on performance of pineapple leaf fiber–polycarbonate composites. *Compos. Part A: Appl. Sci. Manuf.* 40, 628–632, 2009.

28. Panthapulakkal, S., and Sain, M., Injection-molded short hemp fiber/glass fiber-reinforced polypropylene hybrid composites-mechanical, water absorption and thermal properties. *J. Appl. Polym. Sci.* 103, 2432–2441, 2007.

29. Bledzki, A.K., Faruk, O., and Specht, K., Influence of separation and processing systems on morphology and mechanical properties of hemp and wood fibre reinforced polypropylene composites. *J. Nat. Fibers* 4, 37–56, 2007.

30. Naderi, G., Nekoomanesh, M., Mehtarani, R., and Eslami, H., Preparation of a new polypropylene/clay nanocomposites via *in situ* polymerization. *e-Polymers* 16, 1–10, 2011.

31. Zapata, P., and Quijada, R., Polypropylene nanocomposites obtained by *in situ* polymerization using metallocene catalyst: Influence of the nanoparticles on the final polymermorphology. *J. Nanomaterials* 2012, Article ID 194543, 2012.

32. Namazi, H., Mosadegh, M., and Dadkhah, A., New intercalated layer silicate nanocomposites based on synthesized starch-g-PCL prepared via solution intercalation and *in situ* polymerization methods: As a comparative study. *Carbohydr. Polym.* 75, 665–669, 2009.

33. Wu, J., Lin, J., Zhou, M., and Wei, C., Synthesis and properties of starch-graft-polyacrylamide/clay superabsorbent composite. *Macromol. Rapid Commun.* 21, 1032–1044, 2000.

34. Solomon, M.M., and Umoren, S.A., *In-situ* preparation, characterisation, and anticorrosion property of polypropylene glycol/silver nanoparticles composite for mild steel corrosion in acid solution. *J. Colloid Interface Sci.* 462, 29–41, 2016.

35. Solomon, M.M., Umoren, S.A., and Abai, E.J., Preparation and evaluation of the surface protective performance of poly (methacrylic acid)/silver

nanoparticles composites (PMAA/AgNPs) on mild steel in acidic environment. *J. Mol. Liq.* 212, 340–351, 2015.

36. Obasi, H.C., Egeolu, F.C., and Oparaji, O.D., Comparative analysis of the tensile and biodegradable performances of some selected modified starch filled polypropylene blends. *Am. J. Chem. Mater. Sci.* 2(2), 6–13, 2015.

37. Mantia, F.P.L., Morreale, M., and Mohd Ishak, Z.A., Processing and mechanical properties of organic filler–polypropylene composites. *J. Appl. Polym. Sci.* 96, 1906, 2005.

38. Al-Mulla, A., Alfadhel, K., Qambar, G., and Shaban, H., Rheological study of recycled polypropylene–starch blends. *Polym. Bull.* 70, 2599–2618, 2013.

39. Khanna, N.D., Kaur, I., and Kumar, A., Starch-grafted polypropylene: Synthesis and characterization. *J. Appl. Polym. Sci.* 119P, 602–612, 2011.

40. Azhari, C.H., and Wong, S.F., Morphology-mechanical property relationship polypropylene/starch blends. *Pak. J. Biological Sci.* 4(6), 693–695, 2001.

41. Gupta, A.P., Vijay, K., and Sharma, M., Formulation and characterization of biodegradable packaging film derived from potato starch & LDPE grafted with maleic anhydride-LDPE composition. *J. Polym. Environ.* 18, 484–491, 2010.

42. Yao, B., Cheng, G., Wang, X., Cheng, C., and Liu, S., Linear viscoelastic behaviour of thermo-setting epoxy asphalt concrete—Experiments and modeling. *Constr. Build Mater.* 48, 540–547, 2013.

43. Vaidyanathan, T.K., Vaidyanathan, J., and Cherian, Z., Extended creep behavior of dental composites using time-temperature superposition principle. *Dent. Mater.* 19, 46–53, 2003.

44. Hamma, A., Kaci, M., Mohd Ishak, Z.A., Ceccato, R., and Pegoretti, A., Starch-grafted-polypropylene/kenaf fibres composites. Part 2: Thermal stability anddynamic-mechanical response. *J. Reinf. Plast. Compos.* 34(24), 2045–2058, 2015.

45. Haydaruzzaman, A.H.K., Hossain, M.A., Khan, M.A., Khan, R.A., and Hakim, M.A., Fabrication and characterization of jute reinforced polypropylene composite: Effectiveness of coupling agents. *J. Compos. Mater.* 44(16), 1945–1962, 2010.

46. Haydaruzzaman, A.H.K., Hossain, M.A., Khan, M.A., and Khan, R.A., Mechanical properties of the coir fiber-reinforced polypropylene composites: Effect of the incorporation of jute fiber. *J. Compos. Mater.* 44(4), 401–416, 2010.

47. Bagrodia, S., Bash, T., Kelly, W., and Scheer, F., Advanced materials from novel bio-based resins. Cereplast Inc.: Hawthorne, CA, 2008.

48. Bioplastics Magazine, www.bioplasticsmagazie.com/en/online-archive/data/585 (Acessed 24/12/2015).

49. GlobeNewswire, www.globenewswire.com/news-release/2012/10/10(Acessed 24/12/2015).

50. Nie, S., Song, L., Guo, Y., Wu, K., Xing, W., Lu, H., and Hu, Y., Intumescent flame retardation of starch containing polypropylene semibiocomposites:

Flame retardancy and thermal degradation. *Ind. Eng. Chem. Res.* 48(24), 10751–10758, 2009.

51. Matkó, Sz., Toldy, A., Keszei, S., Anna, P., Bertalan, Gy., and Marosi, Gy., Flame retardancy of biodegradable polymers and biocomposites. *Polym. Degrad. Stab.* 88, 138–145, 2005.

52. Prototype Today, www.prototypetoday.com/tag/materials (Accessed 24/12/2015).

53. Inside Indiana Business, www.insideindianabusiness.com/story/29814939/plastics-company-signs (Accessed 24/12/2015).

54. Trellis Earth, www.TrellisEarth.com (Accessed 24/12/2015).

4

Polypropylene (PP)/Polylactic Acid-Based Biocomposites and Bionanocomposites

Xin Wang

University of Science and Technology of China,
Jinzhai Road 96, Hefei, Anhui 230026, P. R. China

Abstract

The production and application of polypropylene/polylactic acid (PP/PLA)-based biocomposites and bionanocomposites have attracted increasing research interest in both scientific and industrial communities over the past few decades. This chapter first introduces the preparation method of PP/PLA-based biocomposites and bionanocomposites, including melt blending, melt spinning and some other new emerging methods like autoclave preparation. Then morphology, compatibility and crystallization of PP/PLA-based biocomposites and bionanocomposites are discussed in order to obtain resultant materials with uniform distribution and good compatibility among different components. We also summarize the rheological, electrical, mechanical, thermal, gas barrier, and flame retardant properties of these biocomposites and bionanocomposites, and how each of these properties is dependent upon the composition of biocomposites and bionanocomposites, the type and amount of compatibilizers as well as the dispersion state of nanofillers in the matrix. Finally, an overview is presented of potential applications for these biocomposites and bionanocomposites along with current challenges in the development of this new class of promising materials.

Keywords: Polypropylene/polylactic acid composite, Compatibility, Thermal properties, Mechanical properties, Flame retardancy

4.1 Introduction

Over the past few decades, polymer composites have received increasing attention due to their advantageous combined properties over

Corresponding author: wxcmx@ustc.edu.cn

Visakh. P. M. and Matheus Poletto. (eds.) Polypropylene-Based Biocomposites and Bionanocomposites, (85–112) 2018 © Scrivener Publishing LLC

single-component polymers [1]. Among the preparation techniques, blending of two or more polymer components has been regarded as a practical and economical method for fabricating polymeric composites with desirable combined properties for specific end uses. With the increasing focus on environmental awareness, considerable efforts have been made to develop polymer composites that contain all or some fraction of sustainable polymers. Polylactic acid (PLA) is such a sustainable aliphatic polyester derived from polymerization of the renewable monomer lactic acid. Besides its well-known biodegradability and renewability, the physical and mechanical properties of PLA are comparable or even better when compared with some petrochemical polymers such as polyethylene, polypropylene (PP), polystyrene and polyethylene terephthalate. PLA shows a tensile strength of 50 ~ 70 MPa with an elongation at break of ca. 4%, a Young's modulus of around 3 GPa, and an impact strength close to 2.5 kJ/m^2 [2]. These excellent properties make PLA a promising alternative to petrochemical polymers in fabricating eco-composites.

Despite of much potential and several advantages, PLA still possesses some drawbacks as well, like brittleness and low ductility [3], poor gas barrier behaviors [4], and poor hydrolysis resistance [5]. Blends of PLA with several synthetic petroleum-based polymers have been prepared in an effort to overcome these shortcomings of PLA. In recent years, a variety of polyolefins, including polyethylene [6], polypropylene [7], poly(trimethylene terephthalate) [8], poly(propylene carbonate) [9], poly(ethylene glycol) [10], and poly(methyl methacrylate) [11] have been adopted for the preparation of PLA blends. Among these composites, blending PP with PLA has been considered to be a facile and effective solution to create a new material with better hydrolysis resistance and lower cost relative to virgin PLA, and better sustainability and faster degradability than PP. However, immiscibility between PLA and PP occurs during the blending process due to their different polarity. Several methods, including chemical and/or physical modifications, have been utilized to improve the compatibility between PLA and PP. In addition to these two phases, various types of nanofillers have also been incorporated into PP/PLA biocomposites as synergists in order to develop bionanocomposites with additional functionalities like reinforcement.

In this chapter, we will review the recent achievements in the realm of PP/PLA-based biocomposites and bionanocomposites. The preparation methods of these biocomposites and bionanocomposites as well as their morphology, crystallization and compatibility will be summarized. The main emphasis is focused on the properties of PP/PLA-based biocomposites and bionanocomposites, involving their rheological and viscoelastic properties, mechanical and thermal properties, gas barrier properties and flame retardant behavior, etc. An overview of potential applications for these biocomposites and bionanocomposites and a brief outlook on the field are provided to guide future development of these promising materials.

4.2   PP/PLA-Based Biocomposites and Bionanocomposites

4.2.1   Preparation of PP/PLA-Based Biocomposites and Bionanocomposites

In recent years, a variety of processing approaches have been developed for PP/PLA-based biocomposites and bionanocomposites. Generally, PP/PLA-based biocomposites and bionanocomposites can be prepared through mixing PP, PLA and/or other additives using melt blending, extrusion or spinning. In addition to conventional methods, this section also describes the latest developments in some of the newly arising techniques such as batch foaming process by autoclave preparation.

4.2.1.1   Melt Blending

The modification of polymers through blending is a popular technology which was first developed in the 1970s or even earlier [12]. Upon mentioning the term melt blending, we understand it to be the simple mixing of different kinds of polymer materials in the melt state under high shear forces. Melt blending is often regarded as one convenient and economical (because no solvent is used) approach to create new materials with the desired combined performances. Blending could be performed by conventional machinery, like vane extruder, twin-screw extruder and internal mixer, which is very suitable for industrial production. Relative to the development of novel monomers and polymerization technology, a very wide variety of polymer blends can be obtained by this route to meet the requirements of the targeted application in relatively short time and for low cost.

The arising biopolymers have caused a revival in blending technology, because their several drawbacks can be overcome by blending, as mentioned above. The amount of literature about the blending of biopolymers is extensive due to the increasing research interest in them. Among these blends of biopolymers, PP/PLA is one of the most often investigated materials. As described above, the goals of PP/PLA blends are various and range from the improvement of toughness and processability to the modification of some other behaviors like gas barrier or anti-flammability characteristics.

Despite of the existence of many advantages, there are still some disadvantages in melt mixing conditions. The physical blending of different kinds of polymeric materials usually leads to the occurrence of miscibility, and the use of compatibilizers is essential during melt blending. Another problem is the inhomogeneous dispersion of nanofillers in the blends when the bionanocomposites are prepared using such a method. Previous studies indicate that, to date, such methods result in agglomerates of nanofillers, as observed from the X-ray diffraction and TEM data [13].

4.2.1.2   Melt Spinning

Up to date, most of the PP/PLA-based biocomposites have been prepared for specimens or films through conventional melt blending, and few studies have been reported on producing PP/PLA-based biocomposite fibers. PP/PLA bicomponent filament can be prepared by melt spinning, which involves the co-extrusion of these two polymers from the same spinneret into a single filament. Fiber production requires similar processing temperatures between the different polymer phases. Obtaining PP/PLA biocomposite fibers turns out to be possible because PLA and PP have similar processing temperatures (200–230 °C) [14]. Wojciechowska and co-authors developed a melt-spinning method for preparing PP/PLA biocomposite fiber [15]. Through optimizing the processing temperatures, PP/PLA biocomposite fibers with good quality were obtained within the whole range of component concentrations. The primary advantage of co-spinning PLA with PP is its full biodegradability, which to some extent can provide PP/PLA biocomposite fibers with suitable biodegradability. Furthermore, adding PLA as a component changes some intrinsic properties of the PP fibers, including the improved mechanical properties and dyeability [16]. Recently, Arvidson et al. synthesized bicomponent core-sheath filaments consisting of PP as core and PLA as sheath via melt-spinning technology [17]. As shown in Figure 4.1, the PP/PLA interface is clearly observed in the cross-sectional TEM image. Up to now the melt-spinning preparation of PP/PLA biocomposite fibers is still in its infant stage, and various cross-sectional geometries, such as segmented pie, side-by-side, or islands-in-the-sea are awaiting exploration.

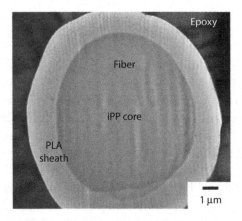

Figure 4.1 Cross-sectional TEM image of filaments in different configurations: a PP core/ PLA sheath bicomponent filament coated in Au and embedded in epoxy (labeled) is shown. (Reproduced with permission from [17]; Copyright © 2012 American Chemical Society)

4.2.1.3   Other Methods for PP/PLA Biocomposite Preparation

Cellular plastics display many advantages over polymer resin such as lower density, excellent thermal insulating behavior, superior energy absorption, and so on. These fascinating performances make cellular plastics suitable for a wide variety of applications in packaging, automotive components, and sporting equipment [18]. The typical commercialized cellular plastics are expanded polystyrene bead foams, expanded polyethylene beads, expanded polypropylene beads and then expanded polylactic acid. Aside from the single-component cellular plastics, considerable interest has shifted to multiphase cellular plastics in order to tune the foam properties. Recently, Tang and coworkers prepared PP/PLA bead foams through autoclave method, with n-pentane as the physical blowing agent [19]. It is found that the incorporation of PLA significantly widens the foaming window of PP. The foaming behavior of PP/PLA is also affected by the n-pentane concentration and the foaming temperature. Under optimal condition, the resultant PP/PLA bead foams displayed high expansion ratio up to 44.4, good ellipse bead foam shape, well-defined cell structure and very high cell density.

4.2.2   Morphology, Compatibility and Crystallization of PP/ PLA-Based Biocomposites and Bionanocomposites

Since property enhancements strongly depend on the microstructure of composite or nanocomposite, morphological characterization is crucial to explore structure-property relationships for these materials. The two most common tools which can provide direct observations of the dispersion state between PP and PLA may be TEM and SEM. Generally, the morphology of polymer blends exhibits either a uniform single phase or separate phases of polymer domains relying on the miscibility. PP/PLA biocomposites are immiscible due to their significant polarity difference, so distinct phase separation morphology between PLA and PP can be clearly observed from the SEM micrographs (Figure 4.2a-4.2f) [20]. The typical structure of islands-in-the-sea is formed: the PLA phase distributed within the PP matrix when the PLA content was not more than 50 wt%; the PP phase distributed within the PLA matrix when PLA content exceeded 50 wt%. Furthermore, the average diameter of the disperse phase particles within the PLA/PP biocomposites is affected by the mass ratios of these two components.

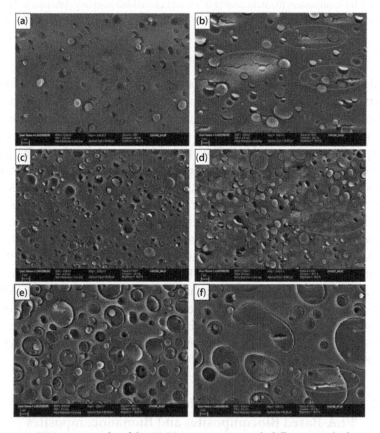

Figure 4.2 SEM micrographs of the PP/PLA composites with different weight fractions: (a) 90:10; (b) 10:90; (c) 75:25; (d) 25:75; (e,f) 50:50. (Reproduced with permission from [20]; Copyright © 2015 Elsevier Ltd.)

Compatibility is a technical term which qualitatively reflects the combined properties of polymer composites in view of a certain application [21]. If the combined properties are comparable or even superior to the expectation, the compatibility of the polymer composites is deduced to be relatively good, and otherwise the components in the composites are incompatible. The most common approaches to determine the compatibility are based on the number of glass transition temperatures measured by DSC or on the phase separation observed on SEM or TEM images. However, these methods just evaluate the compatibility qualitatively, and phase diagrams and the mutual solubility of the components in each other are essential to fully assess the compatibility.

As previously mentioned, the compatibility correlates closely with the combined properties of the resultant composites, so considerable attempts have been made to improve the compatibility between PP and PLA. The compatibility of PP/PLA-based biocomposites is usually enhanced through physical (addition of compatibilizers) or chemical (for example, reactive processing) means. In the first strategy, the compatibilizers are physically incorporated into the PP/PLA-based biocomposites to relieve the interfacial tension between the two components. For instance, the addition of lignin resulted in an increment in the thermal stability and mechanical properties, indicating that lignin acted as an effective compatibilizer in the multiphase PP/PLA-based biocomposites [22]. In the second strategy, premade blocking or grafting copolymers with polar functional groups, which could react with the terminal groups of PLA during melt compounding, were used as compatibilizers in an effort to mediate the polarity at the interface of PP/PLA-based biocomposites. In general, maleic anhydride grafted polyolefins [23–25] and glycidyl methacrylate containing copolymers [26, 27] are utilized as reactive compatibilizers for PP/PLA-based biocomposites. The main compatibilizing mechanism, which involves the facile reaction between anhydride or epoxy and PLA terminal groups at the interface during melt processing, allows the formation of strong interfacial interactions. These types of compatibilizers include a variety of blocking or grafting copolymers, such as polypropylene-g-maleic anhydride (PP-g-MAH) [23, 25, 28], ethylene propylene-g-maleic anhydride (EP-g-MAH) [24], styrene/ethylene/butylene/styrene-g-maleic anhydride (SEBS-g-MAH), ethylene/n-butyl acrylate/glycidyl methacrylate terpolymer (PTW) [26], ethylene–glycidyl methacrylate–methyl acrylate terpolymer (PEGMMA) [27], polyethylene-g-glycidyl methacrylate (PE-g-GMA) [28], and polypropylene-g-acrylic acid (PP-g-AA) [29], which have been demonstrated to be effective in enhancing the compatibility of PP/PLA-based biocomposites. For example, a PEGMMA terpolymer was synthesized and applied

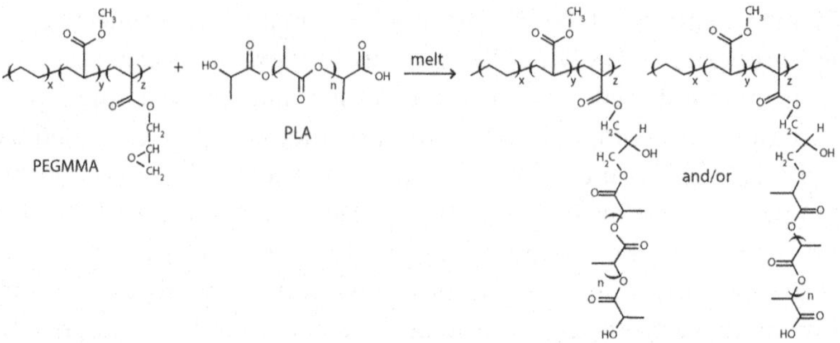

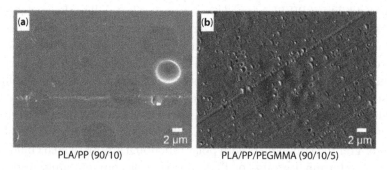

Figure 4.3 Proposed graft copolymer structures formed between PEGMMA and PLA. PEGMMA composition: x:y:z ≈ 44:5:1 with about 25 glycidyl groups per PEGMMA chain on average. (Reproduced with permission from [27]; Copyright © 2015 American Chemical Society)

(a)	(b)
PLA/PP (90/10)	PLA/PP/PEGMMA (90/10/5)

Figure 4.4 SEM images of PP/PLA (10/90) and PP/PLA/PEGMMA (10/90/5). (Reproduced with permission from [27]; Copyright © 2015, American Chemical Society)

for reactive compatibilization of PP/PLA-based biocomposites, as shown in Figure 4.3 [27]. In contrast to the PP/PLA composite (10:90 by weight) (Figure 4.4a), the addition of 5 wt% PEGMMA leads to the smaller droplet sizes and more uniform particle distribution (Figure 4.4b). Table 4.1 summarizes the types and amounts of the compatibilizers used for PP/PLA-based biocomposites from previous studies. Similar phenomenon in terms of more homogeneous distribution and enhanced thermal and mechanical properties has been observed in other compatibilizer investigations [28, 29].

Since PP and PLA are thermodynamically immiscible they have a negative influence on the properties of resultant composites. In order to overcome this disadvantage, nanotechnology has been combined with the immiscible PP/PLA biocomposites, which provides a novel approach to modify their properties. Up to date, introduction of inorganic nanofillers, such as multiwalled carbon nanotubes (MWCNTs) [33], nanoclay [34],

Table 4.1 Overview of a variety of compatibilizers for PP/PLA-based biocomposites.

Matrix	Type and amount of compatibilizer	Comments	Refs.
PP/PLA/ Coffee (8.5/76.5/10)	Lignin (5 wt%)	Increased thermal stability and mechanical properties such as the storage and flexural strength	[22]
PP/PLA (87/10)	EP-g-MAH (3 wt%)	Significant improvement in tensile strength and flexural strength	[24]
PP/PLA (10/90)	PEGMMA (5 wt%)	Substantial improvement in elongation at break and tensile toughness	[27]
PP/PLA (60/30)	PP-g-MAH (3 phr)	Significantly increased tensile strength and flexural strength	[28]
PP/PLA (60/30)	PE-g-GMA (10 phr)	Significant improvement in impact strength	[28]
PP/PLA (70/30)	PP-g-AA (5 wt%)	Enhanced tensile strength and flexural strength	[29]
PP/PLA (50/50)	PP-g-MAH (8 wt%)	Increment in the tensile strength, crystallization property and thermal resistance	[30]
PP/PLA (50/50)	PTW (8 wt%)	Increment in the tensile strength, crystallization property and thermal resistance	[30]
PP/PLA (80/20)	SEBS-g-MAH	Significant improvement in impact strength	[31]
PP/PLA (75/25)	PTW (5 wt%)	A 100% reduction in the PLA droplet size in the PP/PLA composite	[32]

tungsten disulphide nanotubes (INT-WS$_2$) [25], nano-ZnO [35] and sepiolite [36], into PP/PLA-based biocomposites brings a wide range of reinforced performances (strength, stiffness, permeability, thermal stability). However, the property enhancement strongly depends upon the dispersion state of the nanofillers. The dispersion state of the nanofillers can be divided into three categories: (1) a lot of aggregates of nanofillers form within the polymer matrix, also known as a classic microcomposite (Figure 4.5a); (2) polymer chains enter the gallery, expanding the layers of 2D nanomaterials, called intercalated structure (Figure 4.5b); (3) nanomaterials are uniformly dispersed in a continuous polymer matrix, called exfoliated structure (Figure 4.5c) [37]. The last two types of structures can be called nanocomposites. Generally, the exfoliated or intercalated structure is expected, because high aspect ratio of nanofillers can be achieved. Similar to PP/PLA-based composites, most of the PP/PLA nanocomposites were prepared by melt blending, but this method is not effective enough to destroy the aggregates of nanomaterials just through mechanical shearing force. For example, sepiolite agglomerates can still be observed in TEM images (Figure 4.6a,b) for the PP/PLA/sepiolite nanocomposite [36]. A similar phenomenon is also seen in the PP/PLA/clay nanocomposite (Figure 4.6c), and the presence of PTW compatibilizer improves the exfoliation of clays (Figure 4.6d)

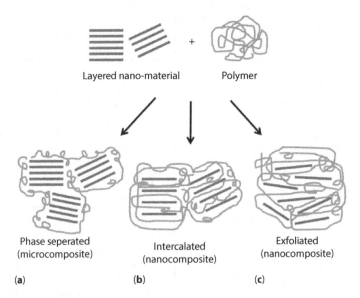

Figure 4.5 Schematic illustration of different structures of composite arising from the interaction of layered nanomaterials and polymers: (a) phase-separated, (b) intercalated, and (c) exfoliated.

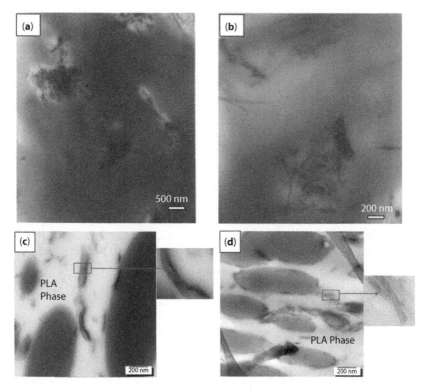

Figure 4.6 TEM images of the PP/PLA composites containing 3.9 wt% sepiolite at (a) low and (b) high magnification, (c) PP/PLA/Clay (25/75/3), and (d) PP/PLA/Clay/PTW (25/75/5/5). (Reproduced with permission from [36] and [38]; Copyright © 2011, 2015 Springer-Verlag)

[38]. According to a previous report [39], it seems that the polar nature of compatibilizer is responsible for a better dispersion state of the nanoclay layers in polymer matrix.

In addition to morphology and compatibility, a large number of studies have focused on the crystallization behaviors of PP/PLA-based biocomposites and bionanocomposites. Differential scanning calorimetry (DSC) and wide-angle X-ray scattering (WAXS) measurements are perhaps the two most common means by which the crystallization behaviors can be evaluated. DSC can give some important crystallization parameters, including the crystallization temperature (T_c), the enthalpy of crystallization (ΔH_c) and the degree of crystallinity (χ_c), while WAXS presents crystallization information regarding the structure of crystals. According to the previous reports, the crystallization behaviors are obviously affected by the PP/PLA ratio [25, 30]. Figure 4.7a displays DSC thermograms of PLA, PP and PP/

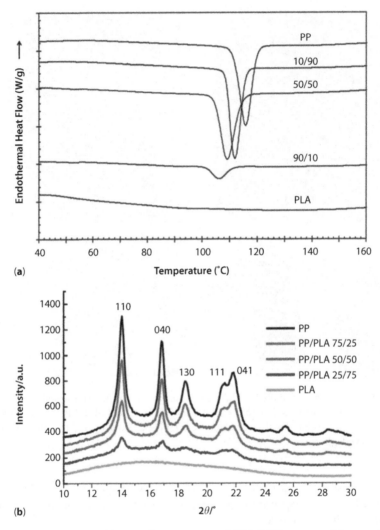

Figure 4.7 (a) DSC crystallization thermograms and (b) WAXS patterns of the pure PLA, PP and PP/PLA composites. (Reproduced with permission from [25], Copyright © 2016 Royal Society of Chemistry; and [38], Copyright © 2015 Springer-Verlag)

PLA composites under the cooling rate of 10 °C/min [25]. As can be seen, virgin PLA shows no crystallization peak due to the relatively low crystallization rate of PLA that may not be able to crystallize when the cooling rate is higher than 5 °C/min [25]. The crystallization exotherms shifted to higher temperatures and exhibited higher enthalpy with the increase of PP concentration, implying that the presence of PLA restricted the crystallization of PP component in PP-rich composites. A similar conclusion can be deduced from WAXS patterns (Figure 4.7b) [38]. The melt-compounded PLA film presented a broad and amorphous WAXS pattern without any diffraction peak, whereas the PP film displayed intense diffraction peaks related to the α-form crystalline phase. This phenomenon can be explained by the different melt-crystallization rates of PP and PLA, i.e., the α-form crystallites were easily formed during the melt-compounding process of the virgin PP film due to its fast crystallization rate, and, in contrast, the amorphous state was formed in the case of the neat PLA film owing to its slow crystallization rate. Meanwhile, with decreasing the PLA content, the intensity of these diffraction peaks increased and the shape was narrowed down, which were associated with increased crystalline size and/or crystalline degree.

Apart from the PP/PLA ratio, the crystallization behaviors are also influenced by the inclusion of the compatibilizer [30], reinforced fibers [23, 24] and/or nanofillers [25, 38]. Chen *et al.* reported that the incorporation of PP-g-MAH improved the ΔH_c and χ_c of the PP component in the PP/PLA biocomposites obviously in comparison with the non-compatibilized samples [30]. This is likely attributed to the stronger interaction between PP and PLA compatibilized by PP-g-MAH, which improved the nucleating activity of PP/PLA biocomposites. The reinforced fibers such as bamboo fiber [23] and oat hull fiber [24] could serve as effective nucleating agents for the crystallization of PP/PLA biocomposites, consequently improving the crystallization rate of PP/PLA biocomposites. For semicrystalline PP/PLA biocomposites, incorporation of nanofiller can also result in an altered crystallization temperature, crystallization rate and degree of crystallinity. A remarkable improvement in degree of crystallinity and crystallization temperature was observed due to the nucleating effect of INT-WS$_2$ [25]. Additionally, clay has been reported to accelerate the crystallization rate of PP/PLA biocomposites [38].

4.2.3 Properties of PP/PLA-Based Biocomposites and Bionanocomposites

4.2.3.1 Rheological and Viscoelastic Properties

The study of composite rheology is crucial to understand the processing operations and it may also provide information about composite microstructure [40]. Figure 4.8 shows the complex viscosity (η) and dynamic storage modulus (G′) curves for PLA, PP and their biocomposites as a function of frequency [26]. As can be clearly seen, the complex viscosity increases with the increase of PP concentration (Figure 4.8a), exhibiting

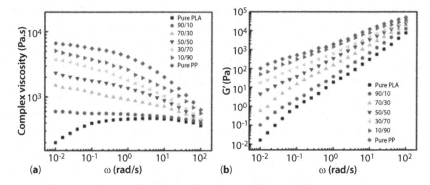

Figure 4.8 (a) Complex viscosity and (b) dynamic storage modulus for the PP/PLA composites obtained in dynamic frequency sweep. (Reproduced with permission from [26]; Copyright © 2015 Elsevier Ltd.)

shear-thinning characteristics in the range of the frequencies studied. The viscosity of PP is greater than that of PLA ($\eta_{PP}/\eta_{PLA} > 1$), which is responsible for the fine dispersion morphology of PLA droplets in the PP-rich biocomposites [32]. Meanwhile, G′ is found to increase with increasing PP concentration across all frequencies (Figure 4.8b). Yoo *et al.* found that the incorporation of compatibilizers (PP-g-MAH and SEBS-g-MAH) improved the complex viscosity of PLA/PP biocomposites in contrast to those without compatibilizers [31]. The increased complex viscosity of the composites usually means the improved compatibility between the components. In another study, the zero shear viscosities (η_0) of PP/PLA/PEGMMA (10/90/5) composite increased by 130% as compared to PP/PLA (10/90) [27], which can be ascribed to the enhanced interfacial interaction, as well as the formation of high molar mass component (e.g., graft copolymers) between PLA and PEGMMA during melt compounding.

4.2.3.2   Electrical Conductivity

As is well known, PP/PLA-based biocomposites are electrically insulating. In order to improve their electrical conductivity, some conductive nanofillers are incorporated into PP/PLA-based biocomposites. Carbon nanotube is such a conductive nanofiller that possesses exceptional electrical conductivity as well as ultrahigh mechanical strength/modulus and outstanding thermal stability/conductivity [41, 42]. However, the electrical conductivity of PP/PLA-based bionanocomposites has rarely been investigated up to now. Lee and Jeong fabricated PP/PLA (50/50 by weight) bionanocomposites filled with a series of pristine multiwalled carbon nanotube (MWCNT)

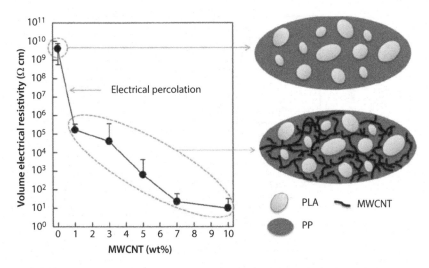

Figure 4.9 Volume electrical resistivity of the PP/PLA/MWCNT composites with different MWCNT contents. (Reproduced with permission from [33]; Copyright © 2014 Elsevier Ltd.)

loadings of 0.0–10.0 wt% [33]. As shown in Figure 4.9, the electrical resistivity of PP/PLA/MWCNT bionanocomposites decreased dramatically as the MWCNT content increased. The electrical resistivity of PP/PLA bionanocomposite with 1 wt% of MWCNT was sharply reduced to ~10^5 Ω·cm, implying that the electrical percolation threshold was obtained at a certain content below 1.0 wt% MWCNT. Such an electrical percolation threshold is even lower than that of PP-based composites, including that of MWCNT reported in earlier literature [43]. The lowest electrical resistivity of ~10^1 Ω·cm was observed for PP/PLA bionanocomposite with 10 wt% of MWCNT, which was attributed to the uniform distribution and the interconnected network formation of the MWCNTs. The interconnected network of the MWCNTs can provide percolated pathways for electron transfer, enabling the resultant nanocomposites conductivity.

4.2.3.3 Thermal Properties

The compounding of polymer composites usually induces an impact on their thermal properties, including the thermal stability and change in glass transition temperature (T_g). The thermal stability can be assessed by the temperatures at 10% (T_{-10}) and/or 50% mass loss (T_{-50}). The typical TGA profiles of PP, PLA and their biocomposites are displayed in Figure 4.10. The thermal stability of the PP/PLA biocomposites decreased with increasing PLA

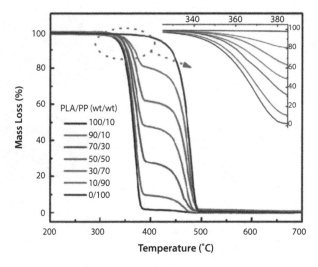

Figure 4.10 Thermogravimetric (TG) curves of the neat PP, PLA and PP/PLA composites. (Reproduced with permission from [26]; Copyright © 2015 Elsevier Ltd.)

concentration, due to the lower thermal stability of PLA [26]. Additionally, the thermal degradation behavior of PP/PLA biocomposites clearly exhibited two mass loss stages, indicating the immiscibility between the two phases. The first stage is ascribed to the thermal decomposition of PLA, and the second one is assigned to the thermal degradation of PP. The addition of compatibilizer, such as PP-g-MAH, can cause an increase in both T_{-10} and T_{-50} of PP/PLA biocomposites [7, 30], due to the improved compatibility between PP and PLA. The inclusion of nanofillers, including MWCNT [33] and INT-WS$_2$ [25], has been reported to obviously increase [25, 33] or slightly [25] change the T_{-10}, depending on the components ratio and the dispersion state of nanofillers. Regarding the glass transition temperature, in some studies the PP/PLA biocomposites usually exhibited two T_gs, as confirmed by DSC and DMA, indicating the formation of immiscible structure [29]. Improving the T_gs by adding nanofillers was demonstrated in PP/PLA/INT-WS$_2$ bionanocomposites in comparison to the values of the PP/PLA biocomposites [25]. This is probably attributed to the presence of nanofillers, which restricts the mobility of both polymer chain segments.

4.2.3.4 Mechanical Properties

As previously mentioned, one of the major objectives of fabricating PP/PLA composites is to obtain desired mechanical properties for engineering applications. The mechanical properties of the resultant PP/PLA

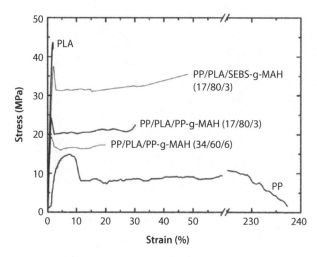

Figure 4.11 Typical stress-strain curves of neat PLA, PP and their composites. (Reproduced with permission from [36]; Copyright © 2011 Springer-Verlag)

composites are affected by several key factors, such as compositions of composites, compatibility between the components, morphology, and addition of nanofillers. Figure 4.11 displays the stress-strain curves of neat PLA, PP and their composites [36]. As a typical brittle material, the neat PLA exhibits high Young's modulus and tensile strength but low elongation at break. Conversely, the pure PP displays ductile material characteristics with low Young's modulus and tensile strength, and high elongation at break. As a result, the resultant PP/PLA composites showed the medium combined tensile properties between PP and PLA. The tensile modulus and strength increased with the increase of PLA content [20, 29]. However, sometimes the tensile strength values were lower than both PP and PLA, which might be caused by the poor interfacial tension of the immiscible polymer blends [44, 45]. The presence of compatibilizers might lead to improvement [24] or deterioration [27] of the tensile properties. The improvement in tensile properties induced by compatibilizers was due to the improved stress transfer efficiency at the interphase boundary of the composites, whereas the deterioration was caused by the plasticizing effect of compatibilizers. Consequently, the optimum tensile properties were observed at a suitable content of compatibilizers [31], due to the balance between the compatibilizing and plasticizing effect. The incorporation of nanoclay (Cloisite 15A) into the PP/PLA composites resulted in a notable improvement in tensile modulus, but the tensile strength and elongation at break values were reduced with the addition of

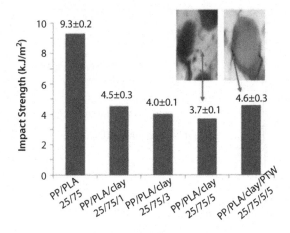

Figure 4.12 Correlation of impact strength with localization of the nanoclays. (Reproduced with permission from [34]; Copyright © 2016 Taylor & Francis)

the nanoclay [20]. This phenomenon can be explained by the fact that the agglomerates of incorporated clays served as stress concentration points, leading to poor mechanical performance.

Impact strength and modulus are other important parameters regarding the mechanical properties of the resultant PP/PLA composites. Similar to tensile property, impact strength and modulus are also closely related to component ratio, compatibility between the components, and addition of nanofillers. The incorporation of PP into PLA led to an increase in the impact strength of the resultant PP/PLA composites compared to neat PLA, due to the toughening effect of PP droplets [20]. The incorporation of compatibilizers, such as PE-g-GMA [28] and PTW [20], can lead to a gradual increase in impact strength with increasing the compatibilizer content, since better homogeneities and stronger interfacial adhesion of two components were achieved with the addition of compatibilizers. Furthermore, the localization of the clay also led to a significant influence on the impact strength of PP/PLA/clay nanocomposites (Figure 4.12). The impact strength of PP/PLA/clay nanocomposites decreased with the increase of clay amount, implying that the aggregates of nanoclays in the PLA phase (see TEM micrographs in Figure 4.12) prevented the motion of the PLA chains and restricted their deformation [34]. In contrast, simultaneous addition of compatibilizer (PTW) and nanoclays made the localization of the clay shift from the PLA to the interfaces. The interface-localized nanoclays could facilitate the deformation of the PP/PLA matrix when exposed to an imposed force, which was responsible for improved impact strength of PP/PLA/clay/PTW nanocomposite.

Furthermore, the influence of accelerated weathering on the mechanical properties of PP/PLA biocomposites has been investigated as well. The addition of modified cellulose [46] or wood [47] into biocomposites successfully avoided the deterioration of the surface of materials, and thus retained the higher tensile and impact strength than those without cellulose or wood.

As discussed above, the change in the mechanical properties of the PP/PLA composites is quite complex and is dependent on the compositions, the type and amount of compatibilizer used, the type and amount of nanofillers as well as the location of nanofillers within the polymer matrix. Regardless, the relationship between these experimental results and theory is essential to establish a further understanding of the relative contributions of each factor on the changes in mechanical properties.

4.2.3.5   Gas Barrier Properties

Another important performance related to PP/PLA biocomposites is gas barrier properties because of their potential applications in packaging films. PP is a host polymer with good barrier to water vapor but high oxygen permeability. Conversely, PLA is very resistant to oxygen and has poor water vapor barrier compared to PP. The gas barrier behavior is highly dependent on the polarity of the host polymer as well as the permeating molecules. Since the polarity of PLA is higher than PP, the dissolving of water vapor (as a polar molecule) into PLA is easier than PP, while the permeation of oxygen (as a nonpolar molecule) into PLA is more difficult than PP. Depending on the difference in morphology and compatibility, the blending of PP with PLA has been reported to cause increases [48], decreases [48] or no significant change [44] in the gas permeability of PP/PLA biocomposites. As shown in Figure 4.13, uniformly distributed and smaller size droplets can improve barrier properties of the PP/PLA biocomposites, which is attributed to the increase in tortuosity of molecular diffusion through the matrix [48]. Moreover, incorporation of layered nanoparticles, such as organoclay, can further improve the gas barrier property of PP/PLA biocomposites [48]. It is found that the oxygen permeability of PP/PLA/clay bionanocomposites is dependent on the clay loading and state of clay dispersion. There is a reduction in oxygen permeability values with increasing nanoclay loading up to 5 wt%, but exceeding 5 wt% nanoclay had only a slight effect on permeation due to the occurrence of nanoclay aggregation at high contents. Crystallization behavior is another key factor

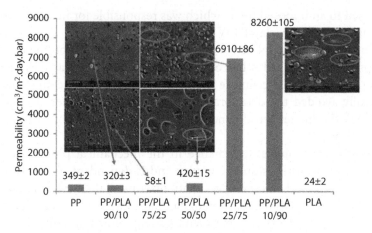

Figure 4.13 Permeability values and morphology of the PP/PLA composites. (Reproduced with permission from [48]; Copyright © 2015 Elsevier Ltd.)

that determines the gas barrier property, since the diffusion of molecules through polymers occurs more easily in amorphous phases compared to crystalline regions. The degree of crystallinity, as determined by DSC, for PP-rich composites is higher than that of PLA-rich ones, in good agreement with oxygen permeability values [48].

4.2.3.6   Fire Retardant Properties

Both PP and PLA are highly flammable synthetic polymeric materials, and thus their composites are still facing the problem of poor fire resistance in construction, electronics and electrical application fields. In order to broaden their applications, flame retardant modification of PP/PLA bio-composites has become an important issue. Over the past few decades, a large amount of literature has been published focusing on flame-retardant modification of either PP [49–51] or PLA [52–55] alone. So far, studies on the flame-retardant modification of PP/PLA biocomposites have been rarely reported. Lin and coworkers prepared PP/PLA biocomposites modified by two kinds of intumescent flame retardants (IFR) systems [56]. IFR1 consisted of ammonium polyphosphate microencapsulated by melamine-formaldehyde resin and dipentaerythritol (3:1 by weight), and IFR2 was comprised of melamine-pyrophosphate microencapsulated by ethoxyline resin and dipentaerythritol (3:1 by weight). Both IFR1 and IFR2 were effective flame-retardant systems for PP/PLA biocomposites (Figure 4.14). The LOI value of PP/PLA biocomposites was only 18% and exhibited no rating in UL-94 vertical burning tests. With the addition of

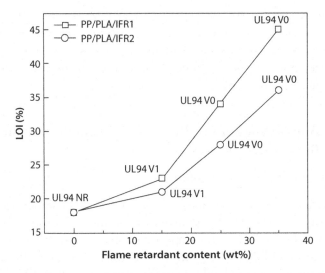

Figure 4.14 The LOI value and UL-94 rating of PP/PLA/IFR1 and PP/PLA/IFR2 biocomposites. (Reproduced with permission from [56]; Copyright © 2012 Taylor & Francis)

Table 4.2 Cone calorimeter results for the PP/PLA/IFR1 and PP/PLA/IFR2 biocomposites with 25 wt% flame retardants. (Reproduced with permission from [56]; Copyright © 2012 Taylor & Francis)

Sample	PHRR (kW/m²)	TPHRR (s)	TTI (s)	THR (MJ/m²)	Residue mass (%)
PP/PLA (45/11.25)	543	340	60	178	6.6
PP/PLA/IFR1 (45/11.25/25)	167	430	42	142	28.8
PP/PLA/IFR2 (45/11.25/25)	207	225	46	153	22.4

Note: PHRR = peak heat release rate; TPHRR = time to peak heat release rate; TTI = time to ignition; THR = total heat release.

25 wt% of IFRs, the LOI value was significantly increased to 34% (PP/PLA/IFR1) and 28% (PP/PLA/IFR2), respectively, and both samples passed the V-0 classification in the UL-94 test. The fire performance of PP/PLA/IFR biocomposites was also evaluated using a cone calorimeter, and the results are listed in Table 4.2. The addition of 25 wt% IFR1 and IFR2 led to a notable reduction in peak heat release rate by 69.2% and 61.9%,

respectively, along with a lower total heat release value. The improved flame resistance of PP/PLA/IFR biocomposites is probably ascribed to the higher char yield that not only insulated the exposure of polymers to heat and oxygen but also restricted the pyrolysis flammable gases entrance into the gas phase as fuel [56].

4.2.4 Applications of PP/PLA-Based Biocomposites and Bionanocomposites

4.2.4.1 Biomedical Applications

One emerging research direction for PP/PLA-based biocomposites and bion-anocomposites is focused on biomedical applications. Tanaka and coworkers used a new PP/PLA (10/90) composite mesh for groin hernia repair [57]. In comparison to PP meshes (commercial product: Prolene*), this new composite biomaterial showed reduced inflammation and cell-mediated immune responses, indicating a better tolerance, which was probably due to the low mesh shrinkage and decreased tissue adhesion. Recently, Zhang *et al.* synthesized PP/PLA composite mesh through treating PP mesh with oxygen plasma followed by grafting PLA onto treated PP mesh [58]. This PP/PLA composite mesh was used for abdominal wall defect repair, exhibiting a better anti-adhesion ability to peritoneal cavity (P < 0.05) and lower inflammation hazard (P > 0.05) compared to PP mesh. Based on these experimental findings, the PP/PLA composite biomaterial is a promising alternative to PP mesh for defect repair, which may overcome chronic pain and high recurrence rate.

4.2.4.2 Packaging Applications

Another notable exception to the PP/PLA-based composite can be found in the development of packaging materials. Ploypetchara *et al.* found that PP/PLA (50:50) composite film displayed a proper tensile strength and modulus as well as a low oxygen permeability [44], which is suitable for packaging application. In order to overcome the low resistance to hydrolysis and poor dyeability of pure PLA, PP/PLA composites were fabricated to create a new material with better resistance to hydrolysis and better dyeability [59]. Future studies of service lifetime and biodegradation ability of these composites are required when applied as packaging materials.

4.2.4.3 Thermal Insulating Applications

Cellular plastics display many advantages over polymer resin such as low density, excellent thermal insulating property, superior energy absorption, and so on. As one representative of cellular plastics, expanded polystyrene bead foams occupy the biggest market share owing to the low cost and high thermal insulation efficiency. However, the caused "white pollution" worldwide demands the development of easy recycling or even biodegradable cellular plastics to replace expanded polystyrene foams. In this context, PP/PLA bead foams were prepared using pilot-scale batch equipment with volume of 50 L [19]. The bead foams obtained showed high expansion ratio (up to 44.4), well-defined cell structure, and very high cell density. These new kinds of bead foams can be considered as substitutes for expanded polystyrene foams due to their excellent combined properties.

4.3 Conclusion

Polypropylene/polylactic acid (PP/PLA)-based biocomposites and bionanocomposites are one kind of the most technologically promising materials with desirable combined properties. Currently, most PP/PLA composites have been prepared through melt-blending technique due to its convenient and economical advantages. However, the melt blending of PP and PLA usually leads to the occurrence of miscibility. Miscibility is often discussed in the literature in regard to polymer blends, but rarely studied properly. Miscibility is usually deduced from the existence of double glass transition temperatures or the phase separation observed by SEM micrographs. Unfortunately, very few studies have dealt with phase diagrams and the mutual solubility of the components in each other. Actually, mutual solubility can be quantitatively determined by the Flory-Huggins interaction parameter. Furthermore, it is essential that miscibility-structure-property correlations be established for PP/PLA composites in the future.

Further property improvements in PP/PLA-based composites will be influenced by incorporating nanofillers. However, up to date, some nanofillers, such as clay [20] or sepiolite [36], have not led to effective property enhancement in PP/PLA-based composites due to the agglomeration of nanofillers. Therefore, how to solve the homogeneous dispersion state of nanofillers in the matrix is still a great challenge.

Despite these challenges, PP/PLA-based biocomposites and bionanocomposites are still one kind of promising composite material in both scientific and industrial communities. The rheological, mechanical, and thermal properties can be optimized through adjusting the composition of composites and controlling their microstructure and morphology. Electrical conductivity enhancement as well as the advantage of improving gas barrier properties can be achieved by carbon nanotubes or graphene. Furthermore, enhanced fire resistance could be obtained through adding flame retardants into PP/PLA composites for special applications. The multifunctional property improvements, associated with their potential for low cost and scalable production, may expedite the applications of these PP/PLA-based biocomposites and bionanocomposites as well as their commercialization.

References

1. Zeng, J.B., Li, K.A., and Du, A.K., Compatibilization strategies in poly(lactic acid)-based blends. *RSC Adv.* 5, 32546–32565, 2015.
2. Anderson, K.S., Schreck, K.M., and Hillmyer, M.A., Toughening polylactide. *Polym. Rev.* 48, 85–108, 2008.
3. Murariu, M., Da Silva Ferreira, A., Alexandre, M., and Dubois, P., Polylactide (PLA) designed with desired end-use properties: 1. PLA compositions with low molecular weight ester-like plasticizers and related performances. *Polym. Adv. Technol.* 19, 636–646, 2008.
4. Nyambo, C., Mohanty, A.K., and Misra, M., Polylactide-based renewable green composites from agricultural residues and their hybrids. *Biomacromolecules* 11, 1654–1660, 2010.
5. Mohd-Adnan, A.-F., Nishida, H., and Shirai, Y., Evaluation of kinetics parameters for poly(l-lactic acid) hydrolysis under high-pressure steam. *Polym. Degrad. Stab.* 93, 1053–1058, 2008.
6. Anderson, K.S., and Hillmyer, M.A., The influence of block copolymer microstructure on the toughness of compatibilized polylactide/polyethylene blends. *Polymer* 45, 8809–8823, 2004.
7. Choudhary, P., Mohanty, S., Nayak, S.K., and Unnikrishnan, L., Poly(L-lactide)/polypropylene blends: Evaluation of mechanical, thermal, and morphological characteristics. *J. Appl. Polym. Sci.* 121, 3223–3237, 2011.
8. Padee, S., Thumsorn, S., On, J.W., Surin, P., Apawet, C., Chaichalermwong, T., Kaabbuathong, N., O-Charoen, N., and Srisawat, N., Preparation of poly(lactic acid) and poly(trimethylene terephthalate) blend fibers for textile application. *Energy Procedia* 34, 534–541, 2013.
9. Ma, X., Yu, J., and Wang, N., Compatibility characterization of poly(lactic acid)/poly(propylene carbonate) blends. *J. Polym. Sci. Part B: Polym. Phys.* 44, 94–101, 2006.
10. Rodriguez-Llamazares, S., Rivas, B.L., Perez, M., and Perrin-Sarazin, F., Poly(ethylene glycol) as a compatibilizer and plasticizer of poly(lactic acid)/clay nanocomposites. *High Perform. Polym.* 24, 254–261, 2012.
11. Bouzouita, A., Samuel, C., Notta-Cuvier, D., Odent, J., Lauro, F., Dubois, P., and Raquez, J.M., Design of highly tough poly(l-lactide)-based ternary blends for automotive applications. *J. Appl. Polym. Sci.* 133, 43402, 2016. doi: 10.1002/app.43402.

12. Imre, B., and Pukanszky, B., Compatibilization in bio-based and biodegradable polymer blends. *Eur. Polym. J.* 49, 1215–1233, 2013.
13. Zhou, K.Q., Jiang, S.H., Shi, Y.Q., Liu, J.J., Wang, B., Hu, Y., and Gui, Z., Multigram-scale fabrication of organic modified MoS_2 nanosheets dispersed in polystyrene with improved thermal stability, fire resistance, and smoke suppression properties. *RSC Adv.* 4, 40170–40180, 2014.
14. Wishman, M., Hagler, G.E., Lewin, M., and Pearce, E.M. (Eds.), *Handbook of Fiber Chemistry*, vol. 4, p. 162, Marcel Dekker Inc.: New York, 1998.
15. Wojciechowska, E., Fabia, J., Slusarczyk, C., Gawlowski, A., Wysocki, M., and Graczyk, T., Processing and supermolecular structure of new iPP/PLA fibres. *Fibres Text. East. Eur.* 13, 126–128, 2005.
16. Tavanaie, M.A., and Mahmudi, A., Green engineered polypropylene biodegradable fibers through blending with recycled poly(lactic) acid plastic wastes. *Polym.-Plast. Technol.* 53, 1506–1517, 2014.
17. Arvidson, S.A., Roskov, K.E., Pate, J.J., Spontak, R.J., Khan, S.A., and Gorga, R.E., Modification of melt-spun isotactic polypropylene and poly(lactic acid) bicomponent filaments with a premade block copolymer. *Macromolecules* 45, 913–925, 2012.
18. Suh, K.W., Park, C.P., Maurer, M.J., Tusim, M.H., De Genova, R., Broos, R., and Sophiea, D.P., Lightweight cellular plastics. *Adv. Mater.* 12, 1779–1789, 2000.
19. Tang, L.Q., Zhai, W.T., and Zheng, W.G., Autoclave preparation of expanded polypropylene/poly(lactic acid) blend bead foams with a batch foaming process. *J. Cell. Plast.* 47, 429–446, 2011.
20. Ebadi-Dehaghani, H., Khonakdar, H.A., Barikani, M., and Jafari, S.H., Experimental and theoretical analyses of mechanical properties of PP/PLA/clay nanocomposites. *Compos. Part B: Eng.* 69, 133–144, 2015.
21. Koning, C., van Duin, M., Pagnoulle, C., and Jerome, R., Strategies for compatibilization of polymer blends. *Prog. Polym. Sci.* 23, 707–757, 1998.
22. Lee, H.J., Lee, H.K., Lim, E., and Song, Y.S., Synergistic effect of lignin/polypropylene as a compatibilizer in multiphase eco-composites. *Compos. Sci. Technol.* 118, 193–197, 2015.
23. Zhang, Y.C., Wu, H.Y., and Qiu, Y.P., Morphology and properties of hybrid composites based on polypropylene/polylactic acid blend and bamboo fiber. *Bioresour. Technol.* 101, 7944–7950, 2010.
24. Reddy, J.P., Misra, M., and Mohanty, A., Injection moulded biocomposites from oat hull and polypropylene/polylactide blend: Fabrication and performance evaluation. *Adv. Mech. Eng.* 2013, Article ID 761840, 2013. doi: https://doi.org/10.1155/2013/761840.
25. Naffakh, M., Diez-Pascual, A.M., and Marco, C., Polymer blend nanocomposites based on poly(L-lactic acid), polypropylene and WS_2 inorganic nanotubes. *RSC Adv.* 6, 40033–40044, 2016.
26. Kang, H.M., Lu, X., and Xu, Y.S., Properties of immiscible and ethylene-butyl acrylate-glycidyl methacrylate terpolymer compatibilized poly (lactic acid) and polypropylene blends. *Polym. Test.* 43, 173–181, 2015.

27. Xu, Y.W., Loi, J., Delgado, P., Topolkaraev, V., McEneany, R.J., Macosko, C.W., and Hillmyer, M.A., Reactive compatibilization of polylactide/polypropylene blends. *Ind. Eng. Chem. Res.* 54, 6108–6114, 2015.

28. Lee, H.S., and Kim, J.D., Effect of a hybrid compatibilizer on the mechanical properties and interfacial tension of a ternary blend with polypropylene, poly(lactic acid), and a toughening modifier. *Polym. Compos.* 33, 1154–1161, 2012.

29. Kim, H.S., and Kim, H.J., Miscibility and performance evaluation of natural-flour-filled PP/PBS and PP/PLA bio-composites. *Fiber Polym.* 14, 793–803, 2013.

30. Chen, R.Y., Zou, W., Zhang, H.C., Zhang, G.Z., Yang, Z.T., and Qu, J.P., Poly(lactic acid)/polypropylene and compatibilized poly(lactic acid)/polypropylene blends prepared by a vane extruder: Analysis of the mechanical properties, morphology and thermal behavior. *J. Polym. Eng.* 35, 753–764, 2015.

31. Yoo, T.W., Yoon, H.G., Choi, S.J., Kim, M.S., Kim, Y.H., and Kim, W.N., Effects of compatibilizers on the mechanical properties and interfacial tension of polypropylene and poly(lactic acid) blends. *Macromol. Res.* 18, 583–588, 2010.

32. Ebadi-Dehaghani, H., Khonakdar, H.A., Barikani, M., Jafari, S.H., Wagenknecht, U., and Heinrich, G., An Investigation on compatibilization threshold in the interface of polypropylene/polylactic acid blends using rheological studies. *J. Vinyl Addit. Techn.* 22, 19–28, 2016.

33. Lee, T.W., and Jeong, Y.G., Enhanced electrical conductivity, mechanical modulus, and thermal stability of immiscible polylactide/polypropylene blends by the selective localization of multi-walled carbon nanotubes. *Compos. Sci. Technol.* 103, 78–84, 2014.

34. Ebadi-Dehaghani, H., Khonakdar, H.A., Barikani, M., Jafari, S.H., Wagenknecht, U., and Heinrich, G., On localization of clay nanoparticles in polypropylene/poly(lactic acid) blend nanocomposites: Correlation with mechanical properties. *J. Macromol. Sci. B*, 55, 344–360, 2016.

35. Zhang, Y.C., Xie, J.F., Wu, H.Y., and Qiu, Y.P., Crystallization and mechanical properties of nano ZnO/PP/PLA composite filaments. *Mater. Sci. Forum* 620–622, 485–488, 2009.

36. Nunez, K., Rosales, C., Perera, R., Villarreal, N., and Pastor, J.M., Nanocomposites of PLA/PP blends based on sepiolite. *Polym. Bull.* 67, 1991–2016, 2011.

37. Alexandre, M., and Dubois, P., Polymer-layered silicate nanocomposites: Preparation, properties and uses of a new class of materials. *Mater. Sci. Eng. R* 28, 1–63, 2000.

38. Ebadi-Dehaghani, H., Barikani, M., Khonakdar, H., and Jafari, S., Microstructure and non-isothermal crystallization behavior of PP/PLA/clay hybrid nanocomposites. *J. Therm. Anal. Calorim.* 121, 1321–1332, 2015.

39. As'habi, L., Jafari, S.H., Khonakdar, H.A., Kretzschmar, B., Wagenknecht, U., ans Heinrich, G., Effect of clay type and polymer matrix on microstructure and tensile properties of PLA/LLDPE/clay nanocomposites. *J. Appl. Polym. Sci.* 130, 749–758, 2013.

40. Zhang, Q.H., Fang, F., Zhao, X., Li, Y.Z., Zhu, M.F., and Chen, D.J., Use of dynamic rheological behavior to estimate the dispersion of carbon nanotubes in carbon nanotube/polymer composites. *J. Phys. Chem. B* 112, 12606–12611, 2008.

41. Yoon, J.T., Lee, S.C., and Jeong, Y.G., Effects of grafted chain length on mechanical and electrical properties of nanocomposites containing polylactide-grafted carbon nanotubes. *Compos. Sci. Technol.* 70, 776–782, 2010.

42. Moniruzzaman, M., and Winey, K.I., Polymer nanocomposites containing carbon nanotubes. *Macromolecules* 39, 5194–5205, 2006.

43. Ezat, G.S., Kelly, A.L., Mitchell, S.C., Youseffi, M., and Coates, P.D., Effect of maleic anhydride grafted polypropylene compatibilizer on the morphology and properties of polypropylene/multiwalled carbon nanotube composite. *Polym. Compos.* 33, 1376–1386, 2012.

44. Yupapin, P.P., Pivsa-Art, S., Ohgaki, H., Ploypetchara, N., Suppakul, P., Atong, D., and Pechyen, C., Blend of polypropylene/poly(lactic acid) for medical packaging application: Physicochemical, thermal, mechanical, and barrier properties. *Energy Procedia* 56, 201–210, 2014.

45. Hamad, K., Kaseem, M., and Deri, F., Rheological and mechanical characterization of poly(lactic acid)/polypropylene polymer blends. *J. Polym. Res.* 18, 1799–1806, 2011.

46. Darie, R.N., Vlad, S., Anghel, N., Doroftei, F., Tamminen, T., and Spiridon, I., New PP/PLA/cellulose composites: Effect of cellulose functionalization on accelerated weathering behavior (accelerated weathering behavior of new PP/PLA/cellulose composites). *Polym. Adv. Technol.* 26, 941–952, 2015.

47. Darie, R.N., Bodirlau, R., Teaca, C.A., Macyszyn, J., Kozlowski, M., and Spiridon, I., Influence of accelerated weathering on the properties of polypropylene/polylactic acid/eucalyptus wood composites. *Int. J. Polym. Anal. Ch.* 18, 315–327, 2013.

48. Ebadi-Dehaghani, H., Barikani, M., Khonakdar, H.A., Jafari, S.H., Wagenknecht, U., and Heinrich, G., On O_2 gas permeability of PP/PLA/clay nanocomposites: A molecular dynamic simulation approach. *Polym. Test.* 45, 139–151, 2015.

49. Wang, Q., Undrell, J.P., Gao, Y., Cai, G., Buffet, J.-C., Wilkie, C.A., and O'Hare, D., Synthesis of flame-retardant polypropylene/LDH-borate nanocomposites. *Macromolecules* 46, 6145–6150, 2013.

50. Wang, X., Sporer, Y., Leuteritz, A., Kuehnert, I., Wagenknecht, U., Heinrich, G., and Wang, D.Y., Comparative study of the synergistic effect of binary and ternary LDH with intumescent flame retardant on the properties of polypropylene composites. *RSC Adv.* 5, 78979–78985, 2015.

51. Yao, D.H., Feng, C.M., Zhang, Y., Lang, D., Liu, S.W., Chi, Z.G., and Xu, J.R., Flame retardant mechanism of a novel intumescent flame retardant polypropylene. *Procedia Eng.* 52, 97–104, 2013.

52. Wang, X., Hu, Y., Song, L., Xuan, S., Xing, W., Bai, Z., and Lu, H., Flame retardancy and thermal degradation of intumescent flame retardant poly(lactic acid)/starch biocomposites. *Ind. Eng. Chem. Res.* 50, 713–720, 2011.

53. Wang, X., Xuan, S., Song, L., Yang, H., Lu, H., and Hu, Y., Synergistic effect of POSS on mechanical properties, flammability, and thermal degradation of intumescent flame retardant polylactide composites. *J. Macromol. Sci. Part B* 51, 255–268, 2012.

54. Ye, L., Ren, J., Cai, S.Y., Wang, Z.G., and Li, J.B., Poly(lactic acid) nanocomposites with improved flame retardancy and impact strength by combining of phosphinates and organoclay. *Chinese J. Polym. Sci.* 34, 785–796, 2016.

55. Tang, G., Wang, X., Xing, W., Zhang, P., Wang, B., Hong, N., Yang, W., Hu, Y., and Song, L., Thermal degradation and flame retardance of biobased polylactide composites based on aluminum hypophosphite. *Ind. Eng. Chem. Res.* 51, 12009–12016, 2012.

56. Lin, Z., Chen, C., Guan, Z., Xu, B., Li, X., and Huang, Z., Polypropylene/poly(lactic acid) semibiocomposites modified with two kinds of intumescent flame retardants. *Polym.-Plast. Technol.* 51, 991–997, 2012.

57. Tanaka, K., Mutter, D., Inoue, H., Lindner, V., Bouras, G., Forgione, A., Leroy, J., Aprahamian, M., and Marescaux, J., *In vivo* evaluation of a new composite mesh (10% polypropylene/90% poly-L-lactic acid) for hernia repair. *J. Mater. Sci.-Mater. M* 18, 991–999, 2007.

58. Zhang, Z.G., Zhang, T.Z., Li, J.S., Ji, Z.L., Zhou, H.M., Zhou, X.F., and Gu, N., Preparation of poly(L-lactic acid)-modified polypropylene mesh and its antiadhesion in experimental abdominal wall defect repair. *J. Biomed. Mater. Res. B* 102, 12–21, 2014.

59. Reddy, N., Nama, D., and Yang, Y.Q., Polylactic acid/polypropylene polyblend fibers for better resistance to degradation. *Polym. Degrad. Stab.* 93, 233–241, 2008.

5

Polypropylene (PP)-Based Hybrid Biocomposites and Bionanocomposites

Svetlana Butylina

Laboratory of Wood and Bionanocomposites, Divison of Material Science, Department of Engineering Sciences and Mathematics, Luleå University of Technology, Luleå, Sweden

Abstract

Hybrid biocomposites and particularly hybrid bionanocomposites are relatively recent additions to the composite family. Biocomposites are composite materials which include parts of biological origin. Hybridization is very useful because it makes it possible to tailor the composite properties according to the desired structure. This chapter deals with the preparation methods applied for PP-based hybrid biocomposites and bionanocomposites. The mechanical and thermal properties as well as the weathering properties and fire performance of various hybrid composites are summarized. The properties of hybrid biocomposites depend on the nature, size and loading of the filler and its interaction with the PP matrix. In most cases, commercial use of biocomposites has been limited to nonstructural or semistructural applications due to low stiffness, impact and thermal properties. Hybridization helps to overcome these problems. Nanofillers can improve the thermal stability, flame retardancy and durability, and therefore give rise to new applications and extend the existing applications of PP-based biocomposites.

Keywords: Hybrid PP-based biocomposites/bionanocomposites, processing, nanofillers, crystallization, durability, water absorption, limited oxygen index, applications

5.1 Introduction

Due to the limited amount of fossil fuel, renewable and low-cost materials, such as biocomposites, are gaining popularity. For the last few years,

---

Corresponding author: svetlana.butylina@ltu.se; sbutylina@gmail.com

Visakh. P. M. and Matheus Poletto. (eds.) Polypropylene-Based Biocomposites and Bionanocomposites, (113–144) 2018 © Scrivener Publishing LLC

thermoplastics as well as thermoset-based natural fiber composites have experienced tremendous growth. This has been driven by their environmental friendliness, the renewability of these fibers, good sound abatement capability, and improved fuel efficiency resulting from the reduced weight of the components [1].

According to the definition, biocomposites are the composite materials, which include parts of biological origin. Biocomposites are divided into those containing non-wood fibers and those containing wood fibers, all of which present cellulose and lignin. The non-wood fibers (also known as natural fibers) are relatively long fibers having a high cellulose content, which delivers a high tensile strength and degree of cellulose crystallinity. Natural fibers are divided into straw, bast, leaf, seed or fruit, and grass fibers. The fibers most widely used in the industry are flax, jute, hemp, kenaf, sisal and coir. Straw fibers, which are found in many parts of the world, are an example of a low-cost reinforcement for biocomposites. Wood fibers have this name because almost 60% of their mass is wood elements. Wood fibers are divided in two large classes: softwood fibers (long and flexible) and hardwood fibers (shorter and stiffer). Besides the aforementioned lignocellulosic fibers, some new sources of biomaterial have been applied to produce PP-based biocomposites. For example, Jang *et al.* [2] have studied PP/green algae (sulfuric acid treated), SGA and PP/SGA/graphite nanoplatelets (GNP) biocomposites. Safwan *et al.* [3] have used palm kernel shell powder to produce hybrid composites with nanosilica.

Thermoplastic composites reinforced with short natural fibers, such as sisal, jute, banana, coir, and pineapple leaf fiber, are gaining increasing attention due to emerging applications in the aerospace, automotive, construction, and textile fields [4]. Exterior nonstructural or semistructural building wood-polymer composite (WPC) products, such as decking, fencing, siding, window framing and roof tiles, are being introduced into the marketplace [5].

Low cost, easy availability, low density, light weight, a high strength-to-weight ratio, less wear and tear in processing machinery, and environmentally friendly characteristics of the natural fiber reinforced composites have been the primary benefits of their commercial application [4]. However, despite the advantages, the use of natural fiber reinforced composites has been restricted due to their high moisture absorption tendency, poor wettability, limited thermal stability during processing, and poor adhesion of the natural with the synthetic counterparts (poor compatibility between hydrophilic fibers and hydrophobic plastic). Natural fibers have some disadvantages because they have hydroxyl groups (OH) in the fiber that can attract water molecules, and thus, the fiber might swell. This results in voids at the interface of the composite, which will affect the mechanical properties and cause a loss of dimensional stability.

The inorganic fillers traditionally used in the plastics industry include glass fibers and mineral fillers such as calcium carbonate, talc and silica. Discontinuous (short) glass fibers were introduced as reinforcement for thermoplastics in the late 1940s/early 1950s, and since then there has been a significant growth in the use of glass fibers in a variety of commodity and engineering thermoplastics [6]. Mineral fillers are used to replace the much more expensive plastics [7]. Due to their unfavorable geometrical features, surface area or surface chemical composition, mineral fillers could only increase the modulus of the polymer moderately, while the strength (tensile, flexural) remained unchanged or even decreased. Depending on the type of filler, other polymer properties could be affected; for example, melt viscosity could be significantly increased through the incorporation of fibrous materials. On the other hand, mold shrinkage and thermal expansion would be reduced, a common effect of most inorganic fillers [8]. The classification of fillers is based on their specific function such as the ability to modify mechanical, electrical or thermal properties, flame retardancy, processing characteristics, solvent permeability, or simply formulation costs. The fillers are, however, multifunctional and may be characterized by a primary function and a plethora of additional functions.

Some exciting new application areas for composites containing nanoparticles such as nanoclays, nanosilicates, carbon nanotubes, ultrafine TiO_2, talc, and synthetic hydroxyapatite, are: structural materials with improved mechanical properties, barrier properties, electrical conductivity, and flame retardancy; high performance materials with improved UV absorption and scratch resistance; barrier packaging for reduced oxygen degradation; and bioactive materials for tissue engineering applications [8]. Nanoparticles are well known for their size advantage, which gives the matrix a high surface contact area. However, the use of nanoparticles in polymer composites is challenging for material engineers, as the homogeneous dispersion of inorganic nanoobjects into a polymer matrix is difficult to ensure. This is the natural behavior of nanoparticles, which have a tendency towards agglomeration [3].

Hybridization allows designers to tailor the composite properties according to the desired structure [4]. Hybrid fillers combining two or more different types of fillers for polymer composites are sometimes very useful because they possess different properties that cannot be obtained with a single type of reinforcement [9]. Several commercial wood plastic composite (WPC) deck boards contain a certain amount of calcium carbonate,

CC (e.g., GeoDeck™ by LDI Composites) and talc (e.g., Timber Tech by TimberTech Company) [7]. The use of mineral fillers in the product can help replace much more expensive plastic, increase the stiffness of the filled products and render the plastic more flame resistant. The development of thermoplastic composites using natural fibers in combination with a small amount of moisture resistant and corrosion resistant synthetic fibers is one technique to enhance the strength and stiffness as well as the moisture resistance of the resultant hybrid composites [1]. So far, very little work has been done on hybrid composites that involves nanosize fillers.

5.2 Polypropylene-Based Hybrid Biocomposites and Bionanocomposites

5.2.1 Preparation

Processing is a critical step in the engineering of biocomposites. In general, plastics processing begins by either mixing or compounding followed by shaping and finishing. Polypropylene composites are processed by many techniques such as extrusion, injection molding, compression molding and rotational molding. Of these, extruders are notably used for mass production in industrial setups [10]. Table 5.1 shows some typical stages and equipment used to produce PP-based hybrid biocomposites and bionanocomposites.

A preprocessing stage is often included in the processing of composites. Washing, drying and grinding are usual preprocessing steps used for natural fibers. For example, the preprocessing of plant fibers, such as bamboo (BF) and sisal (SF), include scouring of the fibers in a hot detergent solution (2%) at 70 °C for 1 hour to remove dirt and core material, followed by washing in distilled water, and drying in a vacuum oven at 70 °C for 3 hours. The dried fibers are cut into the desired length of 4–6 mm by using an electronic fiber cutting machine [4, 11]. The preprocessing stage also includes modification of the fibers and/or other fillers. Gwon *et al.* [12] have used a treatment with a coupling agent (triethoxyvinyl-silane, TEVS) for both wood fibers (Lignocel® C120) and talc, while Kim *et al.* [13] have used silane treatment only for a mineral filler, precipitated calcium carbonate (PCC). The silane pretreatment of the filler was conducted by an immersion method.

Blenders, extruders (single-screw and/or twin-screw), pulverizers, mills (open/two-roll) and mixers are types of equipment widely used for mixing or compounding. Pultrusion is often used for producing long fiber thermoplastic pellets, which are then used for producing relatively long fiber

Table 5.1 Preparation method for hybrid PP-based biocomposites and bionanocomposites.

Composite	Preprocess	Compounding	Shaping/ Forming	Refs.
PP/BF/GF PP/SF/GF BF = bamboo fiber SF = sisal fiber	Washing, drying and grinding of fibers	Counter-rotating twin-screw extruder (CTW 100, Haake, Germany)	Injection molding	[4, 11]
PP/JF/GF JF = jute fiber GF = glass fiber	Twisting of yarns	Pultrusion and re-compounding by using a twin-screw extruder (JSW TEX30HSS)	Injection molding	[14]
recycled PP/PE-BF/PCC BF = bamboo fiber PCC = precipitated $CaCO_3$		Intelli-Torque twin-screw extruder (C.W. Brabender Instruments, NJ)	Extrusion	[15]
PP/BF/GF BF = bamboo fiber		Torque rheometer (Haake Rheocord 90)	Compression molding	[16]
PP/KF/CF/MMT KF = kenaf fiber CF = coir fiber MMT = montmorillonite	Alkaline treatment	Brabender mixer (Model no. 630304-03, GmbH & Co. Ltd)	Compression molding	[17]
PP/WF/talc WF = Lignocel® C120	Triethoxy-vinyl silane (TEVS)	Twin-screw extruder (Bautek Co., Korea)	Injection molding	[12]
PP/WF/recycled mineral wool		Plas Mec TRL 100/FV/W turbomixer with Plas Mec RFV 200 cooler	Extrusion	[18]

reinforced polymer composites by using conventional injection molding and compression molding.

In a study published by Uawongsuwan *et al.* [14], the pultrusion process was used to produce long-fiber pellets of jute fiber/PP and glass fiber/PP. For the making of pellets from the long-fiber jute and PP, four jute yarns were twisted together and passed through the impregnation die. Then, the long-fiber jute/PP pellets were recompounded by using a twin-screw extruder (JSW TEX30HSS) at barrel temperature 200 °C and cut to the length of 4 mm to obtain the recompounded pellets. Long-fiber pellets of glass fiber/polypropylene with 60 wt% glass fiber were also produced by using the long-fiber pellet pultrusion machine. According to Uawongsuwan *et al.* [14], the extrusion compounding process influences the shortening of jute fibers in the recompounded pellets significantly through the shearing effect of the extruder screws.

As a rule, the shaping or molding of PP-based hybrid biocomposites and bionanocomposites takes place at the processing temperatures corresponding to the processing temperature of PP. The temperatures are varied in the range of 160–195 °C, depending on the equipment used for shaping and the composition of the hybrid material. The melt temperatures do not exceed 195 °C to prevent wood degradation.

In addition to the processing which includes a separated compounding stage, direct extrusion is also applied to manufacture hybrid composites. The direct extrusion process has been used to produce a hollow profile of hybrid PP-based composites reinforced with wood fibers and mineral fillers [19–21]. In these studies [19–21], the gravimetric feeding system, which includes the main feeder connected to side feeders for each individual component, was used. All components were fed into the extruder through the main feeder. A Weber CE7.2 conical twin-screw extruder (Hans Weber Maschinenfabrik GmbH, Kronach, Germany) with the L/D ratio of 17 was used.

Coextruded composites are a distinct group of hybrid composites. A number of papers have been published on this topic recently [13, 22–24]. In their study of coextruded WPCs, Kim *et al.* [13] used a pilot-scale coextrusion system consisting of a Leistritz Micro-27 co-rotating parallel twin-screw extruder (Leistritz Corp., Allendale, NJ) for the core and a Brabender 32 mm conical twin-screw extruder (C.W. Brabender Instruments Inc., South Hackensack, NJ) for the shell. A specially designed die with a cross-section area of 13 × 50 mm and a target shell thickness of 1.0 mm was used. A vacuum sizer was used to maintain the targeted size. The coextruded profiles passed through a 2 m water bath with water spraying by using a downstream puller. The manufacturing temperatures were controlled between 165 and 175 °C for the cores and from 150 to 170 °C for the shells with different

Table 5.2 Compositions of coextruded hybrid PP-based biocomposites and bionanocomposites.

Composite	Composition of core layer	Composition of shell layer	Ref.
rPP/HDPE/ WF/TPCC	$(rPP/HPDE)_{40\%}/WF_{50\%}$	$TPCC_{(6-18\%)}/WF_{15\%}/$ $HDPE_{(67-79\%)}$	[13]
rPP/HDPE/ WF/TPCC	$(rPP/HPDE)_{40\%}/WF_{50\%}$	$TPCC_{12\%}/WF_{(0-15\%)}/$ $HDPE_{(63-88\%)}$	[13]
WPC-CB	$WF_{63\%}/PP_{20\%}/MAPP_{5\%}/$ $talc_{8\%}$	$PC_{50\%}/PP_{43\%}/MAPP_{3\%}/$ $CB_{3\%}$	[22]
WPC-EG	$WF_{63\%}/PP_{20\%}/MAPP_{5\%}/$ $talc_{8\%}$	$PC_{50\%}/PP_{43\%}/MAPP_{3\%}/$ $EG_{3\%}$	[22]
WPC-G	$WF_{63\%}/PP_{20\%}/MAPP_{5\%}/$ $talc_{8\%}$	$PC_{50\%}/PP_{43\%}/MAPP_{3\%}/$ $G_{3\%}$	[22]
WPC-CNTs	$WF_{63\%}/PP_{20\%}/MAPP_{5\%}/$ $talc_{8\%}$	$PC_{50\%}/PP_{43\%}/MAPP_{3\%}/$ $CNTs_{3\%}$	[22]
WPC-CFs	$WF_{63\%}/PP_{20\%}/MAPP_{5\%}/$ $talc_{8\%}$	$PC_{50\%}/PP_{43\%}/MAPP_{3\%}/$ $CFs_{3\%}$	[22]

shell formulations. In the works published by Turku and Kärki [22, 23] and Butylina et al. [24], a coextrusion system including a Weber CE 7.2 conical twin-screw extruder and a fiberEX extruder was used to produce the core and shell layer, correspondingly. The processing temperatures in both extrusion processes were between 175 and 200 °C. Examples of coextruded hybrid PP-based biocomposites and bionanocomposites are shown in Table 5.2.

The use of nanoparticles in polymer composites is challenging for material engineers, as the homogeneous dispersion of inorganic nanoobjects into the polymer matrix is difficult to ensure. This is the natural behavior of nanoparticles, which have a tendency towards agglomeration [3]. In their work, Deka and Maji [25, 26] have studied wood polymer nanocomposites prepared using polymer blends (HDPE, LDPE, PP and PVC or HDPE, PP and PVC), wood flour and nanoclay (montmorillonite, MMT) or a combination of MMT and TiO_2 nanopowder. The MMT and TiO_2 were modified using N-cetyl-N,N,N-trimethyl ammonium bromide (CTAB). The polymer (HDPE/LDPE/PP/PVC) blend was prepared by mixing HDPE, LDPE and PP (1:1:1) dissolved in xylene and PVC dissolved in tetrahydrofuran (THF). Wood and clay (and TiO_2) were mixed with the polymer solution.

The composite sheets were obtained from dried and ground material by a compression molding press (Santec Exim Pvt. Ltd., New Delhi) at 150 °C under the pressure of 80 MPa.

5.2.2 Characterization

The following properties of PP-based hybrid biocomposites and bion-anocomposites and the characterization techniques used to study these properties are discussed in this section: mechanical properties, thermal properties (differential scanning calorimetry [DSC], thermogravimetric analysis [TGA], and heat deflection temperature [HDT]), weathering prop-erties (resistance to UV light and moisture) and fire performance (limiting oxygen index [LOI] and cone calorimeter testing).

5.2.2.1 Mechanical Properties of PP-Based Hybrid Biocomposites and Bionanocomposites

Mechanical testing is mainly carried out using ASTM international stan-dards, but in some cases European standards (EN) for composites made from cellulose-based materials and thermoplastics are applied. ASTM D638 and EN-ISO 572-2 standards describe the test procedure for the determina-tion of tensile properties, ASTM D790 and EN 310 are used to determine the flexural (bending) properties of composites, and ASTM D256 notched Izod impact test is used to determine the impact strength of composites having a "V" notch. The ISO 179-1/1fU method is used to determine the Charpy impact strength for unnotched samples tested in flatwise position. The method is based on the recommendation given by TS 15534-1 Wood-plastics composites (WPC)-Part 1: Test methods for characterization of WPC materials and products. Different models of universal testing machines and impact testers have been developed to determine the tensile (tensile strength, TS, and tensile modulus) and flexural (flexural strength, FS, and flexural modulus) properties, and the impact strength (IS) of composites.

The properties of the filler play an important role in the final properties of filler reinforced composites. Generally, the chemical properties of fillers are not varied during melt extrusion processes, but rather physical characteris-tics, such as filler size and size distribution, can be changed [12]. Filler size and size distribution are important for the final material properties of composites. The results of mechanical testing presented in this section are divided into three parts, which correspond to hybrid composites made with glass fibers, hybrid composites made with mineral fillers and hybrid nanocomposites.

Table 5.3 shows the mechanical properties of hybrid PP composites reinforced with glass fibers and natural fibers: sisal fibers (SF), bamboo

Table 5.3 Mechanical properties of hybrid PP composites reinforced with glass fibers and natural fibers: sisal fibers (SF), bamboo fibers (BF), hemp fibers (HF), jute fibers (JF) and wood fibers (WF).

Composite	TS, MPa	Tensile modulus, MPa	FS, MPa	Flexural modulus, MPa	IS, (J/m) or (kJ/m²)ᵃ	Refs.
PP	32.0 (0.6)	586 (1.8)	35.3 (0.9)	1362 (1.56)	32.5 (0.7)	[4, 11]
PP/SF$_{30\%}$	40.4 (0.9)	961 (1.6)	45.2 (0.6)	1885 (4.44)	57.8 (0.9)	[11]
PP/SF$_{25\%}$/GF$_{5\%}$	41.8 (0.6)	970 (1.7)	47.4 (0.9)	1901 (5.6)	59.3 (0.9)	[11]
PP/SF$_{20\%}$/GF$_{10\%}$	42.3 (0.6)	991 (1.6)	52.5 (1.0)	2060 (5.1)	60.2 (0.8)	[11]
PP/SF$_{15\%}$/GF$_{15\%}$	45.4 (0.7)	1095 (1.7)	53.9 (0.8)	2265 (3.4)	63.3 (0.9)	[11]
PP/SF$_{10\%}$/GF$_{20\%}$	41.8 (0.8)	1001 (2.3)	51.5 (0.7)	2045 (2.6)	62.2 (1.0)	[11]
PP/SF$_{5\%}$/GF$_{25\%}$	41.0 (0.7)	980 (3.6)	50.5 (0.8)	1972 (2.2)	59.9 (0.9)	[11]
PP/BF$_{30\%}$	44.0	1240	45.4	1920	53.6	[4]
PP/BF$_{25\%}$/GF$_{5\%}$	45.5	1311	44.2	2100	48.0	[4]
PP/BF$_{20\%}$/GF$_{10\%}$	47.2	1381	51.5	2275	52.2	[4]
PP/BF$_{15\%}$/GF$_{15\%}$	51.0	1426	51.2	2300	61.5	[4]
PP/HF$_{40\%}$			97.5	4500	41	[1]
PP/HF$_{25\%}$/GF$_{15\%}$			101	5400	~ 55	[1]
PP/JF$_{20\%}$	31.9 (1.2)	4320 (440)	54.83 (4.5)	3140 (290)	1.6 (0.1)ᵃ	[14]
PP/JF$_{10\%}$/GF$_{10\%}$	52.3 (5.7)	4660 (370)	95.6 (10.2)	4220 (220)	16.3 (3.8)ᵃ	[14]
PP/WF$_{22\%}$	13.8 (1.7)	4600 (600)			4.1 (0.6)ᵃ	[27]
PP/WF$_{22\%}$/GF$_{10\%}$	16.6 (1.0)	5900 (400)			3.8 (0.3)ᵃ	[27]

ᵃ Impact strength (IS) values are presented in kJ/m²

fibers (BF), hemp fibers (HF), jute fibers (JF) and wood fibers (WF). Uawongsuwan *et al.* [14], who have studied long jute fiber reinforced PP composites, conducted a tensile test (ASTM D638) and a three-point bending test (ASTM D790) by using an Instron universal machine (Instron Model 4206) at a constant speed of 1 mm/min. Notched Izod impact testing was performed on a Digital Impact Tester (Toyoseiki) with 5.5 J pendulum in accordance with ASTM D256. The V-notch shape with 2 mm depth was prepared. Samal *et al.* [4] and Nayak and Mohanty [11] tested flexural and tensile properties of hybrid PP reinforced bamboo/glass fiber and sisal/glass fiber composites using a universal testing machine (LR-100K, Lloyd Instrument Ltd. UK) according to ASTM D638 and ASTM D790 respectively. The Izod impact strength was determined as per ASTM D256 with V-notch depth of 2.54 mm using Impactometer 6545 (Ceast, Italy).

Najak and Mohanty [11] found that at a minimum glass fiber wt% of 5 with a total sisal:glass fiber ratio of 25:5, the tensile strength increased from 40.36 MPa in sisal fiber reinforced PP (PP/SF$_{30\%}$) composites to 41.75 MPa in sisal-glass fiber reinforced (PP/SF$_{25\%}$/GF$_{5\%}$) composites. A similar enhancement in tensile modulus from 960.02 to 975.01 MPa was observed. This behavior was mainly due to the replacement of weak and less stiff sisal fibers with stronger and stiffer glass fibers. The sisal-glass fiber reinforced polypropylene hybrid composites at 15 wt% each of sisal and glass fiber loadings showed optimum tensile strength, as there was an effective transformation of load from the sisal and glass fiber. The same trend has beenfound for bamboo-glass fiber reinforced polypropylene [4]. The bamboo-glass fiber reinforced polypropylene at 15 wt% each of bamboo and glass fiber loading showed optimum tensile strength, as at this composition there was an effective transformation of load from the bamboo fibers [4]. At 15 wt% ratio bamboo:glass fibers, the tensile strength of BGRP (bamboo-glass fiber reinforced polypropylene) increased to 16% in comparison to the pure BFRP (bamboo fiber reinforced polypropylene) at 30 wt% fiber content. At above 15 wt% of glass fiber content, there was a negative hybrid effect due to agglomeration and fiber-fiber interaction [4, 11].

The trend of increased flexural strength and flexural modulus in sisal glass-fiber reinforced polypropylene composites was consistent with the tensile properties in the same manner. The substitution of glass fibers with sisal fibers by an equivalent weight ratio of sisal:glass of 15:15 resulted in an optimal balance of performance, as observed in the case of tensile properties. An increase of approximately 19.26% in flexural strength and 20.18% in flexural modulus as compared to 30 wt% of sisal fiber reinforced polypropylene was obtained. Incorporation of 15 wt% of glass fiber led to a hybrid effect displaying an increase in flexural strength and modulus to 12.73 and 19.7%, respectively, as compared to BFRP with 30 wt% of bamboo fiber [4].

The impact strength at 30 wt% of sisal fiber loading was 57.79 J/m, which increased to 63.27J/m with the addition of 15 wt% glass fiber [11]. The impact strength at 30 wt% of bamboo fiber loading was 53.6 J/m, which increased to 61.5 J/m with the addition of glass fiber to 15 wt% [4]. Panthapulakkal and Sain [1] reported that the impact strength of 40 wt% hemp fiber reinforced PP composite was 41 J/m and it increased by 35% with the increase of glass fiber content from 0 to 15%. This may be attributed to the improved resistance offered by the glass fibers in the composites.

The mechanical properties of injection-molded short hemp fiber/glass fiber reinforced polypropylene hybrid composites were investigated by Panthapulakkal and Sain [1]. The authors observed that hybridization with glass fiber enhanced the performance properties. A value of 101 MPa for flexural strength and 5.5 GPa for flexural modulus was achieved for a hybrid composite containing 25 wt% of hemp and 15 wt% of glass. The notched Izod impact strength of the hybrid composites exhibited great enhancement (34%) compared to unhybridized samples.

According to Uawongsuwan et al. [14], the effectiveness of glass fiber hybridization was highest when combined with recompounded jute fiber/polypropylene (RP-JF/PP) pellet composites for tensile strength (64%), flexural strength (74%) and impact strength (948%), respectively, when comparing at 20 wt% total fiber content. The reason for these occurrences was the better distribution of jute fiber in the hybrid composites. This indicates that the characteristics of the higher strength glass fiber, such as fiber orientation and fiber length, are the major factors controlling the mechanical properties of hybrid composites.

Turku and Kärki [27] have also found that the tensile strength and modulus of hybrid biocomposites made with softwood fiber increased as compared to a biocomposite made without an addition of glass fibers. However, in this case, the addition of glass fiber was found to decrease the impact strength of the biocomposites. The difference may have arisen from the different testing methods used; as in all studies described above, V-notched samples were tested, while Turku and Kärki [27] tested unnotched samples.

Table 5.4 shows the mechanical properties of hybrid PP composites reinforced with wood fibers (WF) and different types of mineral fillers. The addition of talc to the PP-WF composite increased the tensile strength of composites by approximately 29% as compared to a composite made without talc [12]. Talc might simply increase the physical stiffness of the composites.

Table 5.4 Mechanical properties of hybrid PP composites reinforced with wood fibers (WF) and different types of mineral fillers.

Composite	TS, MPa	FS, MPa	IS, kJ/m²	Ref.
$PP_{50\%}/WF_{50\%}$	30.7 (0.7)			[12]
$PP_{47\%}/WF_{50\%}/Talc_{3\%}$	39.6 (0.6)			[12]
$PP_{47\%}/WF_{50\%}$ (3% MAPP)	50.7	76.2		[28]
$PP_{44\%}/WF_{50\%}/Talc_{3\%}$	53.1	76.1		[28]
$PP_{44\%}/WF_{50\%}/Kaolin_{3\%}$	56.5	81.9		[28]
$PP_{44\%}/WF_{50\%}/$ Zinc-borate$_{3\%}$	50.2	70.2		[28]
$PP_{30\%}/WF_{64\%}/MAPP_{3\%}$	17.2 (1.1)	23.4 (1.2)	2.1 (0.3)	[18]
$PP_{30\%}/WF_{44\%}/Mineral$ wool$_{20\%}/MAPP_{3\%}$	13.5 (0.8)	18.7 (3.5)	2.3 (0.3)	[18]
$PP_{30\%}/WF_{44\%}/Talc_{20\%}/$ $MAPP_{3\%}$	16.0 (1.2)	22.9 (1.1)	2.7 (0.3)	[19]
$PP_{30\%}/WF_{44\%}/Calcium$ carbonate$_{20\%}/MAPP_{3\%}$	16.8 (1.6)	23.0 (1.6)	2.9 (0.2)	[19]
$PP_{30\%}/WF_{44\%}/$ Wollastonite$_{20\%}/$ $MAPP_{3\%}$	15.2 (1.5)	22.6 (1.2)	2.9 (0.3)	[19]
$(R\text{-}PP/PE)_{60\%}/Bamboo$ fiber$_{40\%}$		28.1 (0.5)	3.7 (0.2)	[15]
$(R\text{-}PP/PE)_{48\%}/Bamboo$ fiber$_{32\%}/PCC_{18\%}$		29.2 (1.0)	2.4 (0.1)	[15]

However, the addition of talc increased the tensile strength of composites with no alkali treated wood fibers, whereas it reduced slightly the tensile strength of the talc-containing composites made with alkali treated wood fibers.

In their study, Gwon *et al.* [28] compared the tensile and flexural strengths of WPCs filled with different types of inorganic fillers, and the kaolin case kept the highest values of strengths for all filler loadings. Generally, platy fillers have high aspect ratios, and this property increases the wettability of the fillers by the matrix, so fewer microvoids are created between the matrix and the filler. Plate-shaped kaolin and talc have higher aspect ratios than the cube-type zinc borate, and thus they lead to high mechanical strength. The void plays a crucial role in the physical properties of composites because the existence of a void in a composite represents a decrease of interfacial adhesion between the matrix and the fillers. A high void content can reduce the mechanical strength significantly. A cube-shaped structure leads to higher void fractions than other inorganic fillers in composites during the melt extrusion process, and it reduces the interfacial adhesion of the fillers with the polymer matrix. Therefore, the zinc-borate-filled composite showed the worst tensile strength.

Väntsi and Kärki [18] and Huuhilo *et al.* [19] have studied the extruded PP-based hybrid biocomposites containing 20% of mineral filler: mineral wool fibers, talc, calcium carbonate and wollastonite. In these studies the mechanical testing of composites was done according to EN-ISO 572-1, EN 310 and ISO 179-1/1fU standards. The addition of minerals did not lead to improvements of either tensile or flexural strength, but impact strength increased.

Kim *et al.* [15], who studied hybrid bamboo and precipitated calcium carbonate (PCC) fillers in a recycled polypropylene/polyethylene matrix (R-PP/PE), found that the addition of 18 wt% PCC to bamboo fiber-filled composites increased the tensile strength of the composite slightly, but at the same time decreased the impact strength. The impact strength decreased when the PCC content increased as a result of early fractures from a lack of a toughening effect due to the size of the PCC.

Table 5.5 shows the mechanical properties of PP-based hybrid bionano-composites. Clays, also known as layered silicates, are the most frequently used nanosized fillers in the composite design. The results of the studies on PP-based hybrid kenaf (and/or coir) fiber/montmorillonite, MMT [17] and PP-based hybrid wood fiber/MMT [29] composites indicate that the addition of MMT in concentration 1-2 phr (part per hundred of PP resin) enhanced the interfacial interaction and adhesion between the fiber and the polymer matrix and thus improved the mechanical properties of the composites. On the other hand, replacing 10% of wood fiber by organoclay, MMT, decreased the tensile strength by about 12% [27]. The loss in the strength might be explained by the difficulties in achieving a good dispersion of the nanometric particles. In general, nanoparticles having a higher aspect ratio than micro-sized ones can provide better interfacial adhesion. However, as is well known, nanoparticles, having a large aspect ratio and free surface energy, prefer to be agglomerated. Agglomerated particles can reduce the interfacial interaction, thereby deteriorating the interfacial

Table 5.5 Mechanical properties of hybrid PP-based bionanocomposites.

Composite	TS, MPa	Tensile modulus (MPa)	IS kg cm/cm^2 or kJ/m^{2a}	Ref.
PP$_{57\%}$/WF$_{40\%}$/MA$_{3\%}$	33.2 (1.5)	2296 (305)	8.9 (1.3)	[29]
PP$_{57\%}$/WF$_{40\%}$/MA$_{3\%}$/ MMT$_{1\,phr}$	33.7 (1.1)	2408 (306)	9.0 (1.3)	[29]
PP$_{70\%}$/Kenaf$_{15\%}$/ Coir$_{15\%}$	9.8	345		[17]
PP$_{70\%}$/Kenaf$_{15\%}$/ Coir$_{15\%}$/MMT$_{2\,phr}$	10.7	368		[17]
PP$_{22\%}$/WF$_{73\%}$/ MAPP$_{3\%}$	13.8 (1.7)	4600 (600)	4.1 (0.6)^a	[27]
PP$_{22\%}$/WF$_{63\%}$/ MAPP$_{3\%}$/MMT$_{10\%}$	12.2 (0.5)	6100 (300)	2.8 (0.3)^a	[27]

^{a} Impact strength (IS) values are presented in kJ/m^2

stress transfer and inducing a crack between the matrix and the fibers. As opposed to tensile strength, the tensile modulus of a PP-based hybrid wood fiber composite increased with the incorporation of clay. The tensile modulus was less dependent on the interfacial interaction. The improvement of the tensile modulus might be due to increased crystallinity caused by the nanoclay particles.

Microcrystalline cellulose (MCC) is a new class of cellulosic reinforcing agent. Compared to glass fiber, silica and carbon black, MCC in composites has many advantages: low cost, low density, little abrasion to equipment, renewability, and biodegradability [30]. Ashori and Nourbakhsh [31] found that the flexural strength of a PP$_{60\%}$/WF$_{40\%}$ composite (38.9 MPa) increased after the addition of microcrystalline cellulose (MCC) to 48.3 MPa and 50.2 MPa for PP$_{60\%}$/WF$_{36\%}$/MCC$_{4\%}$ and PP$_{60\%}$/WF$_{32\%}$/MCC$_{8\%}$ composites, respectively. Also, the microcrystalline cellulose had a significant effect on the impact strength. The notched specimens made with 4 and 8 wt% of MCC exhibited remarkably higher Izod impact strength than the notched specimen made without MCC (20 and 26 J/m versus 16 J/m). The results showed that microcrystalline cellulose along with wood flour could be used effectively as a reinforcing agent in a thermoplastic matrix.

5.2.2.2   Thermal Properties of PP-Based Hybrid Biocomposites and Bionanocomposites

Two types of thermal analysis are often used to characterize composites: thermogravimetric analysis and differential scanning calorimetry. The rate of decomposition and the thermal stability of composites can be investigated under a nitrogen atmosphere using thermogravimetric analysis (TGA). For example, Nayak and Mohanty [11] analyzed the thermogravimetric properties by using a Pyris 7 thermogravimetric analyzer (PerkinElmer, USA); the samples with mass around 10 mg were heated from 50 to 600 °C at the heating rate of 20 °C/min. The thermal degradation temperature was taken as the minimum of the first derivative of the weight loss with respect to time.

Table 5.6 shows the results of thermogravimetric analysis for three types of PP-based hybrid composites: sisal fiber (SF)/glass fiber (GF), hemp fiber (HF)/ glass fiber and wood fiber (WF)/mineral filler. The values in Table 5.6 show that incorporation of glass fiber into the natural fiber composites shifts the temperature of degradation to a higher value, indicating an increased thermal stability of the hybrid composites. The residue remaining at 500 °C (600 °C) is increased with the glass fiber content because the residue mainly comes from the lignin component of natural fibers in the case

Table 5.6 Thermogravimetric analysis of hybrid PP-based biocomposites.

Composite	Peak T		Residue after 500 °C (%) or 600 °C (%)	Ref.
	1st	2nd		
PP	no peak	448	3.16	[11]
PP/HF$_{40\%}$	354	464	6.6	[1]
PP/ HF$_{25\%}$/ GF$_{15\%}$	433	474	18.5	[1]
PP/SF$_{30\%}$	365	346	4.25	[11]
PP/SF$_{15\%}$/GF$_{15\%}$	368	352	18.57	[11]
PP$_{22\%}$/WF$_{75\%}$/MAPP$_{3\%}$	365	466	14.57	[20]
PP$_{30\%}$/WF$_{44\%}$/talc$_{20\%}$/MAPP$_{3\%}$	365	472	29.32	[20]
PP$_{30\%}$/WF$_{44\%}$/CC$_{20\%}$/MAPP$_{3\%}$	365	471	27.89	[20]
PP$_{30\%}$/WF$_{44\%}$/wollastonite$_{20\%}$/ MAPP$_{3\%}$	366	472	29.07	[20]

of NF/PP biocomposite, whereas glass fiber is also responsible in the case of hybrid composites. In the data shown in Table 5.6, the values presented by Martikka et al. [20] do not make it possible to evaluate the increase in the temperature of degradation of PP/wood fiber composites after the addition of the mineral filler because the total content of the filler for the reference and the hybrid composite was different. As in the case of glass fiber, the addition of the mineral filler increased the residue remaining at 600 °C.

The melting and crystallization behavior of composites can be studied using differential scanning calorimetry (DSC). Usually, analysis is performed under a nitrogen atmosphere. Samples of 5–10 mg are heated from 40 to 200 °C at the heating rate of 10 °C/min. Nayak and Mohanty [11] used a Diamond DSC-7 analyzer (PerkinElmer, USA). Lee and Kim [29] used a different DSC analyzer model (DSC 2910, TA Instruments) and a different heating scheme: from −65 to 200 °C at the heating rate of 5 °C/min. In many cases a procedure consisting of two heating runs is applied, and the first heating run is used to remove the thermal history of the polymer. For example, Samal et al. [4] have used two heating runs: a first heating run from 40 to 200 °C at the heating rate of 10 °C/min, followed by an isothermal step at 200 °C for 1 min and subsequent cooling to 40 °C at the heating rate of 10 °C/min and final second heating step up to 200 °C.

Farhadinejad et al. [32], who studied the thermal properties of PP/wood fiber/nanowollastonite composites on a DCS 200 F3 Maia thermal analyzer (Netzsch-Gerätebau, Germany), applied a heating scheme consisting of two runs conducted with different rates. In the first run the samples were heated from 25 to 250 °C with the heating rate of 50 °C/min to eliminate the thermal history, then kept at 250 °C for 5 min; after that the samples were cooled down to 25 °C with the rate of 5 °C/min, and heated up to 250 °C again after 30 min with the rate of 10 °C/min.

The melting temperature (T_m) and enthalphy of fusion (ΔH_f), also known as melting enthalpy (ΔH_m), are measured from the DSC heating curve, whereas the crystallization temperature (T_c) and enthalpy of crystallization (ΔH_c) are taken from the DSC cooling curve.

The degree of crystallinity (X_c) is determined from the heat of fusion normalized to that of 100% crystalline PP according to the following equation:

$$X_c = \Delta H_m / (\Delta H_m^{(crys)} \times W) \times 100 \qquad (5.1)$$

where ΔH_m and $\Delta H_m^{(crys)}$ are the melting enthalphy of the composite and 100% crystalline PP, and W is the weight fraction of PP in the composites. Different values have been reported by different authors for the melting enthalpy of 100% crystalline PP, e.g., 138 J/g [29], 169.29 J/g [32], 209 J/g [3, 20], and 240.5 J/g [11].

The DSC results for different hybrid PP-based biocomposites and bionanocomposites are shown in Table 5.7. Due to differences in the heating procedures and the characteristics of the PP used, a direct comparison

Table 5.7 Melting and crystallization properties of hybrid PP-based biocomposites and bionanocomposites.

Composite	$T_m[°C]$	$T_c[°C]$	$\Delta H_m[J/g]$	$X_c[\%]$	Refs.
PP (isotactic, Grade M110)	162	116.5	33.8–71.5	23.2–29.8	[4, 11]
PP/BF$_{30\%}$	163.4	118.9	35.52	21.34	[4]
PP/BF$_{15\%}$/GF$_{15\%}$	164	117.0	24.75	17.00	[4]
PP/BF$_{30\%}$/MAPP$_{2\%}$	162.5	121.0	23.03	15.84	[4]
PP/BF$_{15\%}$/GF$_{15\%}$/ MAPP$_{2\%}$	162.6	120	17.56	12.12	[4]
PP/SF$_{30\%}$	162.4	117.0	76.20	31.70	[11]
PP/SF$_{15\%}$/GF$_{15\%}$	163	118.9	79.60	33.00	[11]
PP/SF$_{30\%}$/MAPP$_{2\%}$	163.5	120	68.40	28.44	[11]
PP/SF$_{15\%}$/GF$_{15\%}$/ MAPP$_{2\%}$	164	121.0	70.0	29.00	[11]
PP 5014 (Mw = 180 kg/mol)	163.1	108.7	81.8	58.7	[12]
PP$_{50\%}$/WF$_{50\%}$	164.6	111.1	35.1	50.8	[12]
PP$_{47\%}$/WF$_{50\%}$/talc$_{3\%}$	166.6	118.5	32.8	53.7	[12]
PP$_{30\%}$/WF$_{44\%}$/talc$_{20\%}$/ MAPP$_{3\%}$	163.8	125.4	28.8	51.7	[20]
PP$_{30\%}$/WF$_{44\%}$/CC$_{20\%}$/ MAPP$_{3\%}$	163.1	122.1	27.2	50.4	[20]
PP$_{30\%}$/WF$_{44\%}$/wol- lastonite$_{20\%}$/ MAPP$_{3\%}$	163.4	122.6	27.9	48.9	[20]

(Continued)

Table 5.7 Cont.

Composite	T_m[°C]	T_c[°C]	ΔH_m[J/g]	X_c[%]	Refs.
PP (injection grade homopolymer)			93.95 (ΔH_c)	55.49	[32]
PP/WF$_{25\%}$			56.57 (ΔH_c)	55.69	[32]
PP/WF$_{38\%}$/ Nanowollastonite$_{2\%}$			61.81 (ΔH_c)	60.85	[32]
PP (J-160, Mw = 219,3 kg/mol)	168.9		95.7	69.23	[29]
PP$_{60\%}$/WF$_{40\%}$/MA$_{3\%}$	165.1		57.7	69.7	[29]
PP$_{57\%}$/WF$_{40\%}$/MA$_{3\%}$/ MMT$_1$	165.3		65.5	83.3	[29]
PP (homopolymer, grade PX617)	167.4		77.8	37.22	[3]
PP/PKS (10 wt%) PKS = palm kernel shell	167.4		65.41	31.13	[3]
PP/PKS/ nano silica (10 wt%)	165.7		83.91	40.15	[3]

between data published by different authors is impossible. For the PP/bamboo fiber and PP/sisal fiber composites reinforced with glass fibers, it has been found that the incorporation of fibers and MAPP interrupts the linear crystallizable sequence of the PP and lowers the degree of crystallization [4, 11]. It can also be seen in Table 5.7 that the T_c of PP (116.5 °C) is shifted to a high temperature by adding bamboo (BF) and glass fiber (GF) due to the nucleation effect of fibers. The T_c of 2% MAPP-treated BF/PP and BF/GF/PP composites is higher than that of untreated composites, indicating further enhancement in the nucleation due to the presence of the coupling agent.

According to Gwon *et al.* [12], the addition of talc particles with the size of 10–20 micrometers reduced the crystallininity of PP/wood fiber composite. Martikka *et al.* [20] found that microsized mineral fillers added in a concentration of 20% did not have any noticeable effect on the degree of crystallinity of PP/wood fiber composites. On the other hand, nanowollastonite [32], organoclay [29] and nanosilica [3] act as efficient nucleating agents for the crystallization of the PP matrix. Nanofillers accelerate the crystallization of the PP matrix remarkably. During the absorption of PP chains on the nanoparticle surface, the configurational entropy of the entire chain decreases, forming a nucleus of a certain volume within the adsorbed chains and thus enhancing the crystallization of the composites [3].

Heat deflection temperature (HDT) testing, which is widely used in automotive applications, can be seen as a type of creep where temperature is continually increased. HDT is the temperature at which the material deflects by 0.25 mm at an applied force, when the specimen is placed in a three-point bending mode. Panthapulakkal and Sain [1] measured the HDT for PP, hemp fiber and hybrid PP/hemp/glass composites according to ASTM 648. Birat et al. [33] used the same test procedure to measure HDT hybrid PP-based sisal fiber/glass fiber composites.

An increase of the heat deflection temperature upon the addition of glass fibers to a PP/hemp fiber or a PP/sisal fiber composite was reported by Panthapulakkal and Sain [1] and Birat et al. [33], respectively. According to Panthapulakkal and Sain [1], the HDT value of neat PP was 53 °C and it was increased twice by the incorporation of hemp fibers. Further improvement was observed as a result of hybridization and the maximum value of HDT was exhibited by the hybrid composite with 15 wt% of glass fibers (132 °C). In the case of sisal fibers, the HDT of a homopolymer PP increased from 55 to 102 °C, and hybridization with glass fibers increased the HDT further by 10% for $PP/SF_{20\%}/GF_{10\%}$ and 19.3% for $PP/SF_{10\%}/GF_{20\%}$. The HDT of composites is increased with stiffness. An increased stiffness indicates a reduction in the free volume present in the system, which enhances the dimensional stability of the composites and hence the HDT values.

5.2.2.3   Weathering Properties or Durability of Hybrid PP-Based Biocomposites and Bionanocomposites

In general, the durability of wood-polymer composites exposed to outdoor conditions is determined by their solar radiation and moisture resistance, thermal stability and fungal resistance [34]. Solar radiation is the driving force behind chemical changes of material, leading to its deterioration. Both PP and natural fibers are susceptible to solar radiation. Moisture in combination with radiation is often a key contributor to the weathering of material. The moisture absorbed by WPCs may influence their mechanical properties, dimensional stability and freeze-thaw resistance and intensify their microbiological activity (including decay) [34, 35]. Weathering properties hinder the outdoor applicability of biocomposites.

The weathering of biocomposites can be studied using either natural or accelerated testing procedures. Standards have been developed to study the weathering of biocomposites. A natural weathering test for biocomposites is conducted according to EN ISO 877:1996. The composites are placed on adjacent racks with a tilt angle of 45° facing the equator. Artificial or accelerated weathering is performed according to the ISO 4892-2:2006 standard. Different types of weathering test chambers simulating different natural conditions exist, e.g., the Q-SUN Xe-3 tester (Q-Lab Europe/UK). In the Q-SUN Xe-3 tester, the composites are subjected to a weathering procedure consisting of 102 min of UV irradiation (with an average irradiance of 0.51 W/m^2 at 340 nm) at a temperature of 38 °C and 50% relative humidity, followed by 18 min of water spraying, according to the ISO 4892-2:2006 standard. The irradiance of 32 W/m^2 (in the broad band 290–390 nm) is measured by the Q-SUN Xe-3 tester with the UV light meter UV-340 A.

Ultraviolet radiation (UV) is one of the two major factors affecting the durability of biocomposites. Both lignocellulosic fibers (wood and natural fibers) and PP can undergo degradation while exposed to UV radiation. The photodegradation mechanisms of wood and polymer are complex and well documented [36, 37]. Current approaches to improve the weathering resistance of biocomposites focus on the bulk, i.e., incorporation of additives into the entire product or surface treatment of the wood fiber [38]. The addition of photostabilizers and pigments into the entire WPC provides protection against discoloration caused by ultraviolet radiation. However, weathering primarily occurs at the surface of the material. Thus, a cost-effective means to deal with weathering would involve adding photostabilizer protection only in the surface layer of composite samples.

Coextrusion is one of the methods for providing a protective surface. Coextrusion can produce a multilayered product with different properties at the outer and inner layers, thus offering different properties between the surface and the bulk [39]. In their comparative research on non-coextruded and coextruded WPCs with a pure high density polyethylene (HDPE) or a pure polypropylene (PP) shell, Stark and Matuana [39, 40] showed that the presence of the shell layer significantly reduces the moisture uptake and the addition of nanosized titanium dioxide in the shell layer showed noticeably enhanced color stability.

A series of studies on various types of weathering of hybrid PP/wood fiber/ mineral composites were conducted in the Fiber Composite Laboratory of Lappeenranta University of Technology. Different types of mineral fillers and inorganic pigments were studied as means of improving the weathering

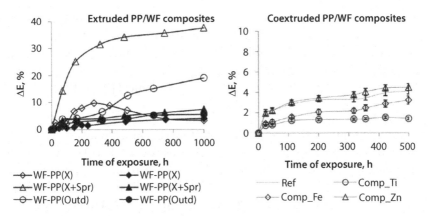

Figure 5.1 Color change of extruded [42] and coextruded PP/wood fiber composites [24] containing inorganic pigments.

behavior of extruded and coextruded PP-based biocomposites [24, 41, 42]. Figure 5.1 combines the results of two studies: in the first one the effect of carbon black on the weathering of an extruded PP-based biocomposite was investigated, while the second study was devoted to the effect of incorporation of three pigments, such as iron oxide, titanium dioxide and zinc oxide, into the shell (surface) layer of coextruded PP-based biocomposites. Carbon black was found to be more efficient in the stabilization of color than other inorganic pigments, e.g., iron oxide [41]. The addition of 2.5% carbon black (CB) to the extruded PP/wood fiber (WF) composite resulted in improved color stability, and dimensional stability, a reduction of surface cracking, and an ability to retain the Charpy impact strength in almost all weathering tests applied. Of all the weathering conditions, only exposure to a high level of irradiation (SUNTEST apparatus) was found capable of decreasing the impact strength of the carbon black-containing composites [42]. Due to their better performance in all weathering tests, the extruded PP/WF and hybrid PP/WF/wollastonite composites stabilized with carbon black were chosen for long-term outdoor exposure.

In the case of coextruded composites, the composite containing white pigment TiO_2 was better in the stabilization of color than the other pigments [24]. Scanning electron microscopy revealed that the density of surface cracking was dramatically reduced in the composites containing inorganic pigments. Serious surface deterioration after weathering was detected only in the case of the reference composite. The tensile test measurements indicated that weathering affected the tensile properties of all the composites adversely. However, the composite containing zinc oxide was capable of retaining its tensile properties. The retention of the tensile properties correlated well with the lowest carbonyl index obtained for the zinc oxide-containing composite. Photodegradation of the surface layer and moisture absorption by the composites in the process of weathering were suggested to be the reasons for the decrease of their mechanical properties.

5.2.2.3.1   Water Absorption

Water absorption studies of hybrid PP-based biocomposites were performed according to the ASTM D570 method or EN 317 method by immersing the samples in distilled water at room temperature. The measurements of weight were performed periodically. The total time of immersion of 28 days was predetermined in EN 317 standard, while the periodicity of intermediate measurements was not determined by this standard. In many cases periods of 24 hours, 7, 4 and 28 days are applied [18–19, 42]. The composites tested according to EN 317 were immersed in water without pre-drying, while the composites tested according to ASTM D570 were immersed in water after drying. The results of studies of water absorption of composites tested as prescribed in ASTM D570 are presented in different ways, e.g., water absorption after 24 hours, 1 week, 10 weeks or at the saturation level (prolongation depends on the type of composite). Nayak and Mohanty [11] reported 1.51% and 0.88% water absorption after 24 hours of immersion in water for $PP/SF_{30\%}$ and $PP/SF_{15\%}/GF_{15\%}$ composites, respectively. Panthapulakkal and Sain [1] have published results on water absorption at the saturation level of PP-based hybrid HF/GF composites; the water absorption of $PP/HF_{40\%}$, $PP/HF_{35\%}/GF_{5\%}$, $PP/HF_{30\%}/GF_{10\%}$ and $PP/HF_{25\%}/GF_{15\%}$ were determined as 8.73, 8.14, 6.78 and 5.49%, respectively. Both these studies showed that incorporation of glass fiber into the PP-based biocomposites (SF or HF) decreased their water uptake significantly. Whereas in the case of glass fiber the trend is clear, the effect of the addition of a mineral filler (micro- and nanosized) in the composite cannot be predicted in advance. Table 5.8 shows the results of water absorption testing for hybrid PP-based composites containing different mineral fillers.

Generally, the water absorption of hybrid biocomposites is affected by the hydrophilicity of the filler surface and the voids between the filler and the matrix. Gwon et al. [28] found that the water absorption of PP/wood fiber composites increased with the addition of a mineral filler, the composite made with zinc borate having the highest water absorption. The increase in the water absorption of the zinc borate-containing composite was attributed to the increased void fractions during the compounding process as well as the high hydrophilicity of the filler. A slight increase of water absorption properties with the addition of PCC (18 wt%) was

Table 5.8 Water absorption of hybrid PP-based biocomposites and bionanocomposites.

Composite	Water absorption %		Refs.
	1 week	28 days	
$PP_{50\%}/WF_{50\%}$	3.9(0.1)		[12]
$PP_{47\%}/WF_{50\%}/Talc_{3\%}$	5.7(0.1)		[12]
$PP_{47\%}/WF_{50\%}$ (3% MAPP)	3.3		[28]
$PP_{44\%}/WF_{50\%}/Talc_{3\%}$	4.7		[28]
$PP_{44\%}/WF_{50\%}/Kaolin_{3\%}$	5.0		[28]
$PP_{44\%}/WF_{50\%}/Zinc\text{-}borate_{3\%}$	6.3		[28]
$PP_{30\%}/WF_{44\%}/talc_{20\%}/MAPP_{3\%}$	4.7 (0.1)	11.8 (0.5)	[19]
$PP_{30\%}/WF_{44\%}/Calcium\ carbonate_{20\%}/MAPP_{3\%}$	5.4 (0.5)	13.5 (1.3)	[19]
$PP_{30\%}/WF_{44\%}/Wollastonite_{20\%}/MAPP_{3\%}$	4.9 (0.3)	16.1 (0.5)	[19]
$PP_{30\%}/WF_{64\%}/MAPP_{3\%}$	3	19 - 22	[18, 42]
$PP_{30\%}/WF_{44\%}/Mineral\ wool_{20\%}/MAPP_{3\%}$	2.5	13	[18]
$PP_{30\%}/WF_{44\%}/Wollastonite_{20\%}/MAPP_{3\%}$		10.5	[42]
$PP_{30\%}/WF_{61.5\%}/CB_{2.5\%}/MAPP_{3\%}$		13.0	[42]
$PP_{30\%}/WF_{41.5\%}/Wollastonite_{20\%}/CB_{2.5\%}/MAPP_{3\%}$		10.5	[42]
$(R\text{-}PP/PE)_{60\%}/Bamboo\ fiber_{40\%}$	2.0 (0.1)	4.8 (0.5)	[15]
$(R\text{-}PP/PE)_{48\%}/Bamboo\ fiber_{32\%}/PCC_{18\%}$	2.9 (0.3)	7.0 (0.9)	[15]

reported by Kim *et al.* [15] for PP/bamboo fiber composites. The increase in the water absorption of PCC-filled composites seems to be mainly related to the interfacial gap caused by the poor compatibility between the hydrophilic PCC and the polymer. The increase in water absorption in the studies mentioned above was due to an increase of the total filler loading (decrease in the content of hydrophobic PP).

On the other hand, the exchange of a hydrophilic wood fiber with a mineral filler leads to a decrease of water absorption of composites [18, 19, 42]. It is interesting to note that the addition of 2.5% of carbon black to a WF-PP composite reduced the water absorption and thickness swelling of this composite almost to the same level as the exchange of 20% of wood fibers with wollastonite [42].

The absorption of water (moisture) by biocomposites usually leads to a weakening of their mechanical performance. According to data published by Panthapulakkal and Sain [1] on the retention of tensile strength of hybrid PP-based hemp fiber/glass fiber composites after 3624 hours of immersion in water, it was 65% for PP/HF$_{40\%}$ and PP/HF/GF composites. These results can indicate that glass fibers did not alter the degradation of hemp fiber composites, despite the reduction in the water absorption in the hybrid composites. The reduction in strength can be explained by the degradation of natural fibers coupled with the dissimilar stress developed at the interface as a result of the prolonged period (3624 h) of immersion of the composite in water. In the case of tensile modulus, the increase in the glass fiber content increased its retention; the percentage retention in modulus increased from 44 to 53% when the glass fiber content increased from 0 to 15%. Panthapulakkal and Sain [1] explain the difference between the retention in strength and modulus by the fact that modulus is less affected by interfacial strength.

5.2.2.4   *Flammability or Fire Performance of PP-Based Hybrid Biocomposites and Bionanocomposites*

The limiting oxygen index (LOI) is an analysis method for measuring the lowest concentration of oxygen which causes the sample to ignite. According to this test, the higher the LOI the lower the flammability of the sample. The LOI test is performed according to ASTM D2863. The sample is placed vertically in the sample holder of the LOI apparatus. The total volume of the gas mixture ($N_2 + O_2$) is kept fixed. The ratio of nitrogen and oxygen at which the sample continued to burn for at least 30 seconds is recorded.

The limiting oxygen index (LOI) values of different PP-based hybrid bionanocomposites are shown in Table 5.9. According to the results reported by Farhadinejad *et al.* [32], the amount of oxygen needed to combust pure PP is 27%, and the addition of 25% wood flour reduced this value to 19%. Wood as flammable material reduced the amount of oxygen

Table 5.9 Limiting oxygen index (LOI) values of different PP-based hybrid bionanocomposites.

Composite	LOI (%)	Char	Ref.
$PP_{75\%}/WF_{25\%}$	19		[32]
$PP_{60\%}/WF_{38\%}/Nanowollastonite_{2\%}$	22		[32]
$PB_1/W_{40\%}/GMA_{10\%}$	42	medium	[25]
$PB_1/W_{40\%}/nMMT_{3\%}/GMA_{10\%}$	51	higher	[25]
$PB_2/W_{40\%}/GMA_{5\%}$	39	little	[26]
$PB_2/W_{40\%}/nMMT_{3\%}/ (TiO_2)_{1\%}/GMA_{5\%}$	62	higher	[26]
$PB_2/W_{40\%}/nMMT_{3\%}/ (TiO_2)_{3\%}/GMA_{5\%}$	67	higher	[26]
$PB_2/W_{40\%}/nMMT_{3\%}/ (TiO_2)_{5\%}/GMA_{5\%}$	64	higher	[26]

PB_1 = (HDPE/PP/PVC); PB_2 = (HDPE/LDPE/PP/PVC); GMA corresponds to glycidyl methacrylate

content required to burn the composite material. The addition of 2 wt% nanowollastonite in $PP_{60\%}/WF_{38}$ increased the LOI index compared to a $PP_{75\%}/WF_{25}$ composite. The great surface area in nanowollastonite particles partially prevents the diffusion of flammable gases from the bulk to the surface, as well as the penetration of oxygen to the bulk of the composite [32]. Also, inorganic materials can protect the composite against burning by forming a charred layer. In their studies on the burning behavior of WPCs made of polymer blends, Deka and Maji found a substantial improvement in LOI values after the addition of nanoclay (nMMT) [25] or a combination of nanoclay (nMMT) and TiO_2 [26]. Nanoclay produced silicate char on the surface of WPC, which decreased the flame propagation property of the composites. TiO_2 nanoparticles similar to clay provided some thermal barrier to the oxygen and heat, leading to an improvement in flame resistant property. The amount of char was higher for WPCs containing nanoparticles compared to composites made without nanoparticles (Table 5.9). Deka and Maji [26] also report that the LOI value decreases at a high percentage (5 phr) of TiO_2. The decrease in the LOI value at high TiO_2 loading has been explained by the agglomeration of oxide nanoparticles.

The effect of minerals in hybrid PP/wood composites on the fire retardancy has been studied by group of researchers in the Fiber Composite Laboratory of Lappeenranta University of Technology [43, 44]. The flammability of the samples was studied with a horizontal burning test using a cone calorimeter (Fire Testing Technology, UK) according to ISO 5660-1. The sample size was $100 \times 100 \times 5$ mm^3. The sides of the samples unexposed to flames were wrapped in aluminum foil. The samples, placed in a retainer frame, were positioned horizontally under the cone heater; the distance between the cone heater and the sample was 25 mm. All samples were tested under irradiation of 50 kW/m^2; the exhaust system flow rate was 24 L/s.

To estimate the improvement of the fire performance of the composites, the cone calorimeter results were evaluated using the European fire classification system. The European fire classification of materials is based on the single burning item (SBI) test. The data from the cone calorimeter were analyzed by tools incorporated into the analyzer for the simulation of the SBI test.

Table 5.10 shows the results of the cone calorimeter test for PP-based hybrid biocomposites containing mineral fillers. Only a slight decrease in heat release rate (HRR) peaks was found for hybrid composites containing 20% of mineral fillers compared to the reference. Moreover, there was no significant improvement in the reaction to fire performance for all samples

Table 5.10 Ignition time (IT), heat release rate (HRR), fire growth rate (FIGRA), and class determined according to the European fire classification system of PP-based hybrid biocomposites [43].

Composite	IT, s	HRR peak, kW/m^2	FIGRA, W/s	Euroclass
$PP_{30\%}/WF_{64\%}/$ $MAPP_{3\%}$	21.6 (1.1)	444.2 (10.7)	1131 (64)	E
$PP_{30\%}/WF_{44\%}/$ $CC_{20\%}/MAPP_{3\%}$	24.6 (1.4)	416.2 (19.4)	879 (40)	E
$PP_{30\%}/WF_{44\%}/W_{20\%}/$ $MAPP_{3\%}$	28.8 (3.3)	432.4 (10.8)	905 (35)	E
$PP_{30\%}/WF_{44\%}/$ $Talc_{20\%}/MAPP_{3\%}$	24.6 (1.7)	415.5 (11.9)	917 (14)	E
$PP_{30\%}/WF_{34\%}/$ $Talc_{30\%}/MAPP_{3\%}$	32.8 (1.8)	320.6 (13.2)	608 (10)	D
$PP_{30\%}/WF_{24\%}/$ $Talc_{40\%}/MAPP_{3\%}$	38.0 (2.2)	314.9 (12.4)	514 (15)	D

containing 20% of mineral fillers. These composites, as well as the reference sample, showed class E. The best improvement could be noticed with the addition of 30% and 40% of talc. The addition of 30% and 40% of talc promoted a decrease of the fire growth rate (FIGRA), sufficient to obtain class D.

The amount of fire retardants required is often so high that they can decrease the mechanical properties of the composite. Similar to weathering, the most destructive processes in composites are surface controlled processes. Thus, modification of the composite surface could improve composite performance as well as save the cost of the final product. Lecouvet *et al.* [45] report that a nanoclay-containing thin layer on top of a polymer is more effective than conventional "bulk" polymer nanocomposites in terms of flammability reduction. The influence of micro- (carbon fibers, CF, expandable graphite, EG, graphite, G) and nanosized (carbon black, CB, and carbon nanotubes, CNTs) carbon-based fillers incorporated into the shell layer on the ignition time, HRR, and the total heat release (THR) of coextruded wood-plastic composites (WPC) was studied by Turku and Kärki [22] (the results of this study are shown in Table 5.11). In general, the time of ignition of the wood-polymer composite did not change with the addition of carbon-based fillers. The composites containing carbon black and carbon nanotubes showed the lowest heat release rate and total heat release. The action of these two fillers as flame retardants can be attributed to their small size or high surface area and interaction with the polymer matrix. The better barrier properties of the nanofillers can be explained by an improved dispersion and generation of a cohesive char layer. The influence of CFs, EG and G on the burning scenario was less

Table 5.11 Fire test results for coextruded composites made with the addition of carbon-based fillers [22].

Coextruded WPCs	IT (s)	HRR (kW/m²)	THR (MJ/m²)	FIGRA (W/s)	Euroclass
WPC	28 (1.4)	395 (21)	162 (5.4)	802 (63)	E
WPC-CB	25 (1.2)	331 (5)	153 (2)	684 (27)	D
WPC-EG	30 (1.7)	371 (9)	156 (7)	649 (30)	D
WPC-G	28 (1)	356 (9)	154 (5)	659 (26)	D
WPC-CNTs	29 (1)	326 (6)	151 (1)	588 (9)	D
WPC-CFs	29 (1)	351 (2)	162 (7)	645 (4)	D

noticeable compared with CNTs and CB. The 3% of carbon-based fillers incorporated into the shell layer of coextruded composites were able to improve their fire class from E (of the reference composite) to D.

5.2.3 Applications

Biocomposites have been gaining much attention for automotive applications and as nonstructural building materials as they can offer environmental and economic benefits by light-weighing of the parts. In today's automotive industry, environmental and regulatory requirements are pushing automakers towards the addition of more sustainable and lighter materials in their future products in order to achieve their emission and fuel economy targets. In most cases, the commercial use of biocomposites has been limited to nonstructural or semistructural applications due to their lower stiffness, impact and thermal properties. Hybridization of biocomposites can help to overcome these problems.

Birat *et al.* [33] compared two types of hybrid biocomposites with the material specification of an automotive battery tray. The comparison showed that both hybrid biocomposites' HDT, melt flow and flexural modulus were similar to the ones of the battery tray, while both exceeded the impact requirements. Furthermore, the density of the hybrid fiber composites was slightly lower compared to the currently used material in the battery tray, suggesting an opportunity for up to 10% weight reduction. The results indicated that the developed hybrid composites have the potential to be used in developing a battery tray prototype. Other potential under-the-hood applications of the developed hybrid biocomposite include air intake integrated parts and extension panel dash that have similar specifications as the battery tray. Due to the significantly higher impact resistance, hybrid biocomposites can be used to develop the underbody shield, belly pan and wheel liner [33].

Hybrid biocomposite materials are a better choice for nonstructural building products subjected to outdoor weathering. Hybridization of biocomposites with mineral fillers or inorganic pigments result in weather-resistant composites. The coextrusion technique allows improvement in the durability of composites exposed to outdoor conditions by hybridization of the thin shell layer. The hybridization of biocomposites with nanoclays, nanosilicates, carbon nanotubes, ultrafine TiO_2, talc, and synthetic hydroxyapatite can lead to some exciting new applications, such as structural materials with improved mechanical properties, barrier properties and flame retardancy, and as high-performance materials with improved UV absorption and scratch resistance.

5.3 Conclusion

This chapter described the processing and characterization of PP-based hybrid biocomposites and bionanocomposites by different methods. Many procedures can be used to prepare PP-based hybrid composites, e.g., extrusion, injection molding and compression molding. The properties of hybrid biocomposites depend on the nature, size (aspect ratio), and loading of the filler and its interaction with the PP matrix. Hybridization with glass fiber enhances the performance properties of the biocomposites: mechanical properties, including stiffness and strength, thermal properties and resistance to water absorption. By applying optimum fiber loading, good mechanical properties of PP-based hybrid composites can be achieved. In general, the hybridization of biocomposites with microsized particulate mineral fillers does not improve the performance of the composites and in some cases can even weaken it. PP-based hybrid bionanocomposites are a new class of composite materials that show improved mechanical properties, thermal stability, flame retardancy and durability at very low loading levels of a nanosized filler. The PP/filler interface, which affects the properties of composites, can be improved by adding a coupling agent. Hybridization of PP-based biocomposites with inorganic/mineral fillers and nanofillers can give rise to new applications and extend the existing applications.

References

1. Panthapulakkal, S., and Sain, M., Injection-molded short hemp fiber/glass fiber-reinforced polypropylene hybrid composites—Mechanical water absorption and thermal properties. *J. Appl. Polym. Sci.* 103, 2432–2441, 2007.
2. Jang, Y.H., Han, S.O., Kim, H.-I., and Sim, I.N. Effect of graphene addition on characteristics of polypropylene biocomposites reinforced with sulfuric acid treated green algae. *Polymer (Korea)* 37(4), 518–525, 2013.
3. Safwan, M.M., Lin, O.H., and Akil, H.M., Preparation and characterization of palm kernel shell/polypropylene biocomposites and their hybrid composites with nanosilica. *Bioresources* 8(2), 1539–1550, 2013.
4. Samal, S.K., Mohanty, S., and Nayak, S.K., Polypropylene-bamboo/glass fiber hybrid composites: Fabrication and analysis of mechanical, morphological, thermal, and dynamic mechanical behavior. *J. Plast. Compos.* 28, 2729–2747, 2009.
5. Stark, N.M., and Matuana, L.M., Surface chemistry and mechanical property changes of wood-flour/high-density-polyethylene composites after accelerated weathering. *J. Appl. Polym. Sci.* 94, 2263–2273, 2004.

6. Dey, S.K., and Xanthos, M., Glass fibers, in: *Functional Fillers for Plastics*, Xanthos, M. (Ed.), pp. 131–147, Wiley-VCH Verlag GmbH & Co. KGaA, 2005.

7. Klyosov, A.A., *Wood-Plastic Composites*, p. 720, John Wiley & Sons, Inc., 2007.

8. Xanthos, M., Polymers and polymer composites, in: *Functional Fillers for Plastics*, Xanthos, M. (Ed.), pp. 3–16, Wiley-VCH Verlag GmbH & Co. KGaA, 2005.

9. Nurdina, A.K., Mariatti, M., and Samayamutthirian, P., Effect of single-mineral filler and hybrid-mineral filler additives on the properties of polypropylene composites. *Vinyl Addit. Technol.* 15, 20–28, 2009.

10. Srikanth, P., Engineering applications of bioplastics and biocomposites— An overview, in: *Handbook of Bioplastics and Biocomposites Engineering Applications*, Srikanth, P. (Ed.), pp. 1–14, John Wiley & Sons, Inc.: Hoboken, NJ, 2011.

11. Nayak, S.K., and Mohanty, S., Sisal glass fiber reinforced PP hybrid composites: Effect of MAPP on the dynamic mechanical and thermal properties. *J. Reinf. Plast. Comp.* 29(10), 1551–1568, 2010.

12. Gwon, J. G., Lee, S.Y., Chun, S.J., Doh, G.H., and Kim, J.H., Effect of chemical treatments of hybrid fillers on the physical and thermal properties of wood plastic composites. *Compos. Part A: Appl. S.* 41, 1491–1497, 2010.

13. Kim, B.J., Yao, F., Han, G., Wang, Q., and Wu, Q., Mechanical and physical properties of core-shell structured wood plastic composites: Effect of shells with hybrid mineral and wood fillers. *Compos. Part B: Eng.* 45, 1040–1048, 2013.

14. Uawongsuwan, P., Yang, Y., and Hamada, H., Long jute fiber-reinforced polypropylene composite: Effects of jute fiber bundle and glass fiber hybridization. *J. Appl. Polym. Sci.* 132(15), 2015.

15. Kim, B.J., Yao, F., Han, G., and Wu, Q., Performance of bamboo plastic composites with hybrid bamboo and precipitated calcium carbonate fillers. *Polym. Composites* 33(1), 68–78, 2012.

16. Thwe, M.M., and Liao, K., Effects of environmental aging on the mechanical properties of bamboo-glass fiber reinforced polymer matrix hybrid composites. *Compos. Part A: Appl. S.* 33, 43–52, 2002.

17. Islam, M.S., Hasbullah, N.A.B., Hasan, M., Talib, Z.A., Jawaid, M., and Haafiz, M.K.M., Physical, mechanical and biodegradable properties of kenaf/coir hybrid fiber reinforced polymer nanocomposites. *Mater. Today Commun.* 4, 69–76, 2015.

18. Väntsi O, and Kärki T., Utilization of recycled mineral wool as filler in wood-polypropylene composites. *Constr. Build. Mater.* 55, 220–226, 2014.

19. Huuhilo, T., Martikka O., Butylina, S., and Kärki, T., Mineral fillers of wood-plastic composites. *Wood Mater. Sci. Eng.* 5, 34–40, 2010.

20. Martikka, O., Huuhilo, T., Butylina, S., and Kärki, T., The effect of mineral fillers on the thermal properties of wood-plastic composites. *Wood Mater. Sci. Eng.* 7, 107–114, 2012.

21. Butylina, S., and Kärki, T., Effect of weathering on properties of wood-poly-propylene composites containing minerals. *Polym. Polym. Compos.* 22(9), 753–760, 2014.

22. Turku, I., and Kärki, T., The influence of carbon-based fillers on the flammability of polypropylene-based co-extruded wood-plastic composite. *Fire Mater.* 40, 498–506, 2016.

23. Turku, I., and Kärki, T., The effect of fire retardants on the flammability, mechanical properties, and wettability of co-extruded PP-based wood-plastic composites. *Bioresources* 9(1), 1539–1551, 2014.

24. Butylina, S., Martikka, O., and Kärki, T., Weathering properties of coextruded polypropylene-based composites containing inorganic pigments. *Polym. Degrad. Stab.* 120, 10–16, 2015.

25. Deka, B.K., and Maji, T.K., Study on the properties of nanocomposite based on high density polyethylene, polypropylene, polyvinyl chloride and wood. *Compos. Part A: Appl. S.* 42, 686–693, 2011.

26. Deka, B.K., and Maji, T.K., Effect of TiO_2 and nanoclay on the properties of wood polymer nanocomposites. *Compos. Part A: Appl. S.* 42, 2117–2125, 2011.

27. Turku, I., and Kärki, T., The effect of carbon fibers, glass fibers and nanoclay of wood flour-polypropylene composite properties. *Eur. J. Wood Prod.* 72, 73–79, 2014.

28. Gwon, J.G., Lee, S.Y., Chun, S.J., Doh, G.H., and Kim, J.H., Physical and mechanical properties of wood-plastic composites hybridized with inorganic fillers. *J. Compos. Mater.* 46(3), 301–309, 2012.

29. Lee, H., and Kim, D.S., Preparation and physical properties of wood/poly-propylene/clay nanocomposites. *J. Appl. Polym. Sci.* 111(6), 2769–2776, 2009.

30. Bai, W., and Li, K., Partial replacement of silica with microcrystalline cellulose in rubber composites. *Compos. Part A: Appl. S.* 40(10), 1597–1605, 2009.

31. Ashori, A., and Nourbakhsh, A., Performance properties of microcrystalline cellulose as a reinforcing agent in wood plastic composites. *Compos. Part B: Eng.* 41, 578–581, 2010.

32. Farhadinejad, Z., Ehsani, M., Khosravian, B., and Ebrahimi, G., Study of thermal properties of wood plastic composite reinforced with cellulose micro fibril and nano inorganic fiber filler. *Eur. J. Wood Prod.* 70, 823–828, 2012.

33. Birat, K.C., Panthapulakkal, S., Kronka, A., Agnelli, J.A.M., Tjong, J., and Sain, M., Hybrid biocomposites with enhanced thermal and mechanical properties for structural applications. *J. Appl. Polym. Sci.* 132, 42452, 2015.

34. Stark, N.M., and Gardner, D.J., Outdoor durability of wood-polymer composites, in: *Wood-Polymer Composites*, Oksman, N.K., and Sain, M. (Eds.), pp. 142–165, CRC Press, 2008.

35. Sobczak, L., Lang, R.W., Reif, M., and Haider, A., Polypropylene-based wood polymer composites-effect of maleated polypropylene coupling agent under dry and wet conditions. *J. Appl. Polym. Sci.* 129(6), 3687–3695, 2013.

36. Williams, R.S., Weathering of wood, in: *Handbook of Wood Chemistry and Wood Composites*, Rowell, G. (Ed.), pp.139–185, CRC Press LLC, 2005.
37. Wypych, G., Data on specific polymers, in: *Handbook of Material Weathering*, Wypych G. (Ed.), pp. 335–512, ChemTec Publishing, 2008.
38. Stark, N.M., and Matuana, L.M., Surface chemistry and mechanical property changes of wood-flour/high-density-polyethylene composites after accelerated weathering. *J. Appl. Polym. Sci.* 94, 2263–2273, 2004.
39. Matuana, L.M., Jin, S., and Stark, N.M., Ultraviolet weathering of HDPE/wood-flour composites coextruded with a clear HDPE cap layer. *Polym. Degrad. Stab.* 96, 97–106, 2011.
40. Stark, N.M., and Matuana, L.M., Co-extrusion of WPCs with a clear cap layer to improve colour stability, in: *Proceedings of 4th Wood Fibre Polymer Composites International Symposium.* p. 1–13, Bordeaux, France, 2009.
41. Butylina, S., Hyvärinen, M., and Kärki, T., Weathering of wood-polypropylene and wood-wollastonite-polypropylene composites containing pigments in Finnish climatic conditions. *Pigm. Resin Technol.* 44(5), 313–321, 2015.
42. Butylina, S., and Kärki T., Resistance to weathering of wood-polypropylene and wood-wollastonite–polypropylene composites made with and without carbon black. *Pigm. Resin Technol.* 43(4), 185–193, 2014.
43. Nikolaeva, M., and Kärki, T., Influence of mineral fillers on the fire retardant properties of wood-polypropylene composites. *Fire Mater.* 37, 612–620, 2013.
44. Väntsi, O., and Kärki, T., Heat build-up and fire performance of wood-polypropylene composites containing recycled mineral wool. *Adv. Mat. Res.* 849, 269–276, 2014.
45. Lecouvet, B., Sckavons, M., Bourbigot, S., and Bailly, C., Highly loaded nanocomposites film as fire protective coating for polymeric substrates. *J. Fire Sci.* 32(2), 145–164, 2014.

6

Biodegradation and Flame Retardancy of Polypropylene-Based Composites and Nanocomposites

S. Butylina[1,*] and I. Turku[2*]

Laboratory of Wood and Bionanocomposites, Divison of Material Science, Department of Engineering Sciences and Mathematics, Luleå University of Technology, Luleå, Sweden
Fiber Composite Laboratory, School of Energy Systems, Lappeenranta University of Technology, Lappeenranta, Finland

Abstract

This chapter summarizes information about the biodegradation and flame retardancy of PP-based biocomposites and nanocomposites obtained from highly ranked journals published during the last two decades. The first part of this chapter deals with the biodegradation of PP-based composites. The second part of this chapter begins with a short description of the specific flammability of PP, followed by an overview of fire retardants and flame testing methods and standards. PP composites reinforced with commonly applied cellulosic fillers, such as wood, flax and others, as well as rarely used wool fibers are considered. The effect of different nanometric fillers alone or in combination with conventional fire retardants on the flammability of composites is also described. Along with comparative analysis of different fire retardants regarding their ability to decrease the flammability of PP matrix composites, the mechanism action and their possible synergy effects are highlighted.

Keywords: PP-based biocomposites/bionanocomposites, soil burial test, water absorption, photo-oxidation, flammability, cone calorimetry, limited oxygen index

Corresponding authors: svetlana.butylina@ltu.se; sbutylina@gmail.com; Irina.Turku@lut.fi

Visakh. P. M. and Matheus Poletto. (eds.) Polypropylene-Based Biocomposites and Bionanocomposites, (145–176) 2018 © Scrivener Publishing LLC

6.1 Biodegradability of PP-Based Biocomposites and Bionanocomposites

The literature in the field of biodegradability of synthetic polymers is almost entirely concerned with the problem of preventing or retarding the attack on plastics by microorganisms and with the susceptibility of plasticizers, etc., to attack [1]. Biodegradable materials can be defined as materials susceptible to being assimilated by microorganisms such as fungi and bacteria. Some non-biodegradable plastics are erroneously believed to be biodegradable because they often contain biodegradable additives, which will support the growth of microorganisms without causing the plastic itself to become assimilated. The term "biodegradable" is often used indiscriminately to refer to various types of environmental degradation, including photodegradation. Because a polymeric material is degraded by sunlight and oxygen, it does not mean that the material will also be assimilated by microorganisms. The term "biodegradable" should be reserved for the type of degradability that is brought about by living organisms, usually microorganisms.

There are four biodegradation environments for polymeric products, namely the soil, aquatic environments, landfills, and composts. Each environment contains different microorganisms and has its special conditions for degradation. In the soil, fungi are mostly responsible for the degradation of organic matter, including cellulosic fibers and polymers [2].

Many different tests have been developed to measure the biodegradation of plastics and plastic-based composite materials. In an earlier study by Potts *et al.* on the biodegradability of commercial plastics, the test procedure described by ASTM D1924-63 was used [1]. In this test, the degradation by fungi was investigated. The samples in the form of strips cut from compression-molded plaques or as finely ground powders were dispersed in or on a solid agar growth medium that was deficient only in carbon. The test fungi consisted of a mixture of *Aspergillus niger*, *Aspergillus flavus*, *Chaetomium globosum* and *Penicillium funiculosum*. After an exposure time of three weeks, the samples were examined and assigned growth ratings as follows: 0, no growth; 1, traces (less than 10% covered); 2, light growth (10–30% covered); 3, medium growth (30–60% covered); and 4, heavy growth (60–100% covered). The biodegradability of commercial plastics was evaluated according to ASTM D1924-63 as follows: 2 for poly(ethylene) household wrap, and 1 for poly(propylene), poly(vinyl chloride) and poly(ethylene terephthalate) [1]. All of the large volume packaging plastics did not show susceptibility to attack.

Strömberg and Karlsson studied the biodegradation of polypropylene, recycled polypropylene and polylactide biocomposites exposed to a mixture of fungi and algae/bacteria in a microenvironment chamber [3]. The fungi in this study were selected according to the standards IEC 68-2-10, ISO 846 and ASTM G21-90. Fungi species (e.g., *Aspergillus niger*, *Aspergillus terreus*, *Aureobasidium pullulans*, *Chaetomium globosum*, *Paecilomyces variotii* and *Scopulariopsis brevicaulis*) and a mixture of algae (*Chlorella* sp.) with different bacterial strains were used. The biodegradation of the samples was evaluated after 84 days. It was found that the composite materials had a larger amount of microbial colonies than the neat PP. The use of cellulose fibers in the composites resulted in more easily colonized surfaces, which were attributed to water uptake by the cellulose fibers. In addition to the agar plate methods, the biodegradability of plastic and plastic-based composites was determined by their weight loss in a soil burial test. The standards for the determination of biodegradability reported included ASTM D6400-99 (standard specification for compostable plastics), which includes D6002-96, D5338-98 and D6340-98; European standards (CEN/TC 261/SC 4N 99 and ISO 14855-1:2012); DIN-Standard draft 54,900, and ISO/CD-standard 15986.2.

The soil is a highly variable entity in terms of its composition and biological activity, both varying with the location, season, pretreatment, and storage. To date it is impossible to standardize soil as a biodegradation test medium. Most test protocols use soils with different texture characteristics, namely sandy, loam and clay textured soil, and under near optimal conditions of moisture, aeration, temperature and pH. The test periods of different tests are quite different. Some variations of soil tests used to evaluate the biodegradability of biocomposites by different researchers are presented in Table 6.1. Optimal conditions, such as temperature, humidity and soil moisture content, were chosen to promote the growth of microorganisms. Biodegradation was evaluated by monitoring changes in weight as a function of burial time. In every specific time interval, the sample was removed and rinsed gently with distilled water to remove the soil, and in some cases dried in an oven at 100 °C for 24 hours. The weight loss (%) due to degradation was calculated by the following equation:

$$\text{weight loss (\%)} = ((w_i - w_f)/w_i) \times 100 \qquad (6.1)$$

where w_i is the weight of the specimen before degradation and w_f is the weight of the specimen after degradation.

According to Arkatkar *et al.*, who have studied the degradation of pure PP films, a 0.43% weight loss was observed after 12 months of exposure to a mixed soil culture; the decrease in tensile strength was estimated to

Table 6.1 Soil test used for PP and PP-based biocomposites.

Composite[a]	Sample dimensions, mm	Conditions			Time		Ref.
		T (°C)	MC_{soil}(%)	Microorg	time interval (days)	total time (days)	
PP	film	30–37	100%	mixed soil culture	6, 9 and 12 months	365	[4]
$PP_{58\%}/RHF_{40\%}/MAPP_{2\%}$ $PP_{58\%}/BF_{40\%}/MAPP_{2\%}$	100 × 100 × 1	27	45–60	n/a	20	120	[2]
$PP_{70\%}/KF_{30\%}$ $PP_{70\%}/CF_{30\%}$ MMT (2 phr)	100 × 150 × 1	25	40–50	n/a	30	90	[5]
$rPP_{65\%}/KF_{15\%}/$ $(nano\text{-}CaCO_3+DAP)_{20\,phr}$	n/a			cellulotic bacteria enriched garbage soil		120	[6]

[a] RHF = rice husk fiber; BF = bagasse fiber; MAPP = maleic anhydride functionalized PP; KF = kenaf fiber; CF = coir fiber; MMT = montmorillonite; and DAP = diammonium phosphate

Table 6.2 Weight loss and water absorption of PP-based biocomposites.

Composite	Weight loss (%)	Water absorption		Ref.
		(%)	test method/time	
rPP$_{80\%}$/KF$_{20\%}$	5.8	1.7	soil burial test/ 4 months	[6]
rPP$_{65\%}$/KF$_{15\%}$/ (nano-CaCO$_3$)	5.3	1.6	soil burial test/ 4 months	[6]
rPP$_{65\%}$/KF$_{15\%}$/(nano-CaCO$_3$+ DAP)$_{20\,prh}$	11.8	1.9	soil burial test/ 4 months	[6]
PP$_{70\%}$/KF$_{30\%}$	19.4	15.8	water absorption test/ saturation point	[5]
PP$_{70\%}$/KF$_{30\%}$/MMT$_{2phr}$	10.5	17.1	water absorption test/ saturation point	[5]
PP$_{70\%}$/CF$_{30\%}$	14.0	19.9	water absorption test/ saturation point	[5]
PP$_{70\%}$/CF$_{30\%}$/MMT$_{2phr}$	17.0	23.0	water absorption test/ saturation point	[5]

be around 25% [4]. Nourbakhsh *et al.* and Suharty *et al.* have reported 0% decrease in weight loss after four months of exposure to soil for recycled PP and for neat PP [2, 6]. Generally, natural fibers increase the biodegradability (weight loss) of composites. For example PP$_{58\%}$/RHF$_{40\%}$/MAPP$_{2\%}$ and PP$_{58\%}$/BF$_{40\%}$/MAPP$_{2\%}$ composites showed 10% and 8% weight loss, respectively, after four months of exposure to the soil.

Table 6.2 shows the results of two studies on composite biodegradability determined in a soil burial test. Due to the difference in the test procedure shown in Table 6.1, a direct comparison of the results is not possible. However, it is clear that a relationship exists between the water absorption capacity of composites and their weight loss in a soil burial test. An increase in the water absorption of a biocomposite after the incorporation of an additive usually results in an increased biodegradability of the composite. The effect of the additive on the water absorption and biodegradability of a biocomposite depends on the nature of the additive and its interaction with other components in the composite. Suharty *et al.* have shown that the addition of diammonium phosphate increased the water absorption and weight loss of a recycled PP/kenaf fiber/nano-CaCO$_3$ composite significantly [6]. Islam *et al.* have shown that the water absorption (measured in a water immersion test) of kenaf/PP and coir/PP composites increased with the addition of montmorillonite (MMT) [5]. This was most probably due to the hydrophilic properties of MMT, which caused the nanocomposite to absorb more water. However, the biodegradability of PP/kenaf fiber composite decreased, while that of the PP/coir fiber composite increased with the MMT addition. Islam *et al.* explain the difference in the effects of MMT on the behavior of PP/kenaf fiber and PP/coir fiber composites by a difference in interfacial interaction and adhesion between the fiber and the polymer matrix.

Normally, degradation processes do not occur by the actions of independent mechanisms, but rather by a combination of several external factors; the most common processes include thermo-oxidation (effect of temperature and oxygen exposure), photo-oxidation (combined effect of light and oxygen exposure, especially interesting in outdoor applications), thermo-mechanical degradation (combination of temperature and shear forces, especially relevant during material processing) and biodegradation (combination of chemical and biological effects) [7]. Commercial equipment and standard tests are readily available for the study of these different processes. Abiotic degradation, which includes water absorption and hydrolysis, and photo- and thermo-oxidation, usually precedes the biological attack and causes an increase in the sensitivity of the material for further degradation, but may also be a factor in every step of the biodegradation process [7, 8]. Abiotic effects complement the biological activity during biodegradation synergistically.

As discussed above in connection with the soil burial test, water absorption is one of the main factors affecting the degradation of biocomposites. Mofokeng *et al.* have studied the degradation through hydrolysis of PLA and PP composites with 1–3 wt% of sisal fibers (SF) [9]. The composite samples were placed in water at 80 °C for 10 days, and the degradation of the samples was tested by monitoring the mass loss. The high temperature was used to shorten the time of exposure in order to speed up the experiment. The results showed that the mass loss of the PP composites was significantly lower than those of the PLA and its composites. No degradation was anticipated for PP during the immersion in water at 80 °C. The neat PP showed no mass loss up to 10 days of immersion, and all the PP/SF composites showed less than 1% mass loss after 10 days, which was probably due to a limited degradation of the fibers on the sample surface. The fibers in the bulk have been protected by the surrounding PP.

The durability or resistance of biocomposites to various types of degradation has been the object of many studies in the field of composites. The question of the durability of the composite is inseparable from the

Table 6.3 Percentage reduction of tensile strength of PP/bamboo fiber (BF) and PP/BF/glass fiber (GF) composites [10].

Composite	% degradation in tensile strength at room temperature	
	502 h	1200 h
$PP_{70\%}/BF_{30\%}$	7.92	13.95
$PP_{70\%}/BF_{20\%}/GF_{10\%}$	5.89	9.11
$PP_{70\%}/BF_{10\%}/GF_{20\%}$	4.5	7.47
$MAPP\text{-}PP_{70\%}/BF_{30\%}$	6.84	11.55
$MAPP\text{-}PP_{70\%}/BF_{20\%}/GF_{10\%}$	5.62	8.9
$MAPP\text{-}PP_{70\%}/BF_{10\%}/GF_{20\%}$	3.54	6.84

question of degradability. Durability determines the usefulness of composite materials in a particular environment. The retention of mechanical properties of biocomposites (natural fiber reinforced composites) during long-term service is crucial in order for them to be utilized in outdoor applications. Thwe and Liao [10] have studied the effect of moisture absorption on retention of the tensile properties of PP/bamboo fiber and PP/bamboo fiber/glass fiber composites by immersing the samples in water for up to 1200 hours at 25 °C; the results of their study are shown in Table 6.3. It can be seen in the table that an exchange of a part of the bamboo fibers with glass fibers resulted in improved retention of tensile strength. The incorporation of a coupling agent (MAPP) also had a positive effect on the mechanical properties, and this was explained by the improved interface between the fiber and the matrix.

It is worth mentioning that the inclusion of more corrosive-resistant fibers, such as glass and carbon fibers, may undermine the environmental impact of natural fiber composites, as these types of synthetic fibers either degrade at a much slower rate or do not degrade at all (under benign conditions) compared to natural fibers [10].

In 2014, the EN 15534 standard for composites made from cellulose-based materials and thermoplastics or natural fiber composites came into practice. The standard describes the test methods for the determination of the durability of composites against biological agents, against artificial and natural weathering, and against moisture. As stated above, photo-oxidation similar to moisture absorption is one of the factors determining the degradation of plastic samples in nature (or otherwise their durability). Moreover, opposite to moisture absorption, which does not usually affect polyolefins because they do not absorb water (or absorb it in very tiny amounts), the commodity of synthetic plastics, including PP, and lignocellulosic fibers are susceptible to photo-oxidation. The photo-oxidation of biocomposites can be tested under both natural and artificial conditions. Artificial weathering is performed by exposure to xenon lamps according to EN-ISO 4892-2, while natural weathering involves direct exposure of the products to solar radiation carried out according to EN ISO 877-2. It requires much longer exposition time to achieve the same degree of degradation under natural conditions than under accelerated weathering test conditions. The degree of degradation can be evaluated by estimating the change of color and appearance of the composites. In a work published by Butylina and Kärki, a comparison was made between natural and artificial weathering of PP biocomposites containing carbon black [11]. Scanning electron microscope (SEM) images of the surface of PP/wood fiber/carbon black composites weathered under natural and artificial conditions for 1000 hours are shown in Figure 6.1. As expected, the degradation of the surface was more pronounced in the case of artificial weathering. According to Butylina and Kärki, an addition of 2.5% of carbon black to the PP/wood fiber composite reduced the water absorption and thickness swelling of the composite.

Very few studies can be found on the weathering of PP-based biocomposites containing fire retardants. Turku *et al.* [12, 13], who have focused their research on the flammability of coextruded PP-based biocomposites containing different fire retardants, have also studied the influence of fire retardants on the durability of these composites [14, 15]. In these studies, 3% of fillers, such as carbon black (CB), graphite (G), expandable graphite (EG), carbon nanotubes (CNTs), and carbon fibers (CFs), and 10% of

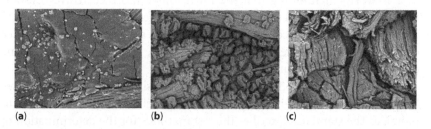

(a)   (b)   (c)

Figure 6.1 SEM images of the surface of a PP/wood fiber/carbon black composite weathered for 1000 hours under (a) outdoor, (b) xenon and (c) xenon and spray conditions (magnification x160) [11].

Table 6.4 Lightness (ΔL) and tensile strength (TS) change of PP-based co-extruded biocomposites after 500 hours of accelerated weathering.

Composite	$\Delta L_{500\ hours}$, %	$\Delta TS_{500\ hours}$, %	Ref.
Reference	4.0	−17	[14,15]
CB$_{3\%}$–WPC	0	−15	[14]
G$_{3\%}$–PC	0.7	−22	[14]
CNTs$_{3\%}$–WPC	0.9	−10	[14]
CFs$_{3\%}$–WPC	2.1	+2	[14]
Melamine$_{10\%}$–WPC	2.0	−14	[15]
G$_{10\%}$–WPC	0	−8	[15]
ATH$_{10\%}$–WPC	2.3	−10	[15]
ZB$_{10\%}$–WPC	3.5	−5	[15]
(TiO$_2$)$_{10\%}$–WPC	1.2	−15	[15]

fillers, such as aluminum trihydrate (ATH), zinc borate (ZB), melamine, graphite, and titanium dioxide (TiO$_2$), were incorporated into the shell (outer) layer of composites. The composites underwent accelerated weathering under a xenon arc lamp source. The changes of tensile strength and modulus were evaluated as well as color (lightness) change, and the surface morphology was also monitored. The results of these studies are summarized in Table 6.4. The fire retardant-loaded samples had a smaller color change compared to the unfilled reference composite. The tensile properties of all composites, except CF-loaded, declined after weathering. No influence of the fire retardants on the mechanism of photo-oxidation of the studied composites was observed [15].

6.1.1   Conclusions

The biodegradation of PP-based biocomposites depends on a number of factors, including the natural fiber content, the biodegradability of each constituent, and the quality of the interface. Natural fibers generally increase the degradation rate of PP-based composites. Additives such as the coupling agent and mineral or synthetic fillers can affect biodegradation.

The rate of biodegradation generally depends on the substrate composition and the existing microorganisms. The susceptibility of composite constituents to photo-oxidation and to water absorption is an indispensable part of the degradation mechanism of composites under natural conditions, and it also affects the biodegradability of the composites. The durability of PP-based biocomposites can be tailored by employing an appropriate amount of synthetic fiber and/or mineral additives, so that in addition to a cost-performance balance, a balance between the environmental impact and performance can also be achieved by the design according to service requirements.

6.2   Flame Retardancy of Polypropylene-Based Composites and Nanocomposites

Natural fiber-containing plastic composites represent an important class of material capable of combining the useful properties of both natural fiber and plastic. Natural fiber composites are used in a great number of applications in different fields, including building, infrastructure, and transportation [16]. Polypropylene (PP) is the most commonly applied polymer in the manufacturing of composites due to its low density, easy processability, high water and chemical resistance, and low cost. A wide range of natural fibers (e.g., wood, flax, sisal, hemp, silk, wool) are used, however, lignocellulosic wood fibers are mostly applied. Natural fibers are ecological and renewable, low cost and low density material with acceptable mechanical properties, making them an attractive alternative to glass and carbon fibers. The main disadvantage of biofiber composites is the poor compatibility between the polymer matrix and filler, resulting in decreased strength, which can be avoided by the addition of a coupling agent (CA). Another disadvantage is their low thermal stability and sensitivity to fire. PP is classified as a highly flammable polymer due to the absence of an aromatic group and double bonds in the polymer chain and, hence, absence of the possibility of char formation. In addition, absence of a crosslinking reaction during thermal degradation, i.e., PP decomposes via random chain scission into smaller fragments, prevents the char-forming reaction [17]. Cellulosic fibers are composed of carbon, hydrogen (fuels) and oxygen (supporter of combustion), and they are thus highly flammable and burn easily [18]. Nevertheless, cellulosic fibers, being char-forming material, can significantly reduce the flammability parameters of PP, including peak of heat release (pHRR), mass loss rate (MLR), total heat release (THR) and others [19–21]. However, natural fiber composites have a lower decomposition temperature and burn faster compared to pure polymer [22–25]. Stronger interaction between the polymer and the reinforcing filler improves the thermal stability of the composite [26–28].

In laboratory studies, composite flammability can be estimated by several tests. In the cone calorimetry test, parameters such as pHRR, THR, total smoke production (TSP), MLR and others can be monitored according to standard EN 16550. The Underwriters Laboratories UL-94 (vertical [V] and horizontal burning [HB]) test measures the flame spread and ignitability of material exposed to a small flame source according to standards ISO 1210 and ASTM D635. The most sited flame retardancy ratings for composites are V0, V1, V2 and HB. A V0 rating is given to material with combustion time less than 10 s after flame application; mean combustion time for 5 specimens should not exceed 5 s and no drips can be observed. The sample is classified as V1 if its combustion time is less than 50 s with mean combustion time for five specimens less than 25 s without combustible drips. A V2 rating is given to material with burning criterion similar to burning criterion of V1 where combustible drips was observed. HB rating indicates that material was tested in a horizontal position and found that flame propagation does not exceed a certain specific velocity. This is the lowest rating of the UL-94 test. The LOI (limited oxygen index) test measures the minimum oxygen concentration required to support candle-like downward flame combustion, according to standards ISO 4589 or ASTM D 2863. Generally, materials with LOI above 26–28% (O_2) are denoted as self-extinguishing. The thermostability of a composite, i.e. the temperature and the rate of decomposition, can be studied by thermogravimetric analysis (TGA). In addition to the tests described above, scanning electron microscopy (SEM) and transmission electron microscopy (TEM) are important techniques for studying char morphology.

The traditional method to reduce the flammability of composites is using fire retardants (FRs). In general, fire retardants can be classified according to their chemical structure and mechanism action. According to mechanism action, FRs are divided as being gas- and condense-phase active. Gas-phase active FRs act through the scavenging of free radicals responsible for the branching of radical chain reactions in the flame, also denoted as chemical mechanism action. Other FRs generate large amounts of noncombustible gases, which dilute flammable gases, and can also dissociate endothermically, and decrease the temperature by absorbing heat. This is the physical mechanism of action in the gas phase. In the condensed phase the most common action is charring, which can be promoted by a fire retardant though chemical interaction with the polymer or physical retention of the polymer in the condensed phase [26]. On the basis of their chemical composition, FRs are divided into phosphorous-containing, metal hydrates and oxides, halogenated, borate, and silicon-containing ones [29]. Halogen-containing FRs are recognized as effective, but they are eliminated from use, e.g. in Europe, due to their high environmental impact [30]. In recent years, nanometric fillers have become popular due to their multifunctionality, influencing the mechanical and physical properties, including flame retardancy, positively.

Phosphorous-based FRs, e.g., ammonium polyphosphate (APP), melamine polyphosphate (MPP), and red phosphorous (RP), are often applied for composite flame retardancy. Upon heating, they decompose into acids or esters which react with the hydroxyl groups of cellulose or other reactive groups containing polymers, facilitating char formation. However, polyolefins, including PP, represent a challenge for the use of phosphorous FRs due to the fact that they do not have reactive groups and burn without leaving char. In order to facilitate the action of phosphorous FRs they are usually used in combination with a char-former agent and blowing agent to produce intumescence. The classic intumescence system for polyolefin consists of APP (acid source and minor blowing agent), pentaerythriol (polyol and char former) and melamine (blowing agent). They participate in char formation during burning which plays role of a physical barrier for heat and mass transfer between the gas and condensed phases [31–33].

Arao et al. have studied the flammability of PP and a PP-based wood plastic composite in the presence of APP, MPP and aluminum trihydrate (ATH) [25]. Cone calorimetry and UL-94 test results are listed in Table 6.5. The cone calorimetry test showed that APP and MPP decreased pHRR and HRR significantly compared to pure PP and the non-retarded composite, whereas ATH had a minor effect. Also, in the UL-94 test, APP had a greater impact than the other FRs used, decreasing the burning rate up to self-extinguishing in horizontal and resulting in a V0 rating in vertical burning test. MPP-loaded composite achieved self-extinguishing in horizontal burning test, but did not achieved even V2 in vertical burning test and was classified as a HB. The mechanism action of APP is its ability to form carbonaceous char where wood flour/cellulose is the charring agent. APP interacts with wood flour and yields a crosslinked ultraphosphate and a polyphosphoric acid with a high crosslinked structure. The char layer prevented the absorption of heat and also blocked the flammable gases produced during polymer/wood decomposition. Accelerated charring by APP was also the reason for shortened ignition time, whereas ATH delayed the starting time of char forming. At a high temperature, ATH decomposes endothermically, releasing water, and thereby diluting flammable

Table 6.5 Cone calorimeter and UL-94 test results for PP and PP composites with 10 wt% of the fire retardants.

Sample	IT (s)	Peak HRR (kW/m^2)	THR (MJ/m^2)	Average HRR (kW/m^2)	Average burning rate (mm/min)	UL-94
PP	31.6	1395	117	434	27.4	HB
WPC(WF: 50 wt%)	21.4	563	93.4	336	32.4	HB
WPC + APP	19.6	312	78.9	136	Self-extinguish	V0
WPC + MPP	20.4	352	98.6	83.7	Self-extinguish	HB
WPC + Al(OH)$_3$	24.8	467	99	96.5	20.9	HB
WPC(WF:30 wt%) + APP	–	–	–	–	21.2	HB

Table 6.6 LOI values and UL-94 rating of PP composite containing various amounts of APP [34].

Sample	LOI (%)	UL-94
PP/WF (10–40%)/APP(10%)	22–25	HB
PP/WF (10–40%)/APP(20%)	24–27	HB
PP/WF (50%)/APP(10%)	30	V0
PP/WF (50%)/APP(20%)	32	V0

gases and absorbing heat. However, the amount of water released was not enough for the self-extinguishing of the sample. Reducing WF amount from 50 wt% to 30 wt% resulted in the worsening of the flame retardancy rating, showing correlation between APP and WF content and synergistic effect [24, 25]. The synergistic effect between APP and wood flakes in a PP-based composite has also been described by Matkó *et al.* [34]. In their study, a silane-based reactive surfactant was used as the compatibilizer. The flammability was characterized by LOI and UL-94 tests. The results are listed in Table 6.6. The LOI of the composite with the amount of wood flakes from 10% to 40% and containing 10% or 20% APP did not change much, and the samples remained highly flammable. Increasing the amount of wood flakes to 50% resulted in an increased LOI up to 30% and 32% for 10% and 20% APP, respectively, and a rating of V0 in the UL-94 test for both. Lignocellulose material is a char source for APP-intumescent FR, and the efficiency of FR action depends on the amount of hydroxyl groups of cellulose available for phosphoric acid formed during thermal decomposition of APP. Thus, it has been observed that 50% of wooden material is the minimum amount for composite to achieve a V0 rating in the presence of 10 or 20% of APP.

Schartel *et al.* have compared the efficiency of APP and expandable graphite (EG) in the flame retardancy of flax/PP composites [35]. The flammability of the composites was tested by LOI, UL-94 and cone calorimetry at different heat fluxes. Thermostability was tested by TGA under a nitrogen atmosphere. Selected results are shown in Table 6.7. As can be seen, the fire retardants increased the char residue in both TGA and cone calorimeter tests. The mechanism action of EG was based on expansion at heating up to 300 times, thereby creating an insulating layer which protected the underlayer material from thermal decomposition. The results showed that EG was more effective than APP. The pHRR and

Table 6.7 Flammability of flax/PP composites containing APP and EG [35].

Sample	TGA residue at 490 °C (%)	Cone Calorimeter residue at flameout (%)	Peak HRR* (kW/m²) at 50 kW/m²	Critical HRR (kW/m²)	LOI (%)	UL-94
PP/flax (70/30)	13.9	10.6	520	167	21	HB
PP/flax/APP (45/30/25)	27.4	29.3	270	131	26	HB
PP/flax/EG (55/30/15)	18.3	28.4	180	73	29	HB
PP/flax/EG (45/30/25)	29.4	> 29.6	160	35	30	V1

*Approximate values taken from a figure in ref. [35]

Table 6.8 Cone calorimeter parameters at external heat flux 50 kW/m² [36].

Sample	IT (s)	pHRR (kW/m²)	UL-94
PP	45	1800	NR
PP/flax (60/40)	30	640	NR
PP/flax/APP (46/31/23)	30	300	NR
PP/flax/(APP/PER/MEL) (46:31:23)	30	270	V0

critical HRR were significantly smaller than in the flax/PP and flax/PP/APP composites. Also, the LOI and UL-94 tests revealed superior protection properties of EG compared to APP. In a similar work, Le Bras *et al.* found that the effect of APP action was reinforced when using an intumescent system [36]. Cone calorimeter and UL-94 test results are shown in Table 6.8. Despite the significant decrease of pHRR, the APP-containing composite had no rating in the UL-94 test. However, the combination of APP with the intumescent system, pentaerythritol (PER) and melamine (MEL), resulted in burning rate being improved up to V0.

The effectiveness of APP, red phosphorous (RP) and expandable graphite and a combination of two of them on wood/PP flammability has been studied by Seefeldt *et al.* [37]. EG loading was found to have the highest flame retardancy properties, reducing pHRR up to 73%. However, the sample cracked completely due to expansion during burning, thereby destroying the thermal barrier and flame retardant effect. The combination of EG with RP or APP prevented cracking due to crosslinking in the residue and resulting in high fire performance of the composite. Also, EG was not only a more effective flame retardant than APP, the composite containing EG was also less biodegradable. APP is a phosphorous- and nitrogen-containing substance, both of which are essential elements for microorganisms growth, and hence their presence in the composite could facilitate the biodegradability of the composite [38].

Yu *et al.* have studied the synergistic effect of APP and EG in wood flour/PP/carbon black composite [39]. They found that the combination of EG (15%) and APP (5%) allowed a decrease of pHRR and THR up to about 50%. The decrease of the values was greater than in the cases where only APP or EG, both at 20%, were used. Char residue increased from 9% to 48% for the non-retarded and APP/EG-retarded composite, respectively. The analysis of the char showed that the sample containing both EG and APP had a dense and swollen structure, which effectively protected the underlayer material from burning. The synergistic effect of EG and an intumescence fire retardant, a combination of APP and triazines-based char-forming agent (CFA), on the fire retardancy of a composite was also observed by Bai *et al.* [40]. The total amount of FR in the composite was 25%. As it was detected, the combination of EG and APP/CFA resulted in the largest decline in the pHRR, THR and MLR, and in the the most significant increase the LOI and time to ignition (Table 6.9).

One of the ways to decrease composite flammability is by using fire-retarded wood flour. Hämäläinen and Kärki modified wood particles with two phosphorous-based and one melamine formaldehyde fire retardant

Table 6.9 Test results for wood fiber/PP composites [40].

Sample	IT (s)	Peak HRR (kW/m²)	THR (MJ/m²)	Mass loss (%)	LOI (%)	UL-94
PP/WF	20	389.6	90.4	79	22.3	NR
PP/WF/EG	24	99.2	39	26	29.4	V0
PP/WF/EG/ APP/CFA	30	89.9	33.6	20	38.8	V0
PP/WF/ APP/CFA	26	243.8	72.3	44	28	V1

[41]. It was shown that composite flammability was decreased significantly compared with a non-modified wood flour-loaded composite. Matkó *et al.* reported that silane-modified cellulose fibers were more thermally stable than non-modified ones [34]. Schirp and Su studied a recycled PP-based wood plastic composite where the wood flour (WF) was fire retarded, and in addition, fire retardants were loaded during composite compounding [33]. The wood flour was residues from particleboard production which were modified with ammonium phosphate. Fire retardants, namely, APP, melamine cyanurate, melamine and phosphorous pentoxide, tris-(2-hydroxyethyl)isocyanurate, expandable graphite (EG), red phosphorous (RP) and their combinations were introduced during composite manufacturing. Thermogravimetric analysis and a LOI test showed that the composites with fire-retarded WF had higher stability than those with non-treated WF. Additional loading of FRs resulted in further improved fire retardancy of the composite, where the combination of EG/RP was found the most effective. The activity of RP was attributed to its ability to act in the gas phase by radical trapping and in the condensed phase, where in combination with oxygen it forms phosphates which are able to crosslink carbonaceous structures. EG acts physically by creating a layer of expanded graphite worms on top of the material. It was also concluded that the effect of the combination of EG and RP appeared to outweigh the effect of fire-retardant action in wood flour.

Metallic hydroxides, which can be characterized as being nontoxic, environmentally friendly, inexpensive and abundant, represent a good alternative to halogen-containing FRs. Aluminum trihydroxide (ATH) and magnesium hydroxide (MH) are most commonly applied in plastics and composites. At enhanced temperatures, starting from 200–220 °C and 300–320 °C, for ATH and MH, respectively, they endothermically decompose to metal oxide and water. Thus, their fire retardancy mechanism is releasing water, diluting combustible gases and cooling down the material at the same time. Also, the metal oxide can form a protective nonflammable layer on the polymer surface, which protects the material from further thermal decomposition and release of toxic gases [42]. ATH and MH are also smoke suppressant in many cases [43]. Similarly to metal hydroxides, zinc borates (ZB) release water vapor during combustion, thereby diluting the combustion volatiles and decreasing the temperature in the system. They also form a glassy layer after the decomposition, which is a barrier for the transfer of heat and volatiles.

Sain *et al.* studied the fire retardancy effect of magnesium hydroxide, boric acid and zinc borate on wood flour and rice husk PP-based composite [22]. The flammability of the composites was monitored by horizontal burning rate and LOI tests. The replacement of natural fibers with MH (25%) reduced the burning rate by about half for both fillers, whereas LOI was significantly higher compared to the composites without MH and pure PP. No synergistic effect of MH with boric acid and zinc borate (in the ratio 20:5) was observed in the experiments. In a similar study, Suppakarn and Jarukumjorn studied the effect of MH and ZB on a sisal/PP composite [23]. The burning rate of sisal/PP was reduced significantly after the incorporation of FRs (15%), where MH had a better effect than ZB. The char residual of the ZB-containing composite, however, was higher than that of the MH-containing blend. No synergistic effect was observed when both MH and ZB were incorporated, which is in consistent with the study of Sain *et al.* [22]. The flammability of the composite was improved without a decline of mechanical properties. Arao *et al.* reported that ATH had no significant effect on WPC flammability (see Table 6.5). This can probably be explained by the low thermal stability of ATH, which starts to decompose at 200–220 °C, whereas fire retardants have to be thermally stable at the processing temperature [44].

Unlike cellulosic fibers, keratin-based wool fibers have low flammability. The fire resistance of wool can be attributed to the relatively high contents of nitrogen and sulphur (3–4 wt%), high ignition temperature (570–600 °C), low heat of combustion (4.9 kcal/g), and high limiting oxygen index (25.2%). The crosslinking and dehydration tendency of the sulphur-containing amino acids in wool can improve char formation under combustion, making wool fibers non-melting and non-dripping [45, 46]. Conzatti *et al.* reported that wool fibers (20–60%) improved the thermal stability of PP, especially in the presence of a compatibilizer [45]. Kim *et al.* studied the fire retardancy of a PP/wool (30%) composite containing 20% of APP [46]. Cone calorimetry results demonstrated a significant decrease in pHRR (~ 82%) and an increase in the time to peak of HRR (~ 170%), compared to those of neat PP. Fire retardancy was attributed to rigid char formation, which protected the underlayer material from combustion. The effective charring was due to cumulative action of APP and wool fibers. At a high temperature, the release of phosphoric acids from APP and hydrolytic scission of keratin polypeptide chains of wool fibers can enhance dehydration and crosslinking with possible formation of chemical bonds with PP, leading to cohesive char formation. Thus, wool and APP together formed rigid, compact and voluminous char, which was an effective barrier for mass and heat transfer in the composite.

Subasinghe *et al.* have shown that fire retadancy of a kenaf/PP composite was improved when wool fibers were applied [47]. The peak of the HRR of the kenaf/PP was decreased from 729 to 330 kW/m^2 when APP (24%)

was loaded, but the sample did not pass the UL-94 test. Additional loading wool fibers (3%) resulted in a slightly decreased pHRR, 311 kW/m², and, most importantly, a sample was self-extinguishing.

6.2.1 Bionanocomposites

In recent decades, nanometric fillers have become an important player in the composite production market. Nanosized fillers are capable of modifying composite properties at very low loadings, usually 1–5%. Important, that unlike to the conventional FRs, nanofillers capable to improve flame retardancy of material without deterioration its mechanical performance. The most widely applied nanofillers are layered silicates and carbon nanotubes. In addition, there is growing interest in nanocomposites reinforced with nano-sized oxides, such as SiO_2, ZnO and TiO_2. The main drawback, however, is the difficulty of dispersing the individual nanoparticles in the composite body. Generally, for better dispersion, layered silicate nanoparticles are modified with a surfactant or/and coupling agent. Besides that, the processing method of nanocomposites has a great effect on the degree of dispersion and the final properties of the composite [48].

The main mechanism of nanofillers is related to the development of a carbonaceous structure on the nanocomposite surface, acting as an insulating layer for heat and mass transfer during combustion and also preventing contact with oxygen. Yet, nanofiller particles act as physical crosslinking sites among polymer chains, forming a network that restricts the mobility of macromolecules and hindering the escape of decomposition products, thereby influencing the HRR [49]. It has also been suggested that clay has the role of a catalytic agent for char formation even in polymers which do not exhibit a char-forming tendency [49–51].

The effect of organomodified clay (OMMT) on PP flammability has been studied by Gilman *et al.* [52]. Maleic anhydride of polypropylene PP (MAPP) was used as the coupling agent. Gilman *et al.* reported that the peak of HRR of 2% and 4% OMMT/PP/MAPP nanocomposite decreased by 70% and 75%, respectively, compared to a polymer without OMMT (see Table 6.10). Char residue also increased slightly compared to PP. However, the heat of combustion, and smoke and carbon monoxide yields were unchanged, suggesting that clay did not have a gas phase effect influencing only condensed phase changes. The char morphology analysis with TEM showed that the clay layers were well dispersed in the matrix, which was attributed to CA action. Thus, the multilayered char formed during burning worked as a mass transfer barrier, slowing

Table 6.10 Comparison of cone calorimeter data for PP and PP/clay nanocomposites at 35 kW/m² heat flux [52].

Sample	Yield residue, %	pHRR, kW/m²	HRR, kW/m²	H_c, MJ/ kg	SEA, m²/kg	CO yield, kg/kg
PP	5 ± 0.5 (0%) ref. [53]	1525	536	39	704	0.02
PP/OMMT 2%	6 ± 0.5	450 ± 70	322 ± 40	44	1028	0.02
PP/OMMT 4%	12 ± 0.5	381 ± 75	275 ± 49	44	968	0.02

H_c = specific heat of combustion; SEA = specific extinction area

the escape of the volatile products generated during decomposition [52, 53]. The role of a coupling agent in the flame retardancy of clay/ PP nanocomposite is described by Gilman [53]. A composite without a MAPP, consisting only of PP and MMT (5%), burned without char leaving. In this case, PP burned completely; only clay residue, white in color, was left. However, in MMT/PP composite where MMT was modified with MAPP (15%) an extensive char formation was observed. Also, the quality of char was improved with the increase of the mass fraction of MMT from 2 to 5%. The coupling agent may have played a dual role in this experiment: it improved clay dispersion and participated in the formation of char. At a high temperature a polymer evolves a products degradation which are collected and form bubbles which are then diffused and liberated as combustible gases. A nanofiller, creating a continuous network in the composite matrix, can shield bubbling and thereby reduce the flammability of the given material [54]. The significance of nanoclay dispersion on the fire retardancy behavior of PP has been reported by Fina et al. [55]. A composite with a better dispersed structure where organophilic MMT was used OMMT/PP, had superior reduction in pHRR compared to microdispersed MMT/PP and pure PP. Qin et al. have shown that the flammability of PP containing 5% MMT, modified or pristine, decreased compared to pure PP [56]. But they also report that degree of dispersion had a minor effect on the thermal-oxidative stability and flammability of the polymer. Moreover, the authors concluded that clay catalyzed the decomposition of the matrix, which was reflected in a shortened ignition time and increased initial HRR.

However, carbonaceous silicate char which was formed due to MMT presence improved fire retardancy of nanocomposite in general. Tang *et al.* used a combination of organo-modified clay and a nickel catalyst (Ni-Cat) to enhance the carbonization of PP during burning [57]. Nickel catalyst is known as a good catalyst for the synthesis of CNTs. This combination was found to increase the fire retardancy of the composite significantly, which was reflected in the decreased peak of HRR and increased residue amount.

It has been shown in several studies that clays exhibit a synergistic effect with conventional FRs, often allowing a reduction in the amount of FRs loading. Marosi *et al.* modified an APP intumescence fire retardant (IFR) system, consisting of 75% APP and 25% polyol, by organoclay which is a potential protective char former in a PP-based nanocomposite [58]. Two types of MMT were applied, one was modified only with a surfactant, OMMT, and one was modified with a borosiloxane elastomer (BSil), additionally. When 1% of OMMT was loaded, no changes in the LOI and UL-94 ratings were observed; however, the pHRR was delayed and decreased (see Table 6.11). The loading of the IFR increased the LOI, delayed and reduced the pHRR and improved the UL-94 rating. The combination of organoclay and intumescence system resulted in a V0 rating and significant delaying and diminishing the peak of HRR. The incorporation of borosiloxane elastomer-treated OMMT resulted in further improvement of the flame retardancy of the composite.

Table 6.11 Flammability test data for PP and PP nanocomposite [58].

Sample	LOI (%)	UL-94	Peak of HRR (kW/m^2)	Peak of HRR delay time (s)
PP	17	NR	2571	110
PP + 1% OMMT	17	NR	1743	150
PP + 34% IFR	29	V2	375	237
PP + 1% OMMT + 33% IFR	31	V0	226	417
PP + 1% OMMT + 31% IFR + 2% BSil	37	V0	201	501

The synergistic effect of APP and organic MMT in the flame retardancy of PP has been reported by Zhu et al. [59]. In those study MMT was used alone or modified with a different FRs, such as melamine, triphenylphonium (TPP) chloride and tetradecyl trihexyl phosphonium (TTP) bromide. TTP was found had a negative effect on the fire retardancy of nanocomposite, whereas combinations of APP with MMT, MMT/melamine or MMT/TPP had a positive effect. The role of the intercalating agent, which plays a significant role in MMT dispersion, was also remarkable in the formation of char as well. Melamine as the blowing agent promoted a uniform distribution of MMT layers. In addition, the gas released during melamine decomposition moved the MMT particles toward the surface of the composite, which resulted in intumescent char formation, whereas TPP, which emits neither gas nor water during combustion, resulted in ceramic-like char with poor protection of the composite. The synergistic effect of nanoclays with MH has been observed by Marosfoi *et al.* [60]. They used two types of clay, sepiolite and MMT, and MH for the flame retardancy of PP. Selected results of a cone calorimeter test are shown in Table 6.12. As can be seen, clays alone had an

Table 6.12 Combustion parameters of PP and PP composites in a cone calorimeter test provided under a heat flux of 50 kW/m² [60].

Sample	IT (s)	Peak HRR (kW/m²)	Peak of HRR delay time (s)	THR (MJ/m²)	Burning rate (mm/min)
PP	37	584	167	75.6	9.7
PP + MH (15%)	31	381	179	61.2	8.2
PP + sep (5%)	24	533	111	68.1	10.2
PP + osep (5%)	23	515	124	66.1	–
PP + MH (15%) + osep (5%)	26	205	255	53.5	–
PP + OMMT (2.5%) + sep (2.5%)	62	417	213	63.7	4.1
PP + MH (15%) + OMMT (2.5%) + sep (2.5%)	50	209	316	50.1	5.6

MMT = montmorillonite; OMMT = organo-MMT; sep = sepiolite; osep = organosepiolite

MH = magnesium hydroxide

insignificant effect on the most parameters, whereas their combination with MH allowed to improve PP stability significantly. Also, the mixture of clays had a synergistic effect, showing maximal protective properties in combination with MH. The action of the clays mixture was attributed to a difference in the physical form of the clays, plate-like MMT and fibrous sepiolite. The mixed clays formed a networked protective layer on the surface during burning, which was more effective than MMT or sepiolite based chars.

Zaikov and Lomakin have shown that PP filled with silica and $SnCl_2$ (3 wt% Si + 2 wt% $SnCl_2$) had significantly reduced flammability compared to pure PP [61]. The ignition time increased from 62 to 91 s, the pHRR decreased from 1378 to 860 kW/m^2, the THR decreased from 332 to 194 MJ/m^2, and char yield increased from 0 to 10 wt%. The mechanism action of the fillers combination was attributed to 1) the formation of a barrier (char) that hindered the supply of oxygen and reduced the thermal conductivity of the material to limit heat transfer, and 2) trapping the active radicals in the vapor and condensed phases. In another work it was reported that silicon powder (1–5%), alone or in combination with MH or APP, improved flame retardancy (decreased CO evolution, smoke production and HRR) of a PP composite significantly. The incorporation of 25% of MH reduced the pHRR by 67%, whereas the combination of 5% of silica with 20–45% of MH resulted in a 73–85% pHRR reduction compared to pure PP. The possible mechanism action of silica was attributed to its silanol groups, (-SiOH), which can form a crosslinked network with the polymer and act as a heat and mass barrier [62].

The influence of OMMT on the flammability of a PP biocomposite was studied by Kord [63]. Organoclay (1–5%) was incorporated into a hemp/PP hybrid. MAPP (2%) was used as the coupling agent. The burning rate decreased monotonically, from approximately 47 to 20 mm/min with organoclay loading of 1–5%, whereas the time to ignition increased from 30 to 80 s. Char residue increased from 15 to 30%. The total smoke production decreased from 300 to 200 m^2/m^2. The data has been taken from figures in the article. The improved parameters were attributed to nanocaly action, which facilitated carbonaceous char formation on the composite surface during burning. The compatibilizer used facilitated the dispersion of organoclay particles in the matrix. Optimal dispersion was achieved at 3 wt% of OMMT, after which the dispersion decreased slightly with increasing of clay portion. Nevertheless, the level of dispersion was good enough for the formation of a functional protective layer. The increased time to ignition with the MMT loading was related to the increased thickness of the protective layer, which requires a higher temperature and longer time period to be destroyed. In addition to reduced flammability, the mechanical properties of the composite were improved.

Kashiwagi *et al.* have studied the influence of CNT on the flammability of PP. They report that the heat release rate was significantly decreased in the presence of multiwalled CNT compared to neat PP [64, 65]. The pHRR of the composites was about 27% and 32% of that of neat PP for 1 vol% and 2 vol% CNT, respectively. The total heat release was not changed significantly after the CNT loading, but the samples of nanocomposites burnt much slower than PP. The ignition time decreased in the CNT-containing composites due to the fact that composites absorb more radiation than pure polymer and hence ignite early. The fire retardancy mechanism of CNT was attributed to its capability to form a protective char layer on the composite surface. Two probable mechanisms of surface protective layer formation are proposed: 1) gaseous bubbles formed during polymer degradation move the carbon nanotubes towards the surface, or 2) the polymer moved from the surface by heat, leaving the CNT on the top. This layer shields the underlayer material from heat and oxygen transfer; and also, the nanotube network acts as a barrier against flammable gases emission formed during the thermal decomposition of the polymer.

Zhang *et al.* combined silica with APP in order to improve a wood fiber/ PP composite. Cone calorimetry and LOI test results showed that the substitution of 10 phr of APP with silica resulted in improved parameters (Table 6.13) [66]. During sample burning, silica tended to accumulate on top of the sample and form a charred layer by combining with APP. The inspection of burned samples also showed that silica prevented cracking of

Table 6.13 Cone calorimeter and LOI results of WF/PP composites [66].

Sample	IT (s)	pHRR (kW/m^2)	Average HRR (kW/m^2)	MLR (g/s)	LOI
PP/WF	12	701	301	0.175	21.4
PP/WF/APP(30)	18	505	199	0.124	27.4
PP/WF/APP(20)/ silica(10)	32	428	156	0.11	28.9

the sample surface. Thus, the silica-based charred layer was more effective in composite protection during burning, than that of without one.

The effect of a combination of CNT (2%) and sepiolite (10%) on the fire retardancy of a PP nanocomposite was studied by Hapuarachichi et al. [67]. The combination of nanofillers resulted in a significant reduction of the phRR, by 82%, compared to neat PP. The reduction in the phRR was attributed to functional char formation where both filler, CNT and clay, were involved. Also, increased melt viscosity in the composite resulted in restrained mobility of polymer chains during combustion, leading to significant improvements in fire retardancy. Char residue increased from 2% to 10% for PP and PP/CNT/sepiolite, respectively. The total smoke release did not change, suggesting that the fire retardancy was due to changes in the condensed phase and not in the gas phase.

Fu et al. studied the effect of CNT on the performance of a wood flour/PP composite [28]. A coupling agent, maleic anhydride grafted PP, was used to improve the dispersion of hydrophilic wood fibers in a hydrophobic PP matrix. A hydroxylated form of CNT, CNT-OH, was used in order to facilitate nanotube dispersion. Hydroxyl groups have an affinity for hydrophilic wood particles, and, on the other hand, CNT-OH reacted with the CA, thereby improving CNT dispersion in the PP matrix. Selected results of the cone calorimeter test are shown in Table 6.14. As can be seen, the coupling agent had no effect on the flammability of the composite, whereas incorporation of CNT reduced the phRR and THR, where hydroxylated CNT had a better effect. On the basis of the char residue study it was concluded that the ability of CNT to create a network in the composite was key to CNT action. CNT-OH had better fire retardancy due to better dispersion

Table 6.14 Cone calorimeter data for composite samples at a heat flux of 35 kW/m² [28].

Sample	IT (s)	phRR (kW/m²)	THR (MJ/m²)
PP/WF	26	383	77
PP/WF/MAPP	25	375	75
PP/WF/MAPP/CNT(1%)	23	318	65
PP/WF/MAPP/CNT-OH(1%)	23	285	56

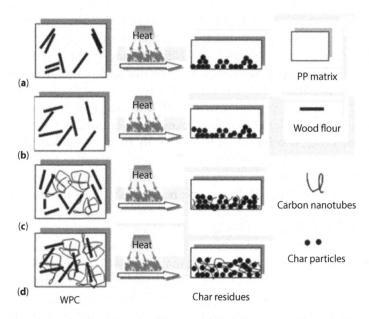

Figure 6.2 Schematic representation of flame retardancy models of WPCs: (a) WF/PP, (b) WF/PP/MAPP, (c) WF/PP/MAPP/CNT(1%) and (d) WF/PP/MAPP/CNT-OH(1%). (Adapted from [28])

and the capability to create more compact char during combustion. The mechanism action of CNT is shown schematically in Figure 6.2.

Ashori and Nourbakhsh reported that partial replacing of wood flour with microcrystalline cellulose did not improve the thermal stability of a wood flour/PP nanocomposite [68]. Lignin is thermally more stable than cellulose or hemicellulose and its removal can be the reason for the decreased stability of the composite.

Taking into account that the most destructive processes in materials, including biodegradation and ignition, are surface-controlled, the improved surface parameters solely may improve the composite performance while reducing cost. Also, reduced the loading amount of conventional fire retardants, can help to avoid deterioration of the mechanical properties of the composite. Turku *et al.* reported that loading different conventional FRs, ATH, melamine, ZB, graphite and TiO_2, in the amount of 10% into the shell layer of a coextruded composite resulted in a pHRR reduction by 8–22%, depending on the FR type [12]. In another study, 3% of carbon-based fillers, CNT, carbon black, graphite, EG and carbon fibers, were applied. The reduction in the pHRR was 6–17% [13].

6.3 Conclusions

This chapter presented an overview of flammability-related studies on PP-based biocomposites and nanocomposites and different methods to protect the composite against thermal decomposition. Conventional fire retardants are effective if they are loaded in large amounts, which generally has a negative influence on the mechanical properties of the composite. The application of nanosized fillers can reduce flammability, usually without a decline of the overall composite performance. Another potential method in the fire retardancy field is using a combination of different fire retardants showing synergism. One of the new approaches to control composite flammability can be using a protective coating, shown by an example of pure polymer [69] or co-extrusion where flame retardants are present on the composite surface.

References

1. Potts, J.E., Clendinning, R.A., Ackart, W.B., and Niegisch, W.D., The biodegradability of synthetic polymers, in: *Polymers and Ecological Problems*, Guillet, J. (Ed.), pp. 61–79, Plenum Press: London, 1973.
2. Nourbakhsh, A., Ashori, A., and Tabrizi, A.K., Characterization and biodegradability of polypropylene composites using agricultural residues and waste fish. *Compos. Part B* 56, 279–283, 2014.
3. Strömberg, E., and Karlsson, S., The effect of biodegradation on surface and bulk property changes of polypropylene, recycled polypropylene and polylactic biocomposites. *Int. Biodeter. Biodegr.* 63, 1045–1053, 2009.
4. Arkatkar, A., Arutchelvi, J., Bhaduri, S., Uppara, P.V., and Doble, M., Degradation of unpretreated and thermally pretreated polypropylene by soil consortia. *Int. Biodeter. Biodegr.* 63, 106–111, 2009.
5. Islam, M.S., Hasbullah, N.A.B., Hasan, M., Talib, Z.A., Jawaid, M., and Haafiz, M.K.M., Physical, mechanical and biodegradable properties of kenaf/coir hydrid fiber reinforced polymer nanocomposites. *Mater. Today Commun.* 4, 69–76, 2015.
6. Suharty, N.S., Almanar, I.P., Sudirman, Dihardjo, K., and Astasari, N., Flammability, biodegradability and mechanical properties of biocomposites waste polypropylene/kenaf fiber containing nano $CaCO_3$ with diammonium phosphate. *Procedia Chem.* 4, 282–287, 2012.
7. Vilaplana, F., Strömberg, E., and Karlsson, S., Environmental and resource aspects of sustainable biocomposites. *Polym. Degrad. Stab.* 95, 2147–2161, 2010.
8. Shah, A.A., Hasan, F., Hameed, A., and Ahmed, S., Biological degradation of plastics: A comprehensive review. *Biotechnol. Adv.* 26, 246–265, 2008.

9. Mofokeng, J.P., Luyt, A.S., Tabi, T., and Kovacs, J., Comparison of injection moulded natural fibre-reinforced composites with PP and PLA as matrices. *J. Thermoplast. Compos. Mater.* 25(8), 927–948, 2011.

10. Thwe, M.M., and Liao, K., Effects of environmental aging on the mechanical properties of bamboo-glass fiber reinforced polymer matrix hybrid composites. *Compos. Part A* 33, 43–52, 2002.

11. Butylina, S., and Kärki, T., Resistance to weathering of wood-polypropylene and wood-wollastonite-polypropylene composites made with and without carbon black. *Pigm. Resin Technol.* 43, 185–193, 2014.

12. Turku, I., Nikolaeva, M., and Kärki, T., The effect of fire retardants on the flammability, mechanical properties, and wettability of co-extruded PP-based wood-plastic composites. *Bioresources* 9, 1539–1551, 2014.

13. Turku, I., and Kärki, T., The influence of carbon-based fillers on flammability of polypropylene-based co-extruded wood-plastic composite. *Fire Mater.* 40, 498–506, 2016.

14. Turku, I., and Kärki, T., Accelerated weathering of wood-polypropylene composite containing carbon fillers. *J. Compos. Mater.* 50, 1387–1393, 2016.

15. Turku, I., and Kärki, T., Accelerated weathering of fire-retarded wood-polypropylene composites. *Compos. Part A* 81, 305–312, 2016.

16. Klysov, A.A., *Wood Plastic Composites*, John Wiley & Sons, Inc., 2007.

17. Wilkie, C.A., and Morgan, A.B. (Eds.), *Fire Retardancy of Polymeric Materials*, 2nd ed., Tailor and Francis Group, 2010.

18. Papaspyrides, C.D., and Kiliaris, P. (Eds.), *Polymer Green Flame Retardants*, Elsevier, 2014.

19. Helwig, M., and Paukszta, D., Flammability of composites based on polypropylene and flax fibers. *Mol. Cryst. Liq. Cryst.* 354(1), 373–380, 2000.

20. Borysiak, S., Paukszta, D., and Helwig, M., Flammability of wood–polypropylene composites. *Polym. Degrad. Stab.* 91, 3339–3343, 2006.

21. Kozlowski, R., and Wladyka-Przybylak, M., Flammability and fire resistance of composites reinforced by natural fibers. *Polym. Adv. Technol.* 19, 446–453, 2008.

22. Sain, M., Park, S.H., Suhara, F., and Law, S., Flame retardants and mechanical properties of natural fibre–PP composites containing magnesium hydroxide. *Polym. Degrad. Stab.* 83, 363–367, 2004.

23. Suppakarn, N., and Jarukumjorn, K., Mechanical properties and flammability of sisal/PP composites: Effect of flame retardant type and content. *Compos. Part B* 40, 613–618, 2009.

24. Unemura, T., Arao, Y., Nakamura, S., Tomita, Y., and Tanaka, T., Synergy effect of wood flour and fire retardants in flammability of wood-plastic composites. *Energy Procedia* 56, 48–56, 2014.

25. Arao, Y., Nakamura, S., Tomita, Y., Takakuwa, K., Umemura, T., and Tanaka, T., Improvement on fire retardancy of wood flour/polypropylene composites using various fire retardants. *Polym. Degrad. Stab.* 100, 79–85, 2014.

26. Morgan, A.B., and Wilkie, C.A., *Flame Retardant Polymer Nanocomposites*, Wiley-Interscience, 2007.

27. Albano, C., Gonźalez, J., Ichazo, M., and Kaiser, D., Thermal stability of blends of polyolefins and sisal fiber. *Polym. Degrad. Stab.* 66, 179–190, 1999.

28. Fu, S., Song, P., Yang, H., Jin, Y., Lu, F., Ye, J., *et al.*, Effect of carbon nanotubes and its functionalization on the thermal and flammability properties of polypropylene/wood flour composites. *J. Mater. Sci.* 45, 3520–3528, 2010.

29. Zhang, S., and Horroks, A.R., A review of flame retardant polypropylene fibers. *Prog. Polym. Sci.* 28, 1517–1538, 2003.

30. Ikonomou, M.G., Rayne, S., and Addison, R.F., Exponential increases of the brominated flame retardants, polybrominated diphenyl ethers, in the Canadian arctic from 1981 to 2000. *Environ. Sci. Technol.* 36, 1886–1892, 2002.

31. Bourbigot, S., Le Bras, M., Duquesne, S., and Rochery, M., Resent advances for intumescent polymers. *Macromol. Mater. Eng.* 289, 499–511, 2004.

32. Horroks, A.R., and Price, D., *Advances in Fire Retardant Materials*, Woodhead Publishing Limited: Cambridge, England, 2008.

33. Schirp, A., and Su, S., Effectiveness of pre-treated wood particles and halogen-free flame retardants used in wood-plastic composites. *Polym. Degrad. Stab.* 126, 81–92, 2016.

34. Matkó, S., Toldy, A., Keszei, S., Anna, P., Bertalan, G., and Marosi, G., Flame retardancy of biodegradable polymers and biocomposites. *Polym. Degrad. Stab.* 88, 138–145, 2005.

35. Schartel, B., Braun, U., Schwarz, U., and Reinemann, S., Fire retardancy of polypropylene/flax blends. *Polymer* 44, 6241–6250, 2003.

36. Le Bras, M., Duquesne, S., Fois, M., Grisel, M., and Poutch, F., Intumescent polypropylene/flax blends: A preliminary study. *Polym. Degrad. Stab.* 88, 80–84, 2005.

37. Seefeldt, H., Broun, U., and Wagner, M., Residue stabilization in the fire retardancy of wood-plastic composites. Combination of ammonium polyphosphate, expandable graphite, and red phosphorus. *Macromol. Chem. Phys.* 213, 2370–2377, 2012.

38. Naumann, A., Seefeldt, H., Stephan, I., Broun, U., and Noll, M., Material resistance of flame retarded wood-plastic composites against fire and fungal decay. *Polym. Degrad. Stab.* 97, 1189–1196, 2012.

39. Yu, F., Xu, F., Song, Y., Fang, Y., Zhang, Z., Wang, Q., and Wang, F., Expandable graphite's versatility and synergy with carbon black and ammonium polyphosphate in improving antistatic and fire-retardant properties of wood flour/polypropylene composites. *Polym. Compos.* 38, 767–773, 2017. doi: 10.1002/pc.23636.

40. Bai, G., Guo, C., and Li, L., Synergistic effect of intumescent flame retardant and expandable graphite on mechanical and flame-retardant properties of wood flour-polypropylene composites. *Constr. Build. Mater.* 50, 148–153, 2014.

41. Hämäläinen, K., and Kärki, T., Effect of wood flour modification on the fire retardancy of wood-plastic composites. *Eur. J. Wood Wood Prod.* 72, 703–711, 2014.

42. Rakotomalala, M., Wagner, S., and Döring, M., Recent developments in halogen free flame retardants for epoxy resins for electrical and electronic applications. *Materials* 3, 4300–4327, 2010.
43. Troitzsch, J., *Plastics Flammability Handbook*, Hanser Publishers, 2004.
44. Mariappan, T., and Wilkie, C.A., Combinations of elements: A new paradigm for fire retardancy. *Macromol. Chem. Phys.* 213, 1987–1995, 2012.
45. Conzatti, L., Giunco, F., Stagnaro, P., Patrucco, A., Marano, C., Rink, M., and Marsano, E., Composites based on polypropylene and short wool fibres. *Compos. Part A* 47, 165–171, 2013.
46. Kim, N.K., Lin, R.J.T., and Bhattacharyya, D., Effects of wool fibres, ammonium polyphosphate and polymer viscosity on the flammability and mechanical performance of PP/wool composites. *Polym. Degrad. Stab.* 119, 167–177, 2015.
47. Subasinghe, A.D.L., Das, R., and Bhattacharyya, D., Parametric analysis of flammability performance of polypropylene/kenaf composites. *J. Mater. Sci.* 51, 2101–2111, 2016.
48. Turku, I., and Kärki, T., Research progress in wood-plastic nanocomposites: A review. *J. Thermoplast. Compos. Mater.* 27, 180–204, 2014.
49. Kiliaris, P., and Papaspyrides, C.D., Polymer/layered silicate (clay) nanocomposites: An overview of flame retardancy. *Prog. Polym. Sci.* 35, 902–958, 2010.
50. Zanetti, M., Camino, G., Reichert, P., and Mülhaupt, R., Thermal behaviour of poly(propylene) layered silicate nanocomposites. *Macromol. Rapid Commun.* 22, 176–180, 2011.
51. Song, R., Wang, Z., Meng, X., Shang, B., and Tang, T., Influences of catalysis and dispersion of organically modified montmorillonite on flame retardancy of polypropylene nanocomposites. *J. Appl. Polym. Sci.* 106, 3488–3494, 2007.
52. Gilman, J.W., Jackson, C.L., Morgan, A.B., and Harris, R.H., Flammability properties of polymer-layered-silicate nanocomposites, polypropylene and polystyrene nanocomposites. *Chem. Mater.* 12, 1866–1873, 2000.
53. Gilman, J.W., Flammability and thermal stability studies of polymer layered-silicate (clay) nanocomposites. *Appl. Clay Sci.* 15, 31–49, 1999.
54. Kashiwagi, T., Mu, M., Winey, K., Cipriano, B., Raghavan, S.R., Pack, S., Rafailovich, M., Yang, Y., Grulke, E., Shields, J., Harris, R., and Douglas, J., Relation between viscoelastic and flammability properties of polymer nanocomposites. *Polymer* 49, 4358–4368, 2008.
55. Fina, A., Cuttica, F., and Camino, G., Ignition of polypropylene/montmorillonite nanocomposite. *Polym. Degrad. Stab.* 97, 2619–2626, 2012.
56. Qin, H., Zhang, S., Zhao, C., Hu, G., and Yang, M., Flame retardant mechanism of polymer/clay nanocomposites based on polypropylene. *Polymer* 46, 8386–8395, 2005.
57. Tang, T., Chen, X., Chen, H., Meng, X., Jiang, Z., and Bi, W., Catalyzing carbonization of polypropylene itself by supported nickel catalyst during combustion of polypropylene/clay nanocomposite for improving fire retardancy. *Chem. Mater.* 17, 2799–2802, 2005.

58. Marosi, G., Marton, A., Szep, A., Csontos, I., Keszei, S., Zimonyi, E., Toth, A., Almeras, X., and Le Bras, M., Fire retardancy effect of migration in polypropylene nanocomposites induced by modified interlayer. *Polym. Degrad. Stab.* 82, 379–385, 2003.

59. Zhu, H., Li, J., Zhu, Y., and Chen, S., Role of organic intercalation agent with flame retardant groups in montmorillonite (MMT) in properties of polypropylene composites. *Polym. Adv. Technol.* 25, 872–880, 2014.

60. Marosfoi, B.B., Garas, S., Bodzay, B., Zubonyai, F., and Marosi, G., Flame retardancy study on magnesium hydroxide associated with clays of different morphology in polypropylene matrix. *Polym. Adv. Technol.* 19, 693–700, 2008.

61. Zaikov, G.I., and Lomakin, S.M., Polymer flame retardancy: A new approach. *J. Appl. Polym. Sci.* 68, 715–725, 1998.

62. Pape, P.G., and Romenesko, T.J., Role of silicon powders in reducing the heat release rate and evolution of smoke in flame retardant thermoplastics. *J. Vinyl Addit. Technol.* 3, 225–232, 1997.

63. Kord, B., Effect of nanoparticles loading on properties of polymeric composite based on hemp fiber/polypropylene. *J. Thermoplast. Compos. Mater.* 25, 793–806, 2011.

64. Kashiwagi, T., Grulke, E., Hilding, J., Harris, R., Awad, W., and Douglas, J., Thermal degradation and flammability properties of poly(propylene)/carbon nanotube composites. *Macromol. Rapid Commun.* 23, 761–765, 2002.

65. Kashiwagi, T., Grulke, E., Hilding, J., Groth, K., Harris, R., Butler, K., Shields, J., Kharchenko, S., and Douglas, J., Thermal and flammability properties of polypropylene/carbon nanotube nanocomposites. *Polymer* 45, 4227–4239, 2004.

66. Zhang, X.X., Lin, Q.J., and Chen, L.H., Effect of flame retardants on mechanical properties, flammability and foamability of PP/wood-fiber composites. *Compos. Part B* 43, 150–158, 2012.

67. Hapuarachichi, T.D., Peijs, T., and Bilotti, E., Thermal degradation and flammability behavior of polypropylene/clay/carbon nanotube composite systems. *Polym. Adv. Technol.* 24, 331–338, 2012.

68. Ashori, A., and Nourbakhsh, A., Performance properties of microcrystalline cellulose as a reinforcing agent in wood plastic composite. *Compos. Part B* 41, 578–581, 2010.

69. Jimenez, M., Duquesne, S., and Bourbigot, S., Fire protection of polypropylene and polycarbonate by intumescence coatings. *Polym. Adv. Technol.* 23, 130–135, 2012.

7

Polypropylene Single-Polymer Composites

Jian Wang

School of Chemical Engineering, Beijing Institute of Technology, China

Abstract

Single-polymer composites (SPCs) refer to the class of composites in which the matrix and reinforcement come from the same polymer; thus SPCs do not have the classical problem of poor interfacial bonding between the reinforcement and the matrix. In addition, SPCs are easily recyclable materials that can be heated to become original materials. Polypropylene (PP) SPCs have specific economic and ecological advantages over traditional PP composites. The reinforcement in PP SPCs is also generated from PP and therefore traditional preparation methods are not suitable. Based on many different preparation principles, such as selective melting, overheating, copolymer, polymorphism, undercooling, etc., many methods have been developed to prepare PP SPCs such as hot compaction, film-stacking, coextrusion and injection molding. High stiffness, high mechanical strength, outstanding impact resistance at low density and easy recyclability are the remarkable properties of PP SPCs; applications include the automotive industry, industrial cladding, building and construction, cold temperature applications, audio products, personal protective equipment and sporting goods, etc. This chapter mainly introduces the preparation principles and methods for PP SPCs and the properties of PP SPCs. The influences of different processing conditions or parameters are also discussed. Finally, the applications of PP SPCs are illustrated.

Keywords: Single-polymer composites, fiber-reinforced polymer composites, polypropylene, hot compaction, film stacking, coextrusion, injection molding

7.1 Introduction

Polypropylene (PP) is one of the most widely used thermoplastics, which is characterized with low density, high tensile strength, and high compressive

Corresponding author: wjj_0107@163.com; wanqian089@qq.com

Visakh. P. M. and Matheus Poletto. (eds.) Polypropylene-Based Biocomposites and Bionanocomposites, (177–246) 2018 © Scrivener Publishing LLC

strength. The main reason for this is the possibility of tailoring its structure and properties for a particular application. The use of fibers as reinforcement in PP matrices has received much academic and commercial interest in the past few decades. Although excellent mechanical properties have been achieved in this way, life cycle assessment does not yield favorable results for PP composites when they are reinforced with glass, carbon, aramid, natrual fibers or fabrics. This is mostly due to the energy-intensive production of the above reinforcing fibers and limited recyclability of the corresponding composites [1]. An important contribution to environment preservation and energy savings may come from our ability to recover, recycle and/or reuse the materials. This has led to a search for alternative recycling-friendly PP composites.

On the other hand, the reinforcement and the matrix come from different materials in traditional composites. Physical or chemical incompatibility between the reinforcement and the polymer matrix in these traditional composites would prevent adhesion. Therefore, the interface problem has restricted the development of traditional PP composites.

Due to the need for environmentally friendly composite materials with good adhesion properties, the study of single-polymer composites (SPCs) is of interest. SPCs are a class of composite materials in which the matrix and the reinforcement come from the same polymer. These materials are also known as one-polymer composites, homocomposites, all-(the same)-polymer composites, or homogeneity composites, self-reinforced composites, mono-materials. The concept of SPCs was introduced by Capiati and Porter in 1975 [2]. The working principle of SPCs is identical with traditional polymer composites; transferring the stress from the weaker matrix to the stronger reinforcement via the interphase. Basic characteristics of the reinforcement differ from the matrix in SPCs; the reinforcement is anisotropic semicrystalline polymers, which always have higher stiffness and strength than amorphous ones. The reinforcements of SPCs, similar to traditional polymer composites, are also fibers, tapes and different textile assemblies. The matrices of the SPCs can be amorphous or semicrystalline polymers. The interphase in SPCs is usually given by crystalline superstructures because excellent fiber/matrix adhesion can be ensured without the help of any coupling agent, since the best adhesion and the highest interfacial shear strength can be obtained between identical materials. This is one of the great advantages of SPCs over traditional polymer composites reinforced with glass, carbon, aramid or natural fibers. It is noted that in traditional polymer composites wetting of fibers and the fiber/matrix adhesion are usually obtained by suitable sizing and coupling agents. Surface treatment of fibers is usually used for an acceptable bonding.

The great advantage of SPCs is recyclability, since the reinforcement and the matrix can be reprocessed together through remelting. Moreover, polymer fiber has lower density than glass, carbon, aramid, and natural fiber, etc., which gives SPCs with reduced weight. For instance, glass fiber has a density of 2.5–2.9 g/cm^3 and carbon fiber has a density of 1.7–1.9 g/cm^3, however, the density of PP fiber is only 0.9 g/cm^3. Therefore, SPCs can compete with traditional polymer composites in various application fields based on their easy recyclability, enhanced interfacial adhesion, reduced weight and cost balance. This is the main reason for the industrial and commercial interest behind the development of SPCs.

The SPCs are classified into two-component/constituent or one-component/constituent SPCs depending on the chemical structure of the reinforcement and the matrix [3, 4]. Two-component SPCs involve the same type of polymer with the same chemical composition but different chain configurations. Examples of these are low-density polyethylene (LDPE)/high-density polyethylene (HDPE) SPCs, LDPE/ultra-high molecular weight polyethylene (UHMWPE) SPCs, and PP copolymer/PP homopolymer SPCs. The advantage for two-component SPCs is that a relatively wide temperature window can be established due to the considerably large melting temperature difference between the matrix and reinforcement. The wide melting point contrast arises from the different chain structures of the two components. However, the different structures result in a compromised compatibility between the two constituents and reduced recyclability.

One-component SPCs are composed of polymers with the same chemical composition and chemical structure. Such polymers include but are not limited to PP, UHMWPE, polyethylene terephthalate (PET), and polyamide 6 (PA6). It was reported that the interfacial shear strength for one-component HDPE SPCs was 17 MPa, 7.5 MPa greater than two-component LDPE/HDPESPCs [10]. Furthermore, when the two constituents are made from the same polymer, it is expected that the compatibility and recyclability can be improved. Loos and Schimanski [5] illustrated this feasibility by means of embedding constrained PP fibers into films made from exactly the same resin as the fibers. Transcrystallization layers were observed along the fiber surface due to the good lattice match between the composite materials and the highly favorable energetics. The related principle will be introduced in the next section.

Topics concerning the material polypropylene (PP) hane been studied in depth. PPs reinforced with oriented PP fibers, so-called PP SPCs, or all-PP composites have specific economic and ecological advantages over traditional PP composites, especially the PP composites based on glass fibers.

This is mainly because upon recycling a PP-blend is obtained that can be reused to make PP SPCs or can be used for other PP-based applications. During the manufacture of PP traditional composites, the reinforcement is little affected in the preparation process. However, because the reinforcement in PP SPCs also comes from PP, the preparation methods for PP traditional composites are not suitable for PP SPCs. Recent studies show the large potential of developed manufacturing routes for SPCs based on different processing methods and conditions.

7.2   Preparation Principles for PP SPCs

In traditional processing methods, the polymer matrix and the reinforcements are usually processed together; however, for SPCs the fibers would be melted and the function of reinforcement would be eliminated because the melting points of the polymer matrix and the polymer fiber are the same or similar. Therefore, the difference between the melting temperatures determines the processing window; accordingly, the basic task for PP SPCs in production should be widening the processing temperature window. Different melting temperatures can be obtained by exploiting the intrinsic features of the corresponding polymers. They can enlarge melting temperature difference in two ways: by decreasing the melting temperature of matrix, or, on the other hand, by increasing the melting temperature of the reinforcement. It can be concluded that there are five different principles to enlarge the processing temperature window to realize the production of PP SPCs: selective melting, overheating, polymorphism, copolymers, supercooling, and thermoplastic elastomer. The first two principles are related to reinforcement aspects, and the others are all related to matrix aspects. Major influencing factors that shift the melting and crystallization temperatures are summarized in Figure 7.1 schematically [4].

7.2.1   Selective Melting

All semicrystalline polymers contain crystallites with different perfections, the crystallites have different melting temperatures and the corresponding polymer has a quite broad melting range. The melting is also characterized by temperatures linked to the onset, maximum and final fusion. Temperatures in the vicinity of melting onset are well suited to produce SPCs [4]. It is the basic principle of direct hot compaction of fibers/tapes [6] which has been used in SPCs based on PE, PP, PET, etc.

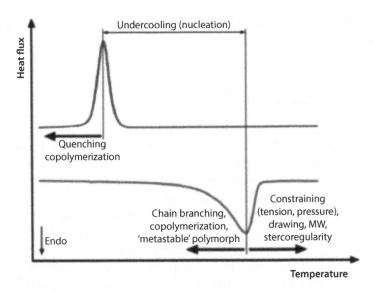

Figure 7.1 Strategies for widening the processing temperature window of SPCs' production [4].

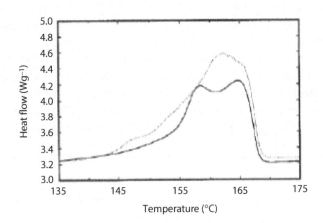

Figure 7.2 DSC endotherms for the (——) uncompacted PP fiber and (......) after melting and rescanning [7].

The method tries to control the temperature which makes the skin of the fibers melt and then become the matrix phase. Figure 7.2 shows the DSC endotherms for the uncompacted PP fiber and after melting and rescanning [7]. It is seen that after complete melting and recrystallization the rescanned sample showed a similar melting peak with the original one but the peak was quite broad, and the original two peaks were substantially overlapped. However, in hot compaction processing, the highest possible compaction temperature, 174 °C, was much higher than the melting peak temperature of the original fiber, which was due to the constraining of the compaction.

7.2.2 Overheating

Physically fixing the fiber ends can prevent shrinkage and molecular reorientation [8]. In the hot compaction method [7, 9], external pressure was applied to aid thermal contact and to restrain the fibers/tapes from shrinking, and interestingly it was found that the optimum compaction temperature was much higher than the melting temperature of the original fiber. The constraining of fibers/tapes played an important role. Compared to the bulk material, drawn fibers can exhibit a shift of the melting temperature and an increased enthalpy of melting. The concept called overheating or superheating can yield a melting temperature increase by constraining. Figure 7.3 shows that the melting temperature of the constrained PP fibers is about 20 °C higher than that of the unconstrained PP fibers [10]. To a certain extent, this principle can enlarge the processing temperature window and thus can be introduced in preparing SPCs [11]. However, constraining cannot always be effective. It is impossible to have a high degree of overheating when the molecular chains are folded or if the chain mobility is high. The overheating of different polymers depends on many different parameters, including the crystallinity level, the crystal size, and the kinetics of melting of these crystals. Barkoula *et al.* [11] investigated the effect of the post-drawing conditions on overheating of iPP. It is concluded that both post-drawing temperature and ultimate draw ratio have a significant

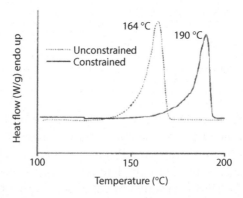

Figure 7.3 DSC curve showing the effect of constraining on the crystalline melting point of an iPP fiber [11].

influence on the degree of overheating. The following prerequisites should be fulfilled for a wide processing window: 1) the drawing temperature should be optimized in order to avoid relaxation processes in the amorphous phase, while at the same time induce orientation and improvement of the crystals in terms of size and perfection; 2) the draw ratio should be high enough (above 7) in order to have chain unfolding which is perfectly oriented (this is not easily applied on less drawable polymers); and, finally, 3) the chain mobility should be relative low for effective constraining.

7.2.3   Copolymers

When two or more different monomers unite together to polymerize, the result is called a copolymer and the process is called copolymerization [12]. Through copolymerization the macromolecular chain becomes less regular. The crystals formed are less perfect and they melt at lower temperature compared to the corresponding homopolymer. The PP copolymer is propylene-based and contains 6% (w/w) ethylene, which results in a significantly lower melting temperature. Researchers have utilized PP copolymers to prepare PP SPCs to enlarge the processing temperature window because the melting temperature of the PP copolymer is significantly lower than that of the PP homopolymer. The processing temperature window was enlarged to greater than 30 °C (see Figure 7.4) [13]. The copolymer is always the matrix giving constituent. It can be incorporated separately (in forms of fibers, films via hot pressing, film stacking) or combined with

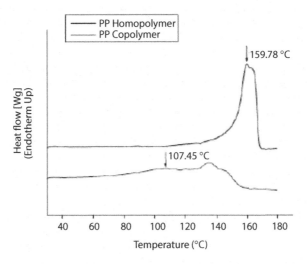

Figure 7.4 DSC melting endotherms of PP homopolymer and PP copolymers [13].

the reinforcement through a suitable preform (coextruded tape, core-shell type bicomponent fiber, cocarding, comingling, etc.) [4].

7.2.4   Polymorphism

Polymorphism is the ability of a solid material to exist in more than one form or crystal structure. PP exists in four different crystalline modifications (α, β, γ, δ) with different crystal unit (monoclinic, hexagonal, triclinic and trigonal lattices, respectively) cell parameters owing to various packing of the chains [4]. The exploitation of the polymorphism of PP can be a very promising possibility to produce PP SPCs. The β-form of isotactic PP homopolymer (β-PP) can be selectively produced by adding suitable nucleating agents [14]. The melting temperature of the β-PP can be up to 25 °C lower than that of the usual α-PP (Figure 7.5). The β/α polymorphs of PP [15] were used mainly in film stacking method to produce PP SPCs. For optimum consolidation temperature, 20–25 °C above the melting peak of the related matrix can be given. Both one-component and two-component PP SPCs are all suitable in this preparation principle.

7.2.5   Undercooling

Undercooling or supercooling refers to cooling a substance below a phase-transition temperature without the transition occurring. Polymers can be even easier to supercool because of their extremely high molecular weight

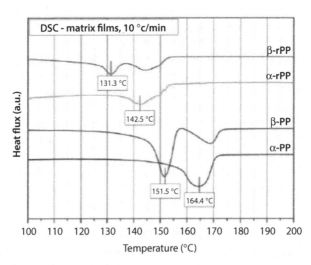

Figure 7.5 DSC curves of the different PP materials with different polymorphs [15].

and long molecular chains. Polymer crystallization can occur over a large temperature range from glass transition temperature to melting temperature. When the processing temperature is below and close to the melting temperature, crystallization can be effectively suppressed. Because there are no nuclei near the melting temperature, the polymer cannot crystallize, although the rate of crystal growth is high. Upon melting, semicrystalline polymers can typically be undercooled to a temperature well below the polymer melting temperature, while crystallization is largely absent. The key idea of applying an undercooled melt in SPC processing is that the fiber can be introduced into a liquid matrix at a temperature well below the matrix melting temperature and the fibers added to the matrix will not melt. The applicability of undercooled melt in SPC processing is expected to be largely dependent on the degree of undercooling that the polymer can undergo without solidification; the processing temperature window could be enlarged. [16].

Dai *et al.* [16] investigated the supercooling properties of PP. Figure 7.6 shows the DSC thermograms of the starting materials. The PP matrix exhibits a large capability of supercooling. It is observed that the PP matrix begins to crystallize at 124 °C, significantly below the melting point. The

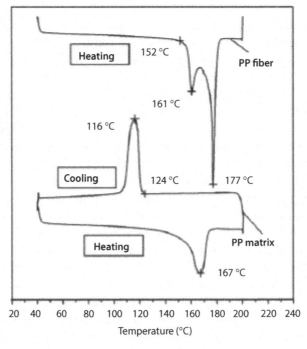

Figure 7.6 DSC thermograms of PP fiber and PP matrix [16].

processing temperature window is not the melting temperature difference between matrix and fibers but the difference between crystallization onset temperature of the matrix and the melting onset temperature of the fibers. It is a very wide processing window, more than 50 °C. However, the supercooled melt is very unstable when the temperature is in the vicinity of crystallization onset. Finally, according to the results of DSC and rheology the supercooling temperature was determined as more than 145 °C, which is much higher than the crystallization onset temperature but much lower than the melting temperature. It also reminds that DSC and rheology studies should be conducted to determine a real processing temperature window.

7.2.6   Thermoplastic Elastomer

Injection molding and extrusion molding are the most popular processes in the industrial production of polymer composites, which are used to create many things. Injection molding is suitable for volume production of products with complex geometries, such as packing, bottle caps, automotive parts and components, toys, etc., while extrusion is suitable for continuous or semicontinuous products such as films, plates, rods, tubes, cases, profiles, etc. In the injection molding or extrusion process for polymer composites, the reinforcements (glass, carbon, natural fibers) and the polymer matrix are usually mixed previously and then added into the barrel, where, under the temperature of the heaters and the shear of screw, a mixture of melted matrix and the fibers are formed. Finally, the mixture is injected into the cavity to mold the products or is extruded through the extrusion die to form the products. However, in order to ensure the melt flowability, the temperature for heating the polymer matrix is usually much higher, 50 °C or more, than its melting point. This processing temperature window cannot be realized by any the above-mentioned principles. Therefore, it is a big challenge to realize the industrial production of SPCs.

According to the copolymer principle, which may be the best one due to its having the widest processing window, Kmetty *et al.* [17] introduced thermoplastic elastomers (TPEs) into the injection molding process of PP SPCs. A significantly wider processing window (about 90 °C) can be obtained. To produce discontinuous natural and man-made fiber-reinforced composites, TPEs have been applied as matrix materials. TPEs can be made by copolymerization (resulting in block or graft copolymers) or blending of a thermoplastic polymer combined with a suitable elastomer material. The thermoplastic polymer forms the continuous phase whereas the elastomer is the dispersed one. Polyolefin-based TPEs with

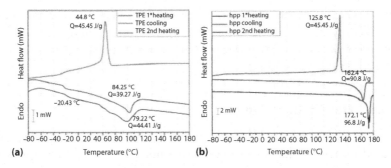

Figure 7.7 DSC curves of the PP-based TPE matrix (a) and the PP homopolymer reinforcement (b) [17].

"dispersed elastomer" phase can also be produced by copolymerization. A wide processing window for SPCs can be produced using TPE as matrix and thermoplastic fibers as reinforcement composites, because the melting temperature of semicrystalline TPE materials is very low (~80–100 °C) compared to semicrystalline thermoplastic fibers (~130–260 °C). The introduction of TPEs opens a new horizon for the industrial production of SPCs [17].

For PP SPCs, the TPE suitable for PP should be used. To compare the selected TPE with PP copolymer, Kmetty *et al.* [17] used Fourier transform infrared spectroscopy (FTIR) tests. Figure 7.7 shows the melting/cooling curves of the PP-based TPE matrix and reinforcement. A significantly wider processing window of about 90 °C can be obtained.

7.3   Processing Methods and Properties of PP SPCs

Besides the effect of temperature, the processing parameters (heating/cooling rates, pressure) are all time dependent, which strongly influences the degree of the property deterioration of the reinforcement and thus the performance of the consolidated SPCs. To realize the production of PP SPCs, adequate processing methods and conditions should be chosen. Abo El-Maaty *et al.* [7] were the first to prepare PP SPCs. Since then, different methods have been developed for preparing PP SPCs, including hot compaction, film stacking, coextrusion, injection molding, and their combination. In fact, the classification of these methods is mainly based on the main key process, since they all include many processes. For example, the hot compaction method includes weaving/winding of fibers/tapes and then hot compression; the film stacking method includes weaving/winding of fibers/tapes, extrusion or hot compression for film manufacture, followed by compression molding; the coextrusion method includes coextrusion of fibers/tapes, winding/weaving of coextruded fibers/tapes, and a final compression molding.

7.3.1   Hot Compaction

The hot compaction method for preparing PP SPCs was initially investigated by Hine *et al.* [7]. Later, this method was successfully applied to woven oriented PP fiber and tapes [9, 18, 19]. Currently, the hot compaction of PP fibers/tapes is the main commercial processing method for preparing PP SPCs.

7.3.1.1   *Hot Compaction of Fibers/Tapes*

In this method, the processing temperature is close to, but below the melting temperature. As shown in Figure 7.8, only the skin of the PP fibers melt and the central part of PP fibers did not melt. The melted section can be fused together. Then, the molten PP recrystallizes upon cooling and acts as the bonding agent among the fibers. The hot compacted PP sheet can be considered as a composite, with an oriented PP reinforcement bound together by an isotropic PP matrix (Figure 7.9). As with PP SPCs, the properties of the final material depend on the properties of the two phases. The role of the oriented reinforcing fiber or tape is to provide high stiffness and strength.

Preparation
In Hine's the investigation of hot compaction of PP fibers by Hine *et al.* [7], they employed a unidirectional arrangement of fibers and used a high-tenacity PP fiber. First, the required number of layers were cut from the PP cloth, then stacked into the matched metal mold. The assembly of PP fibers and the mold were put into a hot press machine for hot press molding. In the process, a compaction temperature between 164 and 174 °C was used and a compaction pressure of 1.1 Mpa; when the temperature was reached, the sample was left for a further 20 min to allow even heating.

Aligned fibers or tapes   Fiber surface melted at the   During cooling the melted part recrystallises
                                 selected temperature and pressure   and forms the matrix of the SRPC

Figure 7.8 Principle sketch of hot compaction on the example of unidirectionally arranged fibers [3].

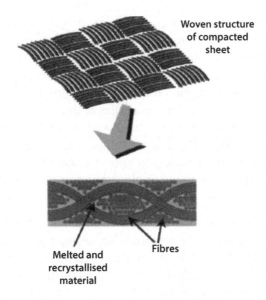

Woven structure
of compacted
sheet

Melted and
recrystallised
material

Fibres

Figure 7.9 A schematic diagram of the woven structure of a compacted sheet [20].

Finally, a compaction pressure of 14 MPa was applied for 10 s before cooling the sample to 100 °C and removing from the hot press. But the transverse strength of all the samples is very low, only reaching 7 MPa at the highest compaction temperature so far achieved of 174 °C. The "selective" surface melting of the fiber surfaces is not achieved to a sufficient degree to give good mechanical properties. The lack of controlled melting could be a consequence of the high shrinkage forces seen with this fiber, resulting in the need to use a high holding pressure during heating to stop the fibers from shrinking. The hot compaction for PP SPCs is mainly based on the principle of selective melting of PP fibers/tapes, but the overheating principle is also considered according to the holding pressure.

Hine *et al.* [9] also prepared PP SPCs by hot compaction of PP tapes (note: not fibers). A woven PP fibrillated tape was used. The weave style was nominally a plain weave. The compacted sheets with 2 mm thickness and 8 layers were manufactured using a hot press. A pressure of 2.8 MPa was applied to the mold assembly. Once the assembly reached the compaction temperature it was left for a further 10 min, after which a higher pressure of 7 MPa was applied for 10 s before cooling the assembly to 100 °C under pressure and then removing it from the hot press. The compaction temperatures from 166 to 194 °C were used to determine an optimum temperature.

Effect of Compaction Temperature

The compaction temperature is always an important parameter in the hot compaction of PP fibers/tapes for preparing PP SPCs. At low compaction temperatures, there is not enough melt to fill the available space between the PP fibers/tapes. Therefore, the mechanical properties of PP SPCs are poor. At high temperatures, the properties of the PP fiber/tape will largely decrease. The in-plane flexural modulus could reach a maximum at around 186 °C [9]. The optimum compaction temperature of 186 °C is obtained as a compromise between developing good mechanical properties through selective melting and losing the original structure of the PP tapes.

Morphology

The morphology of hot compacted PP is shown in Figure 7.10. The sample shown was made at a temperature higher than optimum, making the epitaxially recrystallized material more apparent. It can be seen how the recrystallized matrix has filled in all the gaps in the structure, forming a homogeneous well-bonded structure.

Hine *et al.* [19] established the important parameters that control the hot compaction behavior of five woven oriented PPs. The five materials

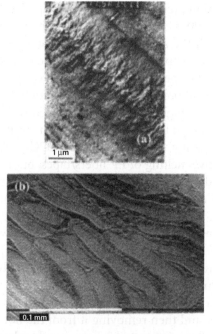

Figure 7.10 Etched SEM micrographs of compacted PP fibrillated tapes: (a) details of epitaxial crystallization, (b) section across a bundle of tapes [6].

studied used different shaped oriented components (fibers and tapes), different molecular weight polymers and various weave styles, allowing the importance of these factors on hot compaction behavior to be studied.

Effect of Molecular Weight

Hine *et al.* [19] found that the level of ductility is of crucial importance in controlling the properties of the hot compacted PP SPC sheets. Unusually for a composite, the failure strain of the oriented reinforcement is higher than the melted and recrystallized matrix phase. It is therefore crucial to have a high strain to failure rate for the matrix in order to allow the composite to retain its integrity up to the failure point of the reinforcement. Higher molecular weight and lower crystallinity (which was produced using a faster cooling rate) was found to improve ductility.

Effect of Reinforcement Geometry (Fiber or Tape)

Fiber geometry is important, in that tapes naturally present a greater surface for compaction than round fibers, which have to be distorted by compression before contact is made. However, the effect in determining the final mechanical properties is less than that of molecular weight distribution.

Effect of Weave Style

The optimum compaction temperature was different for different structures of reinforcement. It can be seen that there is a clear difference in the appearance of the fracture surfaces for pooling in the two directions (Figure 7.11) and a clear geometry (or weave style) effect on the interlayer strength for compacted samples of cloth. The best combination of properties was found for flat tapes, which gave less crimp when woven, and a balanced weave style was also found to be beneficial.

Pros and Cons

The essence of this technique is to take arrays of oriented fibers and tapes, and choose suitable conditions of temperature and pressure such that a thin skin of each fiber or tape is "selectively" melted. On cooling, this molten

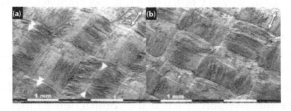

Figure 7.11 SEM micrographs of peel fracture surfaces of PP SPCs: (a) peeled parallel to the weft and (b) peeled parallel to the warp. Arrows at top right indicate peel direction [21].

material recrystallizes to bind the whole structure together. The resulting hot compacted material is therefore composed of a single, and moreover, identical polymeric material, and by virtue of molecular continuity between the phases has excellent fiber/matrix adhesion. Also, by virtue of melting the skin of each fiber or tape, there are no matrix wetting problems. Excellent compatibility between the fiber and matrix could also be realized. But this processing method has a narrow processing temperature window, typically about 10 °C or smaller. Because the skin of PP fibers is melted, heat damage to the strength of PP fiber is unavoidable. The major challenge is the temperature sensitivity of the process compared to the other techniques. The temperature gradient for thick sheet or plate products may be high, leading to non-uniform temperature distribution and incomplete matrix fill region, which will influence the properties of the final products.

7.3.1.2   Hot Compaction Combined with Interleaved Film

Hine *et al.* [22] described a route for manufacturing SPCs by combining the processes of hot compaction and film stacking. The idea is to use an interleaved film, preferably of the same polymer, placed between the layers of woven oriented elements, thereby delivering additional matrix material to the rougher interlayer region. At the same time, it allows the compaction temperature to be reduced away from that where significant melting of the crystalline structure occurs, and this also has the effect of widening the processing window. In this method, the film and the surface part of the fibers/tapes are all melted, thus a much better interfacial property could be obtained.

Preparation
The base material used was a geotextile woven oriented cloth, the film was made with exactly the same chemical composition as the cloth, thus one-component PP SPCs were prepared. The woven PP cloth has a rough surface and so can benefit from additional matrix material in the interlayer region. Compacted samples were made using a single layer of film (15 μm) in between the woven cloth, over a range of compaction temperatures between 175°C (just above the melting point of the film) and 195 °C (the point where significant crystalline melting begins to occur). The chosen assembly using the interleaved PP film between the PP woven cloth layers was placed in a matched metal mold and then put into a hot press set at the appropriate compaction temperature and a pressure of 4.9 MPa was applied. Once the assembly reached the compaction temperature, it was left to dwell for 5 min and then cooled rapidly to 90 °C using water circulation through the heated platens [22].

Effect of the Interleaved Film

The tensile strength and modulus of the optimum PP SPCs (compaction temperature of 191 °C) could be up to 168 MPa and 3.13 GPa respectively. The maximum tensile strength is around 20 MPa higher than that of the hot compacted PP SPCs without film, while the optimum compaction temperature of 191 °C is 2 °C lower than the temperature considered as the optimum for the hot compaction without film. It was found that, in terms of choosing the optimum processing parameters, the addition of an interleaved film allows a lower compaction temperature to be used (hence widening the processing window), while also delivering a significantly higher peel load. Peel load of 12.3 N/10mm could be achieved by the compaction with interleaved film, while the peel load of the hot compacted PP SPCs without film is only 7.5 N/10mm. The peel load increases with increasing compaction temperature from 175 to 195 °C, suggesting that the peel load is roughly related to the sum of the amount of matrix produced by selective melting and from the interleaved film. This increased peel load can be crucial for thermoforming, where a high peel load is of paramount importance to prevent breakup of the structure during thermoforming.

Morphology

Figure 7.12 shows an SEM image of a sample made with a single layer of interleaved film and processed at 191 °C. The image shows the original oriented tapes, and the recrystallized matrix phase formed from a combination of the selective surface melting and the melted interleaved film. It is seen that the melted film layer is able to fill all the large gaps between the woven layers. The melted film material with the same chemical composition as oriented tape is indistinguishable from any surface

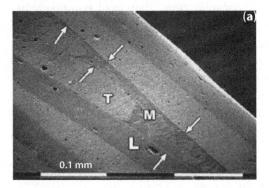

Figure 7.12 Details of the morphology of a compacted PP sample made with an interleaved film: cross section of the sample showing the original tapes (transverse [T] and longitudinal [L]) and the matrix phase [M] [22].

selective matrix melted material, although this is generally a much thinner layer (15 μm). Instances have also been seen where the additional matrix material from the interleaved film can heal gaps in the oriented tapes, which might be too large to be filled just from selectively melting the tape surfaces.

Pros and Cons
It is noted that the use of interlayer film is not absolutely crucial for processing, because PP has a significantly wider melting range. However, it still showed that the combination of film stacking with hot compaction gave a better overall balance of mechanical properties and a wider temperature window for processing compared to a standard hot compaction procedure without a film. The combination process also gave much better wetting of all the oriented elements compared to a traditional film stacking process (where only the film is melted, the film stacking method will be introduced later) due to the partial melting of all the fiber surfaces. The introduction of the interleaved film gives a better overall balance of mechanical properties, and it significantly increases the interlayer peel load and hence improves thermoformability. However, selective surface melting of each oriented element is still crucial in this method, because melting the film only, creates a poor adhesion.

7.3.1.3   Hot Compaction of Fibers with Different Orientation

Izer and Bárány [23] used the principle of overheating to perform hot compaction of textile layers. The textiles consist of two kinds of isotactic polypropylene (iPP) fibers with different orientation. The highly oriented iPP works as reinforcement and the less oriented one fulfills the role of the matrix after hot consolidation. They used two textile assemblies (carded mats, knitted fabrics): a) Plain weft knitted fabric (matrix) with inlaid iPP fiber reinforcement was prepared from the fibers (Figure 7.13), the reinforcing contents were set to be 24 and 38 wt%; b) Chopped (≈ 80 mm) matrix giving and as reinforcement working fibers were carded and needle punched together; the carding process gives a quasi-unidirectional orientation in the reinforcing mat; the reinforcing contents were set at 30, 50 and 70 wt%.

Preparation
The identical textile layers (in the case of knitted 8 and carded 4 plies) were placed on each other by keeping the reinforcement alignment constant and therefore the resulting composite plates are of an isotropic nature. These packages were consolidated by hot pressing at three different temperatures

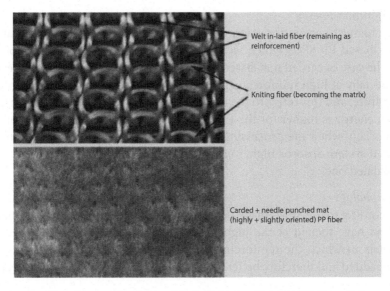

Welt in-laid fiber (remaining as reinforcement)

Kniting fiber (becoming the matrix)

Carded + needle punched mat (highly + slightly oriented) PP fiber

Figure 7.13 The scheme of the plain weft knitted fabric with inlaid iPP fiber reinforcement (hereafter referred to as knitted fabric) and carded and needle-punched mat [23].

(160, 165 and 170 °C) at constant pressure (6 MPa) and for constant holding time (2 min). The thickness of the manufactured composite sheets is approximately 3 mm.

Effect of Fiber Style
Static tensile and dynamic falling weight impact tests were performed on the PP SPC plates. In the case of higher reinforcing contents, the processing (hot consolidation) temperature influences the tensile strength and perforation energy the most significantly for both textile structures. The highest tensile strength of 139.7 MPa was obtained by the PP SPCs with knitted fabric matrix (fiber content 38.2 wt%); its tensile modulus could be up to about 2.5 GPa. The knitted fabric-based PP SPCs possess better mechanical properties than the carded mat-based version, even if the reinforcement content of the latter is much higher. The cause of the difference in tensile strength can be the fact that the knitted fabrics are composed of continuous fibers while carded mats contain chopped fibers.

Effect of Compaction Temperature on Tensile Strength
The tensile strength increases with increasing processing temperature and with increasing reinforcing fiber content. The highest temperature of 170 °C used in this process is the optimum temperature for high tensile properties.

Perforation Energy

The highest perforation energy of 25.1 J/mm was obtained by the carded mat-based PP SPCs which were made at the lowest temperature, 160 °C. In the case of carded mat-based PP SPCs the perforation energy decreases, especially at high reinforcement content, with increasing processing temperature, and this refers to the improvement of consolidation. The perforation energy is higher for the PP SPCs with higher reinforcement content, especially when the processing temperature is lower. The poorly consolidated system absorbs higher energy up to the failure than the better consolidated one.

Morphology

According to the SEM images (Figure 7.14), it can be stated that the matrix fibers have melted at the highest temperature (170 °C) and formed the matrix in which the reinforcing fibers or roving (for knitted fabrics) are embedded and therefore better adhesion can be supposed. With increasing

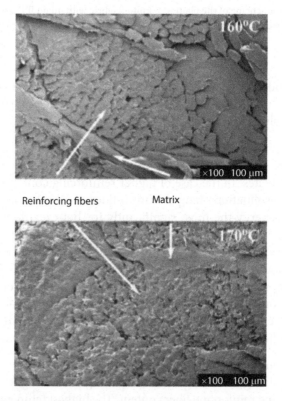

Figure 7.14 SEM images of the cut cross section of the hot consolidated (at 160 and 170 °C) knitted fabric (nominal reinforcing content: 38.2 wt%) [23].

processing temperature, the matrix fibers melted to a greater extent, wetting the reinforcing fibers more, and better adhesion can be supposed, hence consolidation improved.

Pros and Cons

This method could be an extended version of the hot compaction method. By using the different orientation properties, it enlarges the processing temperature window, although it is still 10 °C. The less oriented fibers undertake the effect of matrix, which ensures the distribution and content of matrix in the SPCs, and the effect of temperature distribution could also be suppressed. So the advantage of this technique is that the matrix phase is produced around each oriented element, giving excellent wetting and infiltration. Compared with hot compaction combined with interleaved film, this method can obtain better matrix distribution inside the SPCs.

7.3.2   Film Stacking

The film stacking method uses the melting point difference between PP fibers and PP films to prepare PP SPCs. All the principles mentioned in Section 7.2 are available in this method.

7.3.2.1   Film Stacking Based on Copolymer

Based on the principle of copolymer, Kitayama *et al.* [24] investigated the feasibility of the film stacking method for manufacturing PP SPCs. Figure 7.15 shows a schematic illustration of the film stacking process of PP SPCs. The homoisotactic PP fibers are sandwiched between the PP-PE random copolymer films. The stacking material was introduced between the female and the male molds. Then, the fiber bundles of polymer are put between the polymer films at a given pressure and temperature. The

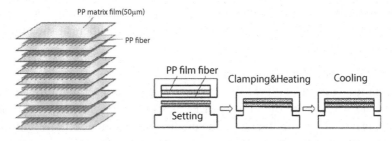

Figure 7.15 Schematic of stacking sequence and the hot pressing process for preparing PP SPCs [24].

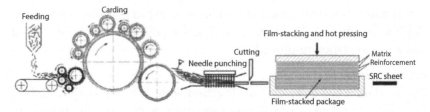

Figure 7.16 Schematic of film stacking process for preparing PP SPCs [15].

temperature should be above the melting point of the polymer films but below the melting point of the PP fiber, and PP SPCs are obtained upon cooling and release of pressure.

Preparation

The investigation of Kitayama *et al.* [24] was mainly on the interfacial properties based on morphology, and their research object was also only a model of single-fiber reinforced PP. Bárány *et al.* [15] prepared the real PP SPCs by film stacking method based on the principle of copolymer. They used carded PP as the reinforcement and random PP copolymer as the matrix to manufacture PP SPCs with film stacking. Figure 7.16 shows a schematic illustration of the film stacking process of PP SPCs. The PP fibers were carded and needle punched, the carding process results in a quasi-unidirectional orientation in the reinforcing mat. Films with a thickness of 400 μm were extruded from the granules. Composite plates with a thickness of 3 mm were produced using film stacking method by hot pressing at different processing temperatures (150–175 °C) and holding time of 90 s under a constant pressure of 5.5 MPa with a nominal reinforcement content of 50 wt%. The reinforcing layers were placed on each other in an identical direction and therefore the resulting composite plates are mainly anisotropic. The film-stacked packages were inserted between preheated plates, held for 30 s at a contact pressure of 0 MPa, pressed for 60 s at 5.5 MPa and then cooled under pressure. The holding time was kept short to prevent shrinkage and molecular relaxation of the fibers at the highest processing temperature.

Effect of Consolidation Temperature

For manufacturing of PP SPCs, Bárány *et al.* used a temperature range of 150–175 °C, where the fluidity of PP is still appropriate to wet out the highly oriented reinforcing fiber. The quality of consolidation can improve significantly in this temperature range. Increasing the temperature can significantly improve the quality of the consolidation, but the relaxation

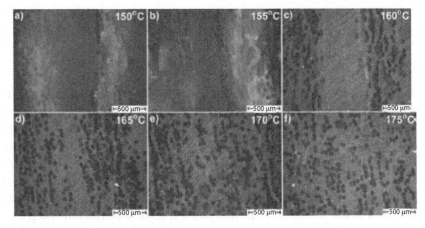

Figure 7.17 Light microscopic images taken from the polished cross section of PP SPCs in the range of 150–175°C [15].

of oriented reinforcement accelerates exponentially as soon as it melts. So the relaxation rate and the melting point of the reinforcing fibers determine the upper limit of the processing window. Figure 7.17 presents the light microscopic images of the composites with a matrix hot pressed in the temperature range of 150–175 °C. The laminate structure is preserved at lower temperatures and the transformation (the quasi-homogeneous distribution of the reinforcing fibers along the cross section) occurs at 165–170 °C for the matrix. The perforation energy decreases with increasing temperature. This refers to the improvement of consolidation, which corresponds with the other methods.

Effect of Melt Flow Indices

Two kinds of PP copolymers with different melt flow indices (MFR, 12 and 45 g/10 min) were used. By comparing the mechanical results as a function of the fluidity of the matrix, it can be stated that the tensile parameters for the matrix of MFR 45 g/10 min are a bit higher, but the perforation energy is lower than for that of MFR 12 g/10 min. The higher MFR of the material means that the length of the molecular chains should be shorter (lower molecular mass). This is reflected by the values of perforation energy determined from out-of-plane dynamic impact tests. Moreover, the poorer consolidation increases the out-of-plane fracture toughness irrespective of the matrix fluidity, since the inappropriate wetting of fibers and the higher void content increase the energy dissipation ability of the material. It is worth considering its positive effect from the viewpoint of applicability and the requirements when the main loading is out of plane, and even the dynamic impact. If the load of the material is more complex, the processing temperature of 165 °C was found to be optimum [15].

7.3.2.2   Film Stacking Based on Polymorphism

The β-form of isotactic PP homopolymer (β-PP) can be selectively produced (by adding suitable nucleants) and its melting temperature is up to 25 °C lower than the usual α-form (α-PP). Based on this principle, Bárány et al. [25] investigated the film stacking of PP SPCs by using α-isotactic PP fibers as reinforcement and β-isotactic PP homopolymer film as matrices. The carded and needle-punched PP fibers (Section 7.3.2.1) were still used. For the matrix a β polymorph rich isotactic PP (β-PP) was selected and used in the film form. This film had a thickness of 100 μm and β content of 50%. The composite plates were produced using the film stacking method by hot pressing at different processing temperature (a lower temperature range of 150–170 °C) and holding time under high pressure of 7 MPa with a nominal reinforcing content of 50 wt%. The reinforcing layers were placed on each other in identical directions and therefore the resulting composite plates are mainly anisotropic.

Consolidation Degree
Based on the work of Bárány et al. [25], it is found that the processing temperature is a more suitable control parameter than the holding time when the consolidation pressure is invariable. The consolidation degree is well characterized by the density and fairly by the peel strength data. This is due to the fact that the laminate-type lay-up progressively disintegrates with increasing hot pressing temperature. Changes in the consolidation degree are far better reflected by the tensile strength than by the stiffness data under static in-plane loading. The specific perforation impact energy under dynamic out-of-plane loading changed adversely with the density or tensile strength. This was assigned to the hindered delamination owing to a good consolidation achieved at high processing temperature. Accordingly, poor fiber/matrix bonding reduces the tensile strength (lacking stress transfer possibility via the interphase) and enhances the impact performance (triggering multiple delaminations).

Abraham and Karger-Kocsis [26] also used the film stacking method based on β and α polymorphic forms to prepare PP SPCs, but PP homopolymer tapes (note: not fibers) were used in their study.

Extrusion, Winding and Consolidation
The PP tapes used for the manufacture of laminates were produced using a twin-screw extruder. The temperature zones in the extruder were

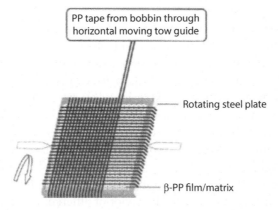

Figure 7.18 Scheme of the tape winding process for the fabrication of PP SPCs [26].

maintained at 190, 200, 210 and 220 °C from the feeder to the nozzle. The die used had a dimension of 10 × 9 × 2 mm². A thin film of β polymorph rich isotactic PP (β-PP) served as the matrix. The thin film (120 μm thickness) was obtained by compression molding β-nucleated PP sheet at 180 °C for 5 min. The manufacture of the PP laminates involved a two-stage process: winding of the PP tapes in a cross-ply manner and consolidation of the related tape consisting of fabric using hot compaction. The schematic of the tape winding process is shown in Figure 7.18. Before winding the PP tapes a thin β-PP film layer was placed on the surface of a thin steel plate. The PP tapes were wound from a bobbin onto the same steel plate rotating at a constant speed. After laying one layer of PP tape, another layer of β-PP film was placed on the surface and the winding direction on the steel plate was changed. The same process continued and the total number of layers of PP tapes and β-PP film was kept as four and five, respectively. Laminates of PP SPC were produced in a hot press at a temperature of 160 °C and a holding time of 5 min under high pressure of 7 MPa. The flexural strength of the β-PP matrix (44 MPa) is increased to 60 MPa by the effective reinforcement of α-PP tape. The flexural modulus of the final PP SPCs is 2.3 GPa, higher than the modulus of the β-PP film (1.2 GPa).

Bárány *et al.* [27] also used α-PP woven fabric as reinforcement, whereas β crystal forms of isotactic PP homopolymer and random PP copolymer (with ethylene) were used as matrix materials in the film stacking method. A plain woven fabric composed of highly stretched split PP tapes was selected and used as reinforcement; it has a melting temperature of 178 °C. Three kinds of PP were used as matrix materials: (i) β form of isotactic PP homopolymer; (ii) random PP copolymer, and (iii) β form of random PP copolymer. The non-nucleated PPs exhibited a melt flow index of 8 g/10 min.

β-Nucleation Process
For the β-nucleation of the PP, first a masterbatch was produced in an extruder with 1.5 wt% of calcium salt of suberic acid (Ca-sub) as a selective β-nucleating agent. The extrusion temperature was 220 °C and the screw rotation speed was 5 rpm. The matrix films having a thickness of 180 μm was manufactured by extrusion. The extrusion temperature was 230 °C and the screw rotation speed was 60 rpm. The final β-PP with 0.15 wt% Ca-sub content was produced by mixing the neat PP with the masterbatch in the film extrusion. To promote the β-crystallization of the PP, the rolls were heated close to 100 °C, because the preferred crystallization temperature range of the β-modification of PP is between 100 and 140 °C [14]. Figure 7.5 compares the thermal behavior of different PP matrix materials. The melting temperature value of the β modification was clearly below that of the corresponding α version, as expected.

Effect of Holding Time
It is noteworthy that the holding time at processing temperature was kept as short and low, respectively, as possible to prevent shrinkage (relaxation) of the fibers.

Effect of β-Nucleation on Homo- and Copolymers
The static mechanical and impact (perforation and tensile impact) performance of PP SPCs were studied. It has been found that the effect of β-nucleation of the matrix giving material was markedly larger for the homo- than for the random PP copolymer.

Effect of Consolidation Temperature
Both static tensile and dynamic impact response were strongly affected by the consolidation temperature. Stiffness and strength values were mostly enhanced whereas the toughness was reduced with increasing consolidation temperature. This adverse tendency was more prominent under dynamic than under static testing conditions. Figure 7.19 demonstrates the typical failure behavior of specimens of β-PP-based SPC after static tensile and dynamic impact tests manufactured at 5 and 15 °C above the relevant matrix melting temperature respectively. At low consolidation temperature, the failure occurs typically by delamination and fiber pullout. At high consolidation temperatures, the delamination was restricted and the specimens broke in the usual way.

Izer *et al.* [28] continued to investigate the film stacking of PP SPCs with β-nucleated homo- and copolymer matrices. Woven PP fabric (woven from highly stretched split PP homopolymer yarns) with melting temperature of 172.4 °C was selected as reinforcement and incorporated into about

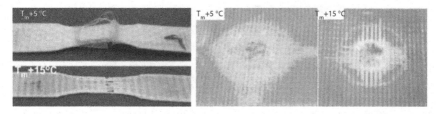

Figure 7.19 Typical failure behavior of specimens of β-PP-based SPC after static tensile and dynamic impact tests manufactured at 5 and 15 °C above the relevant matrix melting temperature respectively [27].

50% of the corresponding PP SPCs. Different crystal forms (β forms of both isotactic PP homopolymer, α and β forms of random PP copolymer) which had the same melt flow index served as matrix materials. The layers (eight reinforcing plies between nine matrix films) were placed on each other in a cross-ply arrangement to obtain orthotropic composite plates. The manufacturing process was detailed in the above section. Composite sheets with a thickness of ca. 2.5 mm and a nominal reinforcement (i.e., α-PP fabric) content of ca. 50 wt% were produced by compression molding of a film-stacked package at seven different processing temperatures. They are selected at 5, 10, 15, 20, 25, 30 and 35 °C above the relevant matrix melting temperature. The consolidation process took place as follows: after heating up the molds, the film-stacked package was inserted and held for 30 s without pressure and for 90 s under a pressure of 7 Mpa; then it was cooled to 50 °C with a cooling rate of 7.5 °C/min and demolded. The holding time at processing temperature was also kept as short and low.

Effect of Consolidation Temperature
Izer *et al.* [28] found that increasing processing temperature improved the consolidation degree based on density and peel-strength results. Moreover, this was accompanied with the appearance of significant transcrystallization in the interphase. Static tensile and flexural responses were strongly affected by the consolidation temperature. With increasing processing temperature, modulus and strength values increased and passed through a maximum. The energy absorption ability (perforation energy) decreases with increasing consolidation temperature. For optimum consolidation temperature, 20–25 °C above the DSC melting peak of the related matrix can be given. The PP SPCs produced at this small temperature range are well consolidated, have excellent interlaminar strength and the presence of transcrystalline layer (as can be seen in Figures 7.20 and 7.21), and have good mechanical performance with relatively good energy absorption at the same time.

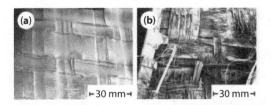

Figure 7.20 Fracture surface of peel specimens of β-PP based SPCs manufactured at (a) 157 °C and (b) 187 °C after testing [27].

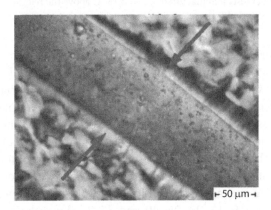

Figure 7.21 PLM image of a thin section of α-rPP-based SPCs compression molded at 172 °C [28].

Effect of β-Nucleation
The β-rPP-based SPCs provide higher perforation energy in a broader temperature range than the α-rPP- and β-PP-based ones. β-modified PP homopolymer-based one-component PP SPCs gave similarly good mechanical properties as the α-random PP copolymer-based two-component ones.

7.3.2.3 Film Stacking Based on Overheating, Copolymer and Polymorphism

According to the overheating principle, Loos and Schimanski [5] prepared PP SPCs by embedding constrained high-modulus isotactic PP (iPP) fibers in thin films of a matrix material based on the same iPP grade. In the hot press, the PP films melt, but the PP fibers do not. It illustrated that the method is feasible. However, the study was only conducted on a model of single-fiber reinforced PP and there were not any results about PP SPC products and their mechanical properties. Abraham *et al.* [29] investigated the tensile mechanical and perforation impact behavior of PP SPCs containing random PP copolymer as matrix and stretched PP homopolymer

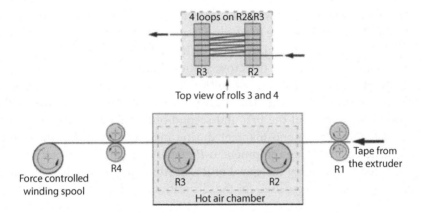

Figure 7.22 Scheme of the α-PP tape production by twin-screw extruder followed by stretching in a hot air chamber. This figure also shows the top view of the hot air chamber along with the winding pattern of the tape within [29].

as reinforcement. The β polymorph rich version of the random PP compolymer (β-rPP) was also used. The manufacture processing of the matrix film is the same as in the above investigation (Section 7.3.2.2). The final thin films of 120 μm thickness were obtained by compression molding at 180 °C for 5 min.

Stretching Process
The α-PP tapes used as reinforcement of the laminates were produced from homopolymer. According to the principle of overheating, Abraham *et al.* [29] constructed and built a stretching unit with a hot air chamber in which stretching of the extruded tape occurred and its axial orientation could be controlled. The schematic of the stretching unit is shown in Figure 7.22. The PP tape coming out from the extruder is passed through a water bath and entered into the stretching unit through the roller (R1). The temperature inside the stretching unit chamber is kept at 100 °C by blowing hot air. Then the tape is wound four times between rollers R2 and R3, which were rotating inside the hot air chamber. The number of windings was kept to four to ensure that the tape is heated enough and softened properly before stretching. The tape was heated up fast with the help of not only the hot air but also due to the direct contact with the hot steel rollers in the chamber. From the roller R2, the tape went directly to R4, which was outside of the hot air chamber, and thereafter to the force-controlled winding machine. The speed of the rollers and the temperature inside the hot air chamber were computer-controlled and monitored. The speed of R4 was changed while the speed of R1, R2 and R3 was the same. The actual

stretching took place between the last winding (beginning of fifth loop) of R2 and R4. In general, the draw ratio is a measure of the degree of stretching during the orientation of a tape or filament. The draw ratio is usually expressed as the ratio of the cross-sectional area of the undrawn to that of the drawn material. This definition assumes no volume (no density) change during stretching. The draw ratio can also be considered as the ratio of the speeds of the first and second pull-roll stands used to orient the flat polyolefin mono-filament during manufacturing. On the home-constructed stretching equipment, α-PP tapes of different draw ratios were produced by changing the speed of R4. In Abraham's research, α-PP tapes of draw ratio of 8 were used for the fabrication of PP SPC laminates [29].

Winding and Consolidation

The manufacture of the PP SPC laminates involved a two-stage process: (i) winding of the α-PP tapes (both unidirectional (0) (UD) and cross-ply (0/90) (CP)) using a thin steel plate and inserting between them films of α-rPP and β-rPP, respectively. This occurred by a typical winding machine. Accordingly, α-PP tapes were wound from a bobbin onto the steel plate rotating at a constant speed. After laying one layer of PP tapes, a layer of matrix was placed on the surface and the next layer of tapes laid. The winding direction on the steel plate was changed for CP lay-up. The same process continued and the total number of layers of α-PP tapes and matrix were kept at 10 and 11, respectively. (ii) Afterwards, PP SPC laminates were consolidated in a hot press at a temperature of 155 °C and 145 °C for the matrices of α-rPP and β-rPP, respectively. For both (UP and CP) laminates a holding time of 8 min and a pressure of 7 MPa were applied. Based on the fact that two different temperatures were used for the α- and β-rPP matrix-based composites, their properties may not be directly compared. The related processing temperature was selected 10 °C above the melting temperature of the related matrices.

Effect of β-Nucleation

PP SPCs with UD and CP lay-ups were successfully produced from α-PP tapes as reinforcement and β-rPP and α-rPP films as matrix-giving materials by using the film-stacking method. The properties of PP SPCs were investigated as a function of the matrix modifications. Though the temperature of the hot pressing conditions differed for the PP SPCs with α- and β-rPP polymorphs as matrices, their reinforcement content, density and void content were similar. Their mechanical performance under both static and dynamic conditions was compared. It turned out that the β-rPP matrix-based composites outperformed the α-rPP-based ones in respect to their static flexural and dynamic impact properties. Transcrystallization

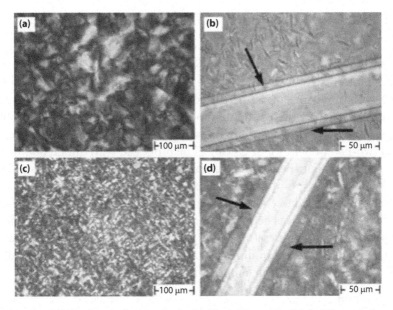

Figure 7.23 Polarized optical micrographs for (a) α-rPP matrix, (b) β-rPP matrix-based single tape model composite, (c) β-rPP matrix and (d) β-rPP matrix-based single tape model composite. Note: arrows represent the transcrystalline regions [29].

phenomenon was traced for the efficient stress transfer from the matrix to the α-PP tape reinforcement (Figure 7.23). It was more pronounced in the β-rPP than in the α-rPP matrix. The perforation energy of the β-rPP-based PP SPC laminates was found to be superior to the α-rPP-based ones irrespective to their lay-up (UD or CP). This was attributed to the higher inherent toughness of β-rPP, and to the development of a wider transcrystalline layer in β-rPP matrix compared to the α-rPP version.

Pros and Cons
The PP SPCs could be produced from α-PP fibers or α-PP tapes as the reinforcement and α-PP copolymer or β-PP copolymer or β-PP homopolymer film as the matrix by the film stacking method. All of these above studies have proved that, with the help of copolymer, polymorphism and overheating, α-PP fiber/tapes (high melt temperature) can be used for reinforcing the β-PP matrix (low melt temperature). By exploiting the melting temperature difference, it can be concluded that the processing temperature window could be enlarged according to different requirements; it is also up to the mechanical properties of the final PP SPC products. Although the adding of films leads to low void content, it may decrease the fiber volume fraction, and the interface adhesion is needed to regulate by temperature and pressure.

7.3.2.4   Film Stacking Based on Undercooling Melt (Undercooling)

Dai *et al.* [16] investigated the supercooling properties of PP and developed the undercooling melt film stacking method. Undercooling melt should be created first, and then the compaction of the undercooling melt film with highly oriented fibers is conducted. The undercooling temperature is lower than the melting temperatures of both the matrix and fiber, which will not influence the reinforcing effect of fibers. One-component and two-component PP SPCs can all be produced with undercooling melt film stacking. For the creation of undercooled melt, a temperature higher than the melting point of the matrix should be used first, and then a fast cooling rate is used to cool the melt into undercooling melt afterwards, the undercooling melt was held at an undercooling temperature which is lower than the melting temperature but higher than the crystalline temperature. For a good interfacial bonding, the undercooling temperature is usually in the vicinity of the melting temperature of the matrix.

Preparation

Thin PP sheets were prepared by compression molding the PP granules at 200 °C and 1 MPa for 5 min, followed by quenching at room temperature. The molded PP sheets were then melted and consolidated with the high-strength PP fabric to form SPC by using a customized two-station compression molding process, as schematically illustrated in Figure 7.24. The two-station process allows the PP sheets to be heated and melted at

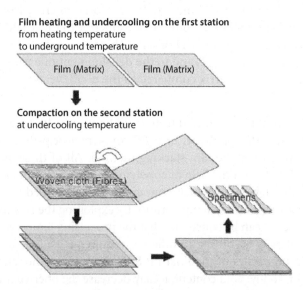

Film heating and undercooling on the first station
from heating temperature
to underground temperature

Film (Matrix)   Film (Matrix)

Compaction on the second station
at undercooling temperature

Woven cloth (Fibres)   Specimens

Figure 7.24 Schematic of undercooling melt film stacking for PP SPC manufacturing [31].

one temperature and undercooled to a second temperature within a short period of time. Specifically, two pieces of PP sheets were first heated to 200 °C for 10 min on the first station to obtain two layers of molten PP sheets. The molten PP sheets were then quickly transferred to the second station set at a lower temperature, where the molten PP sheets were supercooled. After the undercooled PP melt sheets were stabilized on the second station, a PP fabric was inserted in between them and the lamination was immediately compressed under a pressure of 9 MPa for 10 min. Then the lamination was removed and cooled to room temperature. The PP fabric was preheated to the same temperature as that of the second station before it was introduced to the undercooled PP melt layers. The compaction temperatures from 125 °C to 150 °C were used in the second compaction procedure. They are below the melting onset temperature of the fibers (152 °C) but above the crystallization onset temperature of the matrix (124 °C), so the matrix can keep its liquid state while the fibers will not melt. The compaction pressure (6–12 MPa) and compaction time (5–20 min) were used in the investigation of Wang et al. [30].

Mechanical Properties
The PP SPCs with 53.8 wt% fiber fraction could be successfully prepared by the undercooling melt film stacking method. The processing at the compaction temperature of 150 °C, holding time of 10 min and compaction pressure of 9 MPa could obtain the optimum tensile strength, 176 MPa in warp direction and 215 MPa in weft direction. The optimum tensile modulus could be up to 3.5 GPa in warp direction and 3.9 GPa in weft direction, but the maximum modulus emerged at the intermediate temperature of 135 °C. The subsequent decline in the modulus with increasing temperature is primarily due to the orientation loss of the specimen. The peeling strength of the specimen compacted at 135 °C could be up to 15.8 N/cm.

Effects of Different Processing Parameters
The influences of undercooling compaction temperature, holding time and compaction pressure were investigated. The experimental results showed that the undercooling compaction temperature is the most important factor, and that the tensile strength and peeling strength could be increased by increasing compaction temperature. Since the undercooled polymer has different holding time-dependent behavior at different compaction temperature, optimum holding time depends on the compaction temperature. Tensile strength can be promoted by raising the compaction pressure. The application of high pressure encourages polymer flow and prevents shrinkage, but the pressure cannot be set too high so as to avoid an adverse effect on wetting.

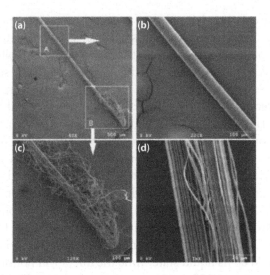

Figure 7.25 Scanning electron micrograph of the pulled-out fiber from the second single-fiber specimen [29].

Morphology

The effects of the compaction temperature were confirmed by the results of the morphological observation; the undercooling compacted composites showed good interfacial adhesion between fibers and matrix. Morphological analysis of the single-fiber specimens gave further confirmation of these conclusions (Figure 7.25).

Pros and Cons

With the aid of DSC and parallel-plate rheometry, a processing temperature window of at least 25 °C, from 125 °C to 150 °C, was established for processing PP SPCs. Within this processing temperature window, high fluidity of the matrix PP can be obtained without significantly reducing the fiber properties. This method still has the limitations of the film stacking method. Moreover, through the process of heating (to a higher temperature than the melting point) the processing temperature should experience cooling (to the undercooling temperature) and heating (at the undercooling temperature); DSC and rheometry measurements are needed in order to control an exact temperature.

7.3.3 Coextrusion

The existing technologies have some inherent limitations which reduce their viability, such as small temperature processing windows or low volume fractions of reinforcement, limiting the ultimate mechanical properties of the composites. Highly oriented, high modulus fibers or tapes can be effectively welded together by melting the surface of the tapes and applying pressure to achieve a good bonding and fill any voids between tapes.

Alcock *et al.* [13] developed a route to produce PP SPCs by using a combination of constraining/overheating and copolymer. Coextrusion technology uses the coextrusion of two types of polymer tapes (e.g., random PP copolymer/PP homopolymer) of different melting temperatures, cold drawing of the tapes to increase the mechanical properties, and finally consolidation of the tapes. The final PP SPC products can compete with glass fiber reinforced PP. They prepared unidirectional PP SPCs [13], PP SPCs with woven fabric [32–34], and PP SPC panels [35] by using hot compaction of coextruded PP tapes/fabrics.

7.3.3.1   Preparation

Coextrusion
Figure 7.26 shows that the use of tapes with a thin coextruded skin layer can lead to the creation of composites with extremely high volume fractions of reinforcement phase (oriented homopolymer). The tape direction is into/out of the plane of the paper. The coextruded tape has a skin-core structure consisting of a core of PP homopolymer surrounded by a thin skin of PP copolymer. The tape with skin-core structure was formed by coextruding in a high-viscosity melt phase and subsequently drawn in a two-stage solid-state drawing process (Figure 7.26) through hot air ovens to obtain tapes with highly oriented and high modulus. Typically these tapes have dimensions of 2–4.5 mm width and 60–125 μm thickness, depending on the draw ratio. When the draw ratio of 17 is used, it has approximate dimensions of 2.15 mm wide and 0.65 μm thick.

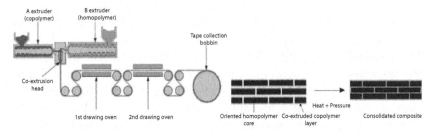

Figure 7.26 Schematic of continuous tape coextrusion and solid-state drawing for the production of highly drawn coextruded tapes [36].

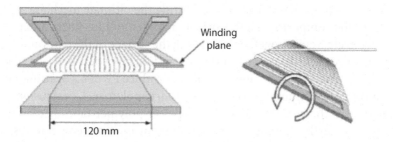

Figure 7.27 Schematic of filament winding plate and hot press to manufacture PP SPCs [13].

Winding and Consolidation

The coextruded PP tapes were regularly wound on the winding plate (Figure 7.27). Then the assembly of PP tapes and the mold was put into a hot press machine for hot press molding. In the process, a holding pressure was applied to the assembly of coextruded PP tapes and the mold. When the temperature reached the required compaction temperature (130–170 °C), the pressure was increased rapidly (1–12 MPa) and held for a certain time (10 min). Finally, the pressure was released and the mold was cooled to room temperature. The coextruded tapes can be consolidated into a composite material by the application of heat and pressure, either by filament winding for unidirectional specimens or by stacking plies of woven tape fabrics. The application of pressure also causes a physical constraining effect which raises the melting temperature of highly oriented polymers, allowing them to be "overheated."

7.3.3.2   Properties

Mechanical Properties

The drawing process results in a high degree of molecular orientation and the drawn tapes possess a high tensile strength (~ 450 MPa) and the maximum modulus (18–20 GPa) in PP tapes, while glass fibers possess a tensile strength of 3.5–5 GPa and modulus of 70–90 GPa, so for PP SPCs to compete with glass fiber reinforced PP, the volume fraction of PP tapes in the PP SPCs must be maximized. The high volume fraction of reinforcement achievable in PP SPCs (> 90%) allows PP SPCs to have competitive mechanical properties with conventional PP matrix composites.

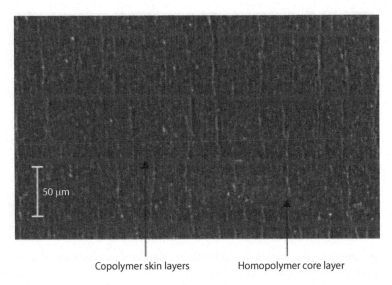

Copolymer skin layers   Homopolymer core layer

Figure 7.28 Optical micrograph of a cross section of coextruded unidirectional PP SPCs [36].

Interfacial Properties

Figure 7.28 shows the skin-core morphology of stacked tapes in a consolidated composite laminate; the tapes are oriented horizontally with tape thickness oriented in the vertical direction and tape width oriented out of the plane of the paper. The copolymer layer is clearly visible in between the highly oriented homopolymer core even within the consolidated composite. Alcock *et al.* [36] investigated the interfacial properties of the highly oriented coextruded PP tapes, the T-peel strength of PP SPCs for a given homopolymer/compolymer combination is determined by the tape draw ratio, the compaction temperature, and the drawing temperature of the tape. The temperature at which bonding occurs is dictated by the melting temperature of the copolymer skin layer. It is also affected by the temperature applied during solid-state tape drawing since drawing at too low temperature results in greater orientation of the copolymer layer. Greater temperatures are required during composite consolidation to achieve autohesion between neighboring tapes. The temperature processing window for PP SPCs is determined by the difference between the onset of adhesion of adjacent copolymer skin layers at lower consolidation temperatures and the loss of tape properties due to molecular relaxation at higher consolidation temperatures. In the research of Alcock *et al.*, the temperature

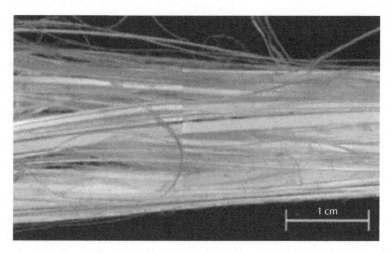

Figure 7.29 Composite optical micrograph of the fibrillated surface of a unidirectional all-PP composite loaded in the tape direction [13].

processing window was seen to be > 30 °C, allowing PP SPCs to be consolidated at a range of temperatures. Thus the interfacial properties of the composite can be tailored during production to suit the final application.

PP SPCs with Unidirectional Tapes
The optimum tensile strength of the unidirectional PP SPCs is 385 MPa and the tensile modulus is 13 GPa, which was obtained at the compaction temperature of 160 °C. Figure 7.29 shows a composite optical micrograph of a unidirectional all-PP specimen, in which the complete fibrillation of the composite can be seen. The compression strength is very close to that of bulk PP (~ 50 MPa); it is likely that no benefit is gained by using the highly oriented PP rather than bulk PP for pure compression applications. In comparison with the mechanical properties of the commercial unidirectional glass fiber reinforced PP containing 75% weight fraction of glass fiber (tensile strength of 1025 MPa), the longitudinal mechanical properties of unidirectional PP SPCs are less, but the specific tensile strength is comparable due to the low density of PP tapes.

PP SPCs with Woven Fabrics
Alcock *et al.* [32–34] made PP SPCs with woven fabrics (Figure 7.30). The coextruded tapes are woven into a plain weave fabric. The fabric is used for subsequent composite laminate production. Plies of the fabric are cut and stacked in a close-fitting mold, which is then subjected to heat and

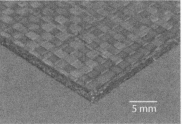

Figure 7.30 Photographs of the woven PP SPC tape fabric (left) and the consolidated PP SPC laminate (right) [34].

pressure in the hot press. In Figure 7.30, the plain weave of the fabric is clearly visible on the upper surface of the laminate; although manufactured from woven highly oriented PP tapes, these tapes are easily fibrillated and the fibrillar nature of the tapes is clearly visible on the edge of the laminate. The mechanical properties of the PP SPCs with 90% high reinforcement fraction can compete or outperform a popular type of glass fiber reinforced PP. The highest tensile strength of the PP SPCs (prepared at compaction temperature of 160 °C, pressure of 12.4 MPa) can be up to 232 MPa, which is higher than 145 MPa of the random glass reinforced PP (50 wt%) and 55 MPa of the random flax reinforced PP (40 wt%), and shows only slightly inferior tensile strength to 350 MPa of the woven glass reinforced PP (60 wt%). And the specific tensile strength could be higher than all the compared PP composites.

Processing Window
The compaction temperature and pressure can affect the mechanical properties of these composites by altering both the mesostructure of the composite (i.e., the structural composition), and the microstructure (i.e., the molecular orientation) of the tape material; but by choosing suitable parameters, loss of mechanical properties can be minimized. Figure 7.31 shows the theoretical processing window. As consolidation temperature is increased, viscosity of the copolymer decreases and so a lower minimum pressure would be predicted for adequate consolidation. However, as temperature is increased, shrinkage forces in the tapes increase and so greater pressure is required to prevent tape shrinkage. As pressure is increased, the maximum applicable temperature decreases, since at high pressures, flow and hence relaxation, is encouraged; and so a combination of high temperature and pressure aids relaxation by lateral flow of the composite in the mold.

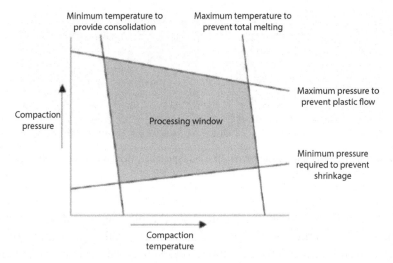

Figure 7.31 Temperature–pressure processing window for PP SPCs consolidation [33].

Impact Performance

Falling weight penetrative impact and ballistic impact tests have been conducted to analyze the impact performance of the PP SPCs. The PP SPC plates possess excellent resistance to falling weight impact penetration and can compete or outperform glass or natural fiber reinforced PP. The highest impact energy absorption of the PP SPCs compacted at a low temperature and low pressure (140 °C, 0.1 MPa) can be the highest value, 45 J/mm. The penetrative energies for woven glass reinforced PP (40 wt%), glass mat reinforced PP (23 wt%) and flax fiber reinforced PP (40 wt%) are 28.4, 9.8 and 4.5 J/mm, respectively. The dominant failure modes of PP SPCs in impact are delamination and tape fracture. Impact performance increases with decreasing interfacial strength, with penetration energy increasing with decreasing compaction temperature and pressure. The normal glass transition temperature which results in a significant decrease in impact resistance of isotropic PP at low temperatures (< 0 °C), is absent in all-PP composites, leading to high impact energy absorption even below glass transition temperature [33].

Pros and Cons

A large temperature processing window (> 30 °C) and a high fiber volume fraction (> 90%) possessing high tensile moduli and strengths can be successfully created by using a combination of constraining and coextrusion. Despite the high temperatures involved during the compaction process, the excellent mechanical properties of the oriented tapes are retained in the resulting PP SPCs. Furthermore, the composite is not sensitive to deviations in the process temperature since mechanical properties proved approximately constant within the processing temperature window. The specific mechanical properties are comparable to those reported for a commercial glass fiber reinforced PP, while the PP SPCs clearly have great advantages in terms of recyclability. Complicated procedures and long cycle time are disadvantages of this method, and shape or geometry of PP SPC products are limited in this method.

7.3.4 Injection Molding

The weaknesses of the above-mentioned methods more or less include long cycle time on the order of minutes, complicated procedure, expensive, etc., and they cannot seem to escape compression molding, greatly limiting the economics and capabilities of SPCs. Injection molding has been widely used in the field of polymer processing, and can also be used in the volume-production of SPCs. In the traditional injection molding process for polymer composites, the polymer matrix and the fiber reinforcements were usually added into the barrel of the injection molding machine together. Traditional injection molding cannot be used to produce SPCs because high temperature and intensive mixing with high shear will lead to extensive fiber damage. The key issue for injection molding of SPCs is how to establish a wide processing temperature window.

7.3.4.1 Injection-Compression Molding

Khondker *et al.* [37] used injection-compression molding method to realize the production of PP SPCs. Although the injection-compression molding method is also limited in compression process, it takes some benefits of injection molding such as short cycle time. Before the injection-compression, weft-knitting technique was adopted to produce plain knitted textile fabric (Figure 7.32) using homo-PP fibers due to its excellent drapability, stretchability and high productivity. Homo-PP and block-PP resin pellets

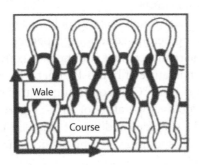

Figure 7.32 Plain knitted fabric structure [37].

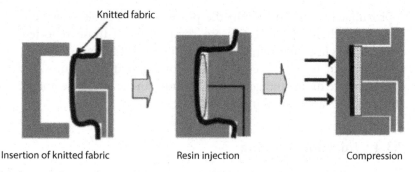

Figure 7.33 Schematic of the injection-compression molding process [37].

were used as the matrices respectively. Although they did not demonstrate the advantages of injection-compression molding, it can definitely have a shorter cycle time compared with compaction processing (hot compaction, film stacking, coextrusion, etc.), and the volume production of PP SPCs can also be realized.

Preparation

A schematic diagram of the injection-compression molding process is shown in Figure 7.33. The plain knitted PP fabric used as reinforcement was inserted into the mold, then the melted matrix was injected into the mold cavity, after which compression was conducted. In order to select an optimum processing temperature for the injection-compression molding, specimens were produced using three different combinations of resin and die temperatures (200 °C /30 °C, 200 °C /80 °C and 260 °C /80 °C). Injection speed was kept at 1 mm/s, while injection-compression load was maintained at 300 kN with a holding time of 40 s. Up to three layers of knitted fabrics were used in the fabrication of injection-compression molded composite panels, with the fiber volume fractions ranging only between 2.3% and 6.9%

Effect of Molding Temperature

Khondker *et al.* [37] have found that the molding temperature is of extreme importance for strengthening adhesion properties and keeping the fiber shape because of the reinforcement and the matrix being the same materials. The melting temperature of homo-PP fiber was 174 °C and that of the homo-PP matrix was 166.6 °C. An optimum resin temperature was chosen at 200 °C with the mold temperature of 30 °C. Figure 7.34 shows the cross section of the PP SPCs molded at different temperatures. The observations clearly reflect the corresponding melting phenomena.

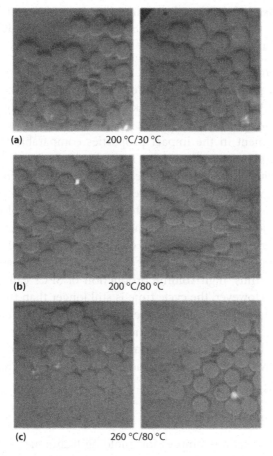

Figure 7.34 Photographs of polished cross section of the PP SPC specimens under different molding conditions [37].

Mechanical Properties

All the PPSPC specimens did not exhibit any improvement or deterioration in the tensile and 3-point bending properties as compared to those of the virgin specimens. Tensile and 3-point bending performances of virgin homo-PP and block-PP matrix materials were not all significantly influenced by the introduction of PP knitted layers over a narrow range of fiber volume fractions (less than 6.9%). Whereas it indicated that the homo-PP-based SPCs exhibited superior energy absorption capability, static and dynamic impact performances as compared to their virgin counterparts. The considerable amount of improvement was noted in both static (between 39% and 64%) and dynamic (between 63% and 345%) impact behaviors of the homo-based PP SPCs in comparison with the virgin homo-PP matrix materials.

Pros and Cons

Although virgin block-PP material exhibited better impact performances than block-PP SPCs with homo-PP knitted fabric, a notably small increase in the reinforcement fiber content (less than 6.9%) revealed considerable improvement in the impact properties comparable to those of the virgin block-PP matrix materials. These homo-PP/block-PP SPCs have clearly indicated that they have the potential to out-perform the block-PP materials via modification and/or manipulation of the reinforcement knit structural/geometrical parameters and the content of reinforcement fibers. The injection-compression molding method can have a very short cycle time to realize the production of PP SPCs, less than 1 minute can be deduced from the injection speed of 1 mm/s and the compression holding time of 40 s. Thus, high-volume production of SPCs can be realized by this method. However, the cycle time is still longer than the conventional injection molding; furthermore, the small fiber volume fraction ranging between 2.3% and 6.9% limits the improvement of mechanical properties. Additionally, like in the compaction method, the shape of the product is also limited by this method. Many more investigations still need to be done in this promising method.

7.3.4.2   Insert Injection Molding

In the above injection-compression molding technique, the compression process needs a holding time of 40 s, which is still longer than that of the conventional injection molding. Insert injection molding offers the potential for high-volume production of SPCs. In this method, PP fabric is preplaced and affixed onto the mold cavity wall by using double-sided tape or by clamping force, and then PP matrix is injected into the cavity and infiltrates the fabric by pressure. In insert injection molding, the preplaced fibers will be partially melted or unmelted. Because the injected matrix cools down in the mold cavity in several seconds, it is feasible to form SPCs even though the injection temperature is much higher than the melting temperature of the fiber. Figure 7.35 shows the temperature distribution of the injected melt in the mold cavity. The temperature of the injected melt 0.2 mm away from the cavity wall is around 160 °C with an injection temperature of 200 °C, 180 °C with an injection temperature of 240 °C, and 210 °C with an injection temperature of 280 °C. The temperature of the melt is lower closer to the mold wall due to heat conduction (Figure 7.35a). The

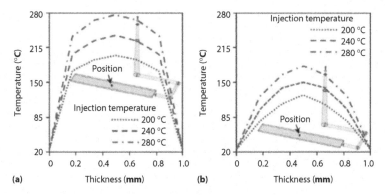

Figure 7.35 Temperature distribution of the melt in the mold cavity for injection temperatures of 200, 240 and 280 °C: (a) 1 s after injection and (b) 2 s after injection [38].

melt temperature drops sharply after about 2 s (Figure 7.35b). It is noticed that the temperature of the melt 0.1 mm away from the cavity wall can drop to around 170 °C, suggesting part of the fibers in the fabric are able to maintain maximum fiber properties. Furthermore, the undercooling melt may appear during the injection process. In addition, the melting temperature of constrained fibers shifts to a higher value due to the overheating principle. Therefore, in this method a wide temperature window can be obtained in which both one-component and two-component SPCs can all be prepared.

Wang *et al.* [38–40] realized the preparation of PP SPCs by using insert injection molding. As first, they set the two layers of fabric on both sides of the mold cavity (Figure 7.36), but a bad appearance existed due to incomplete infiltrating which was caused by the very low temperature at the cavity wall. Then, PP SPCs with sandwiched fabric were prepared by the insert injection molding, in which the inserted woven fabric was fixed by clamping force (Figure 7.37). Thus, both improved mechanical properties and good appearance were obtained.

Insert-Injection Molded One-Constituent PP SPCs
For the one-constituent PP SPCs, PP pellets were used to manufacture PP fibers first, and then they were also used as the matrix in the insert-injection molding. In order to obtain fiber which not only has higher melting temperature but also higher mechanical strength, fiber spinning and heat treatment processing conditions were investigated first. The fiber with a high melting temperature of 170.5 °C (about 5 °C higher than the melting temperature of the matrix) and a high tensile strength of 540 MPa was prepared by the pretreatment temperature of 150 °C. The fibers were further

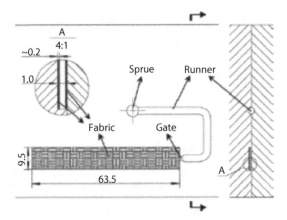

Figure 7.36 Schematic of the mold section of the insert injection molding for PP SPCs [38].

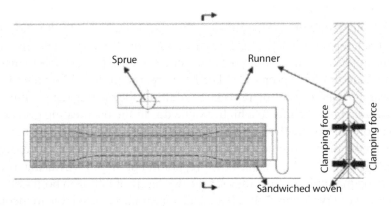

Figure 7.37 Schematic of the mold section of the insert injection molding for PP SPCs with sandwiched woven fabric [38].

woven into fabric. The fabric was affixed onto the cavity wall along the melt flow direction using double-sided tape. The preparation time for the pre-placing of fabric or fibers for each SPC sample was less than 60 s. The molten PP was injected into the cavity after the mold was closed. Then the material in the cavity was cooled and solidified. After demolding, the double-sided tape was immediately removed from the sample surfaces. For both types of PP SPCs, the injection time and holding time were fixed at 1 and 10 s, respectively. Injection and holding pressures were 207 and 167 kPa, respectively. The cooling time was 10 s. Injection temperatures from 200 to 280 °C were chosen as the major variable to investigate the influence of temperature on the properties of SPCs.

Properties
The PP SPCs containing 36 wt% lab-made fabric achieved a tensile strength of 70 MPa, 2.3 times that of neat PP. The tensile strength increased with increasing injection temperature and achieved a maximum value of 70 ± 3.9 MPa for the SPCs samples made at 260–265 °C. The increase in tensile strength may be ascribed to the lower melt viscosity at higher injection temperature and therefore improved permeability into the fabric. The tensile strength of SPCs sample decreased significantly at an injection temperature of 280 °C due to partial melting of the fabric. It was noted that the SPCs made at 280 °C still maintained a tensile strength 39 MPa greater than that of neat PP, suggesting that the partially melted fabric can retain structural integrity and play a role in reinforcing the SPCs even at an injection temperature as high as 280 °C.

Morphology
One-component PP SPCs with uniaxial fibers were prepared for the morphologic observation to examine the fiber-matrix interface (Figures 7.38 and 7.39). The presence of both fibers and matrix were clearly demonstrated in these figures, indicating that fibers can still retain their physical forms during injection molding even at a high injection temperature of 280 °C. Fiber diameters were observed to be smaller than that of the original PP fiber, showing that fibers would partially melt when they were touched with the injected melt. In addition, a significantly reduced fiber diameter

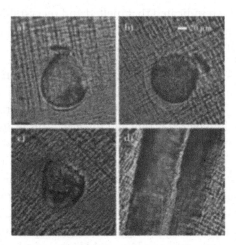

Figure 7.38 Polarized images of sections of PP SPCs molded at different injection temperatures: (a) 200 °C (cross section); (b) 240 °C (cross section); (c) 280 °C (cross section); and (d) 280 °C (longitudinal section) [38].

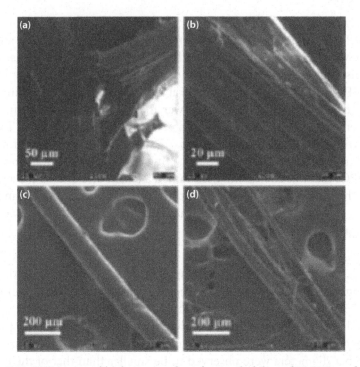

Figure 7.39 SEM images of (a) fracture surface after tensile failure of PP SPCs made at 280 °C; (b) fiber at fracture surface; (c) original fiber; and (d) pulled-out fiber after tensile failure made at 280 °C [38].

can be found, which indicated that melted and recrystallized PP may act as a "bridge" to combine the fiber and the matrix. The SEM images all suggest good interfacial bonding in the PP SPCs.

Insert-Injection Molding with Interleaved/Sandwiched Fabric
Incomplete permeability occurred in the above insert-injection molding for one-component PP SPCs due to the very low temperature at the cavity wall and the melt cooling before it touched the wall. This leads to a bad appearance of the sample surfaces. Therefore, in order to suppress the incomplete permeating issue, the woven fabric can be inserted in the middle of the mold cavity. The clamping force can be used to fix the fabric instead of double-sided tape. Thus, both improved mechanical properties and good appearance could be obtained. Two-component PP SPCs can also be prepared by insert injection molding. By using the principle of copolymer, the processing temperature window could be enlarged. However, when the fabric is set in the middle of the cavity, the injection temperature should be lower because the temperature in the middle of the cavity is much higher

than the temperature in the vicinity of the cavity wall and the high temperature can remain for a longer time. Thus, the copolymer principle can compensate the temperature difference.

For the two-component PP SPCs, random copolymer PP granules were used as the matrix (melting temperature 142.74 °C) and a plain woven fabric with tensile strength of 560 MPa and modulus of 6.6 GPa was used as the reinforcement (melting temperature 172 °C). The PP SPCs with sandwiched woven fabric were prepared under different processing conditions using an injection molding machine. The PP granules were melted under the effects of heat and shear force provided by the screw and barrel, then the melt was injected forward. The temperature of the barrel frontier was 5 °C higher than the nozzle temperature. The nozzle temperature was changed from 160 to 200 °C. According to insert injection molding, the woven fabric was preplaced like an insert on the half cavity of the moving mold plate. After the mold was closed, the woven fabric was fixed in the middle of the whole mold cavity, and the clamping force could press the sandwiched woven fabric tightly (see Figure 7.37). During the injection molding process, the PP melt was injected into the cavity within 2 s under an injection pressure, following by the packing phase for a holding time (5, 15, 25 s) under a holding pressure (32, 48, 64 MPa). The holding pressure was 80% of the injection pressure. The back pressure was 2 MPa and the mold was at room temperature (around 25 °C). Finally, the material in the cavity was cooled for a cooling time (10, 20, 30 s) and solidified, then the SPCs products could be removed by opening the mold. The unnecessary woven fabric was cut off by knife.

Insert-Injection Molded Two-Constituent PP SPCs

The tensile strength could be up to a maximum value of 38.22 MPa, 45% higher than that of the non-reinforced PP. It was obtained by using nozzle temperature of 190 °C, holding time of 15 s, holding pressure of 64 MPa and cooling time of 20 s. However, the modulus of the PP SPCs was improved a little compared to the pure PP. This is due to the fact that the injection temperatures (180–200 °C) are all higher than the melting temperature of the fiber, the high temperature causes the fabric to soften and molecular interdiffusion occurs. The influence of nozzle temperature, holding time, injection pressure and holding pressure on the properties of PP SPCs was analyzed. Suitable parameters should be chosen to ensure good mechanical properties. According to the orthogonal analysis, the nozzle temperature and holding time were found to play a profound role in influencing the properties of PP SPCs. Lower nozzle temperature could be used to prevent the fiber melting. At the lower nozzle temperature, a longer holding

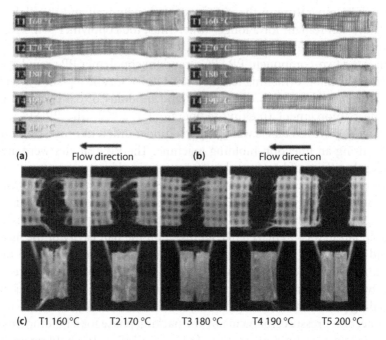

Figure 7.40 Photograph of the PP SPC samples made at different nozzle temperatures of 160, 170, 180, 190 and 200 °C: (a) before tensile tests, (b) after tensile tests, (c) fracture sections [40].

time could be used to improve the permeability of matrix. The injection pressure plays a main role in permeation, and the holding pressure plays a role in improvement of matrix-fiber adhesion.

Morphology
Morphological properties of the PP SPC samples were characterized. The morphology of the samples reflects the effects of different processing parameters, especially the temperature. Figure 7.40 shows the samples made at different nozzle temperatures from 160 to 200 °C. Many more fibers could retain their form, but weak adhesion might exist at much lower temperatures from 160 to 180 °C. As the temperature increased to 190 and 200 °C, the woven structure disappeared and only a little of the short fibers stayed on the edge of the samples, indicating that most of the fibers were melted. Compared with samples before tensile tests (Figure 7.40a), the woven structure in the samples (Figure 7.40b) emerged after the tests since the tensile force destroyed the matrix-fiber adhesive bond. The fibers were melted at high injection temperature and afterwards recrystallized in the

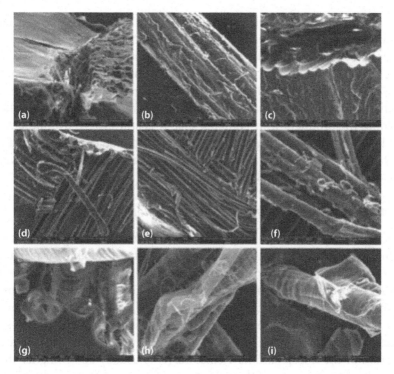

Figure 7.41 Micrographs of the fracture surface of the broken PP SPC sample [40].

matrix, which improved the adhesion and obscured the interface between the fibers and matrix. Even though a higher temperature of 200 °C was used, the woven structures can still be clearly seen after tensile testing. It proves that the reinforcing still existed in the composites made at higher temperature. It furthermore confirms the processing window from 160 to 200 °C. In fact, in a later study by Wang *et al.* [40] it was found that the reinforcing effect of the fibers remained even at 240 °C, although more fibers were melted. The micrographs of the PP SPCs by SEM also illustrated good bonding and compatibility between the fibers and matrix (Figure 7.41).

Pros and Cons

In the insert injection molding method, the processing window can be up to 80 °C, which is much wider than the other SPC processing methods. Another very valuable benefit is that a short cycle time of less than 30 seconds could be realized despite the time for insert setting. The improvement of mechanical strength is determined by the fiber volume/weight fraction. The highest fiber weight fraction was found to be only 36 wt% in insert injection molding of one-component PP SPCs where two layers of fabric were used, but incomplete permeation and bad appearance occurred. Insert injection molding for PP SPCs with sandwiched woven fabric was developed, in which much better interfacial properties can be obtained. However, it limits the setting of higher temperature and the content of the fiber volume/weight fraction. Furthermore, the use of insert injection molding for products with complex shapes is still expected to be developed. Setting and dragging of the inserted fabric should be noted because the fabric is soft. Additionally, large warpage may be a possible disadvantage of insert injection molding.

7.3.4.3 Injection Molding Based on Thermoplastic Elastomers

Injection-compression molding and insert-injection molding for SPCs are all limited by the fabric setting, and they are still not suitable for products with complex shapes. Three-dimensional parts with complex geometry cannot be produced and thus the most design-friendly and versatile processing advantages of injection molding cannot be adapted. A real injection molding, like traditional processing for polymer composites, is still necessary but challenging. In injection molding process, the heating temperature of the nozzle and its vicinal barrel is usually 50 °C higher than the melting temperature of the polymer melt. The 50 °C temperature is too high to be obtained by any principle for the production of SPCs. According to the copolymer principle, which is the best one to have a wide processing window, Kmetty *et al.* introduced thermoplastic elastomers (TPEs) in the injection molding of PP SPCs. They described the preparation of injection moldable PP SPCs by using PP-based TPE as the matrix material and high-tenacity PP fiber as the reinforcement, and reported on their processing-structure-property relationships. A significantly wide processing window (about 90°C) can be obtained.

Preparation
Highly oriented homo-PP multifilament was used as reinforcement. It had a melting temperature of 173 °C. As matrix material, PP-based TPE was selected and used. The TPE was a propylene-ethylene copolymer. From the TPE pellets a 50 μm thick foil was prepared by extrusion film blowing. The matrix film and the reinforcing PP multifilament were laminated to an aluminium core using filament winding combined with film stacking. The orientation of the filament was unidirectional in between the TPE foil layers. Afterward, TPE/hPP sheets with a thickness of 1.6 mm and 70 wt% nominal reinforcement content were produced by compression molding

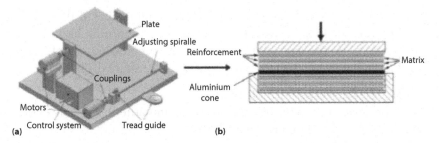

Figure 7.42 Schematic of preparation of unidirectional PP SPC by filament winding (a) followed by compression molding (b) [17].

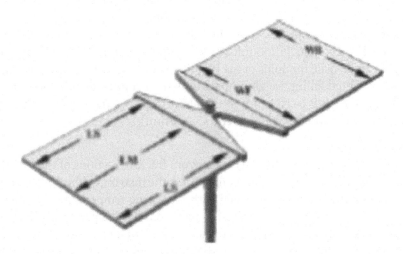

Figure 7.43 Injection-molded PP SPC specimen [17].

(Figure 7.42). The consolidation process took place as follows: after heating up the mold to 140 °C, the wound, film-stacked package was inserted and held for 240 s without pressure and for 480 s under pressure of 5.26 MPa, and then it was cooled to 45 °C. The consolidated sheets were chopped into small pellets having a dimension of 5 × 5 mm which were used for injection molding. From the pre-impregnated pellets, the specimens (Figure 7.43) can be injection molded using an injection molding machine. The injection molding parameters were as follows: injection volume of 44 cm³, injection rate of 50 cm³, injection pressure of 800 ± 200 bar, switch-over point of 10 cm³, holding pressure of 400 bar, holding time of 10 s, residual cooling time of 15 s, screw rotational speed of 15 m/min, back pressure of 20 bar, decompression volume of 5 cm³, decompression rate of 5 cm³/s, melt temperature of 120/140/160 °C, mold temperature of 20 °C.

Properties of the TPE-Based PP SPCs
The shrinkage of the TPE-based PP SPCs is similar to that of conventional PP. The static tests demonstrated that the yield stress and tensile modulus of the PP SPCs agreed with the mechanical parameters of the conventional homo-PP grades (yield strength of 25–30 MPa, E modulus of 950–1000 MPa). The yield stress of the all-PP composites depended on the processing temperature and on the analyzed area of the plaque. With increasing processing temperature, the yield stress values of the composites slightly increased. The PP SPC specimens injection-molded at 160 °C showed a yield stress of about 30 Mpa, which is very close to that of conventional homo-PP. The tensile modulus can be up to 900–1000 MPa. The yield stress and tensile modulus of the PP SPCs showed orientation dependence. Instrumented falling weight impact tests were conducted. The perforation energy of the PP 6–10 J/mm SPCs (about) exceeds the values of the conventional random PP copolymer (0.5–1 J/mm).

Morphology
The SEM images (Figure 7.44) taken from fracture sections of the PP SPCs confirm good interfacial adhesion between the matrix and the fibers. The PP SPCs became better consolidated with increasing temperature. There is no visible skin-core effect in contrast to injection-molded parts with rigid discontinuous fibers such as glass fibers. The high fiber content (70 wt%) and the flexible reinforcing fibers played important roles. A quasi-homogeneous distribution of individual fibers can be found in the cross section.

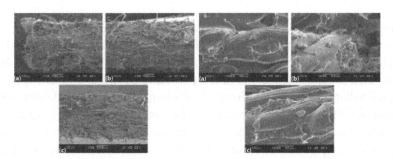

Figure 7.44 Fracture surfaces of the PP SPCs injection molded at (a) 120 °C, (b) 140 °C, and (c) 160 °C, respectively [17].

Pros and Cons
The introduction of TPEs ensures a wide processing window (~ 90 °C). This processing window is 50–70 °C higher than that of the conventionally applied technologies (hot compaction, coextrusion and film-stacking method). The production of PP SPCs with complex shapes can be realized. Actually, not only injection molding, but also extrusion for industrial production of PP SPC products, can be realized by using the PP SPC pellets based on TPEs. However, since the components of PP SPCs belong to the same polymer family, the TPE itself should also be based on PP. To compare the selected TPE with the PP copolymer, previous tests, such as Fourier transform infrared spectroscopy (FTIR), have to be performed. Although a high fiber content (70 wt%) can be obtained, the preparation of the pre-impregnated material includes many complicated processing, such as film blown extrusion for TPE film, filament winding combined with film stacking, compression molding and afterwards pelleting. Although the thermoplastic PP homopolymer worked as efficient reinforcement and significantly increased the yield stress and tensile modulus of the corresponding composites, the tensile properties of the final PP SPCs based on TPE can only be comparable with conventional homo-PP grades. In fact, the TPE is still not the same with PP, and has much lower mechanical properties which also influence the mechanical properties of the PP SPCs. Therefore, considerable effort is still needed to produce injection-moldable PP SPCs which have comparable mechanical properties with standard PP SPCs.

7.3.4.4   Coextrusion and Injection Molding

Traditional compression molding methods that have been used so far as the main processing methods should be replaced by more flexible processing methods. There is ongoing research to determine the usefulness of a two-stage extrusion/injection molding process. As the most popular plastics processing technologies, extrusion as the preparation technique and injection molding as the final shaping technology can be used for basic processing of SPCs. Andrzejewski *et al.* [41] used the principle of copolymer in the extrusion/injection molding method. A temperature difference of 33 °C was established by their materials. Two-component PP SPCs were prepared.

Preparation
The composite pellets were prepared using PP fibers (melting point at about 170 °C) and low-melting PP copolymer (melting point at 137 °C) as the input materials. The coextrusion of PP fibers was performed using an angular extrusion head (see Figure 7.45). Optimal coextrusion parameters

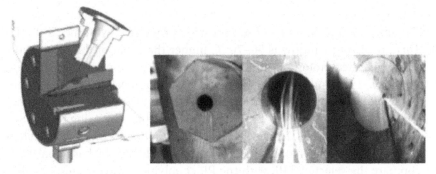

Figure 7.45 The coextrusion die head used for PP SPC pellets preparation (left) and the inlet and outlet of the coextrusion die with visible PP fibers (right) [41].

were set to achieve best filling of space between fibers and matrix and not to overheat and break fibers. The extrusion process and the coextruded fibers of the PP SPC with low reinforcement fibers content (5%) are first investigated to determine the processing temperatures. The low content of PP fibers allowed avoiding the problems usually observed by coextrusion such as breakage of fibers, molten polymer flow instabilities, etc. Then the production of PP SPC pellets with the highest possible content of reinforcement (40%) was investigated. Due to the complexity of coextrusion die arming with PP yarns, the amount of fibers in the yarn was changed. During preliminary tests there were 24 fibers in the yarn, one yarn having a maximum of 460 single fibers. A high screw speed was needed to make the coextrusion process faster, and a higher processing temperature was also needed to obtain a low viscosity of matrix for better filling of the space between fibers. The shortening of the thermal exhibition time of the fibers in the die was also key to stabilize the process. Finally, the extruded PP SPC fibers were wound on the rotating barrel and then were cut using the pelletizing machine to small pieces. The pellets can be injection-molded into SPC products with different, even complex shapes. The injection molding was performed for PP SPC pellets using a small piston injection molding machine and standard heating system. The barrel temperature was set to 155 °C and the mold temperature was 80 °C.

Morphology

The prepared PP SPC pellets were processed and shaped using injection molding technology. Obtained samples were subjected to mechanical testing in the static tensile test and dynamic mechanical analysis. Because during coextrusion the material undergoes phase change and is subjected

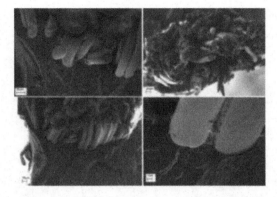

Figure 7.46 SEM images of the coextruded pellets, PP SPC with 5% fibers [41].

to high temperature and pressure, morphological and physical character-ization were needed to assess the presence of fibers and the rheological properties of the compound. Figure 7.46 shows the SEM images of the coex-truded PP SPC pellets with 5% fibers. There were no melted fibers found in the fracture sections, a local poor adhesion of matrix to the fibers can be observed. Figure 7.47 presents a SEM micrograph comparison of neat PP, PP SPC with 5% fibers, and PP SPC with 40% fibers. The two-component structure with a very sharp boundary can be observed, and there were also no melted fibers. Some small holes indicated that fibers were pulled out from the matrix, but no breakage at the surface of the main fracture of the matrix was found, suggesting the lack of joint penetration. For the PP SPC with 40% fibers, fiber aggregation was found, which might result in uneven distribution of fibers, thus reducing the strengthening effect.

Properties of the Injection-Molded PP SPCs
The DMA analysis performed before the mechanical tests showed a decrease in mechanical properties for the PP SPC samples; however, it suggested that the increase of fiber content is correlated with increasing reinforcing effect. The results of tensile strength tests can be considered as satisfactory. When the content of fibers in the composite reached 40% the tensile strength was equal to 30 Mpa, which was about 11% higher in comparison to neat matrix material, while the tensile modulus was about 1.5 GPa, which was about 15% higher than that of the neat matrix. The probable reason for the low improvement of mechanical properties is fibers aggregations, which can be considered as inside notches which reduce the strength parameters. The fiber aggregation can be caused by the point-shaped die used during injection molding process and high shear rate during flow of the polymer.

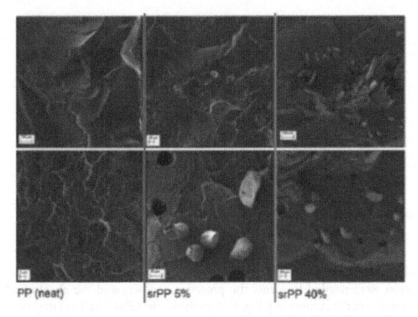

PP (neat)   srPP 5%   srPP 40%

Figure 7.47 SEM images of injection-molded samples: neat PP (left), PP SPC/5% fibers (center), PP SPC/40% fibers (right) [41].

The second reason for the insufficient changes in mechanical properties may be the loss of fibers due to melting; it is noted that the shear rate may increase heating of the fibers. Nevertheless, the SEM images proved that the fibers are still visible in the samples without drastic shape change and still need to be discussed.

Pros and Cons
Compression molding methods limit the product shape, extend the processing time, and require the use of specialty tool/machine equipment. Most of the compression molding drawbacks can be eliminated using extrusion/injection molding techniques. The presented methodology is not the best solution in terms of possibilities to achieve maximum mechanical properties but this approach has been studied as the simplest method that does not require specialized instrumentation. The presented results of the injection-molded PP SPCs characterization confirmed the reinforcing effect of the PP fibers. However, the comparison of the mechanical properties to the standard SPCs shows the relatively low growth of the properties for injection-molded PP SPCs. Much higher injection temperature and more fiber content are needed to investigate optimum processing conditions and improve the mechanical properties.

7.4   Applications

High stiffness, high tensile strength, outstanding impact resistance at low density and easy recyclability are the remarkable properties of PP SPCs. The applications of PP SPCs fall within a broad range of industries, including automotive industry, industrial cladding, building and construction, cold temperature applications, audio products, personal protective equipment, sporting goods, etc. The product is also being tested for numerous other applications such as underbody panels, roof lining and flooring for low-volume cars, anti-riot protection and luggage.

In the automotive industry, various parts are made from PP or its composites, and PP SPCs can be much better alternative materials. Because of the ELV (end-of-life of vehicles) directive (vehicles must be made of 95% recyclable materials, of which 85% can be recovered through reuse or mechanical recycling and 10% through energy recovery or thermal recycling) [40], the development of fully recyclable PP SPCs is especially attractive for this market sector and it is expected that the first applications will be to replace current glass mat reinforced thermoplastics (GMTs)-based products such as body panels, underbody structures, seating components, and bumper bars. PP SPCs have the toughness of regular plastics with levels of rigidity normally associated with certain grades of glass reinforced polymer (GRP), while at the same time being up to 50% lighter than the latter. Based on the properties of PP SPCs, the first commercial applications of these materials are likely to be in strength and impact critical applications like underbody shields or as an alternative to natural fiber reinforced plastics for inner-trim parts [42]. Besides that, PP SPCs can also be used in sporting goods, safety helmets, covers, shells, luggage, etc. Moreover, a promising application field of PP SPCs is based on their excellent sound damping properties (e.g., audio equipment).

At present, the commercially available SPC materials being marketed under the brand names CURV, PURE, Armordon, Para-Lite, Compomeer, Kaypla, and Comfil, are all commercially realized outcomes of the SPC concepts presented in the previous sections. The first five brands are all related to PP SPCs. In particular, CURV is a single-constituent multistep product manufactured from 2D fabric reinforcement. PURE, Armordon and Para-Lite all use a two-constituent multistep manufacturing method with a 1D reinforcement in the form of stretched tapes comprised of different grades of the same polymer in their core and surface layers. Kaypla is mainly known for their polyethylene (PE)-based SPCs and Comfil for injection-moldable PET SPC pellets. Unfortunately, injection-moldable PP SPC pellets are still in further development to be commercially available [43].

Figure 7.48 Some examples of CURV products [44].

7.4.1 CURV

CURV® is an SPC product manufactured by Propex Operating Company, LLC [44]. It consists of PP fibers as the reinforcement material and PP as matrix. Hot compaction of woven PP fabrics is their processing method. Some examples include suitcases and housings for sport articles such as shin guards and ice hockey skates and others (Figure 7.48). When combined with other materials, CURV can be used as an enabling material for new composites and applications. There are two main applications—foam and honeycomb laminates—that maintain as well as extend the concept of a 100% PP composite product (Figure 7.48). CURV/expanded PP (EPP) foam laminates can be produced without adhesives during the compression cycle where the foam automatically bonds to the CURV sheets. Applications for this product include architectural panels, automotive components and insulation. In CURV/PP honeycomb laminates, the CURV sheets are bonded to honeycomb using any of a variety of adhesives.

7.4.2 PURE®

Lankhorst Pure Composites, part of the Royal Lankhorst Euronete Group, produces and sells PURE [45]. This product is made of different grades of PP that are combined using the coextrusion process. The core is a highly oriented, high modulus PP tape which is skinned with a specially formulated PP that allows welding tapes together. PURE is available as tape, fabric or sheet and can be used to mold or construct various composite applications and components. The fabric is ideally used for thermoforming process. PURE composite material, for example, has excellent impact properties; the material is therefore validated and used within anti-ballistic and blast protection products. In addition, PURE composites can be used in a variety of markets such as automotive, construction, consumer (sports) parts and flat panels. Their application possibilities include: underbody

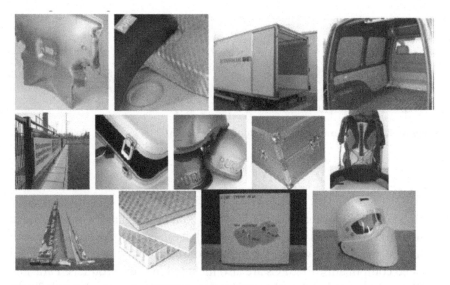

Figure 7.49 Some examples of PURE products [45].

shields, interior panels, scuff plates, impact panels for minivans, advertising panels, suitcases, helmets, flight cases, sports gear, sail reinforcement, architectural panels, sandwich panels, anti-ballistic panels, de-mining masks (Figure 7.49).

7.4.3   Armordon®

Armordon is also a PP-based self-reinforced plastic manufactured by Don & Low Ltd. [46]. The production process is similar to the production of PURE and combines two different types of PP, ensuring a strong affinity between the two constituents of the tape. Further processing steps may use the tapes directly for winding or tape placement processes or they can be woven into fabrics. Armordon product exhibits high impact strength combined with low density, 100% recyclability, significant environmental benefits and has no glass reinforcement. It holds an advantage over the materials more traditionally used in these market areas. Typical applications can be found in the security industry as armor, since Armordon offers excellent cost-effective performance in areas of ballistic and blast protection, helping armoring specialists meet the increasing need for budget-friendly protective solutions. Additionally, applications include the automotive sector, consumer products such as passenger luggage, sportswear, orthotics and packaging (Figure 7.50).

Figure 7.50 Some examples of Armordon products [46].

7.4.4   Para-Lite®

Para-Lite® is a register trademark of Von Roll Deutschland GmbH (one of the oldest companies in Switzerland) [47]. It especially offers products with the benefits of lightweight, flexibility, non-corrosivity, and approved and tested components. PP SPCs are also included in their products. Para-Lite PP is a PP-based lightweight laminate enabling partial or full aramid substitution without weight increase that can be covered with various customized decorative skins or with a ready-to-glue layer. It is used for ballistic producta with very lightweight laminate for backing and linear applications, and is fully recyclable and water resistant.

7.4.5   Compomeer

Compomeer B.V. is a company that was founded in late 2012 and is located in The Netherlands [48]. Compomeer B.V. produces innovative, super strong, high-performing, ultra-light and recyclable thermoplastic composite materials for a variety of applications in markets like automotive, air and sea freight, mobile homes and hurricane shields. The products include suitcases, flight cases, air freight containers, flooring, sidewalls, scaffolding, etc. Compomeer specializes in customized press-consolidated sheets, laminates and sandwich products for use in a wide range of industrial applications.

Flight cases in particular must become lighter to save cost, and Compo2 is a unique PP SPC honeycomb perfectly suited for this purpose. The Compo2 panels are composed out of two CURV skins (consisting of a unique woven structure of stretched PP fibers) combined with a PP SPC honeycomb core, joined together by adhesive bonding (Figure 7.51). The panels consist of 100% PP and are consequently very cost effective. These ultra-lightweight (a density of 80 kg/m³) panels are designed to meet the highest performance characteristics upon impact and are abrasion resistant.

Figure 7.51 Compomeer Compo2 panels [48].

Besides that, the specific properties of the CURV-based skins allow a temperature application of –40°C without any loss of mechanical properties. The Compo2 panels are particularly suited for applications in transport, packaging, temporary installations, partitioning systems and doors.

For the recently developed nanofibrillar SPCs, in which the nanofibrillar reinforcing structure is "surface bonded," various medical applications can be forecast (scaffolds for bone repair/healing) when using in body resolvable polymers. Their nanoporous structure could well be exploited in different pervaporation membranes.

7.5   Summary

Designing for "recycling" or "eco-design" is becoming a philosophy that is applied to more and more materials and products. SPCs offer the promise of superior mechanical properties, reduced weight, and enhanced recyclability. PP is a thermoplastic with a wide range of applications. Over the past 30 years, the concept of SPCs has been extensively investigated with PP. In this chapter, different processing methods for preparing PP SPCs, such as hot compaction, film stacking, coextrusion, and injection molding, were introduced respectively. The properties of the produced PP SPCs using different methods were illustrated. As a review of the published articles on PP SPCs, the process routes of existing methods for PP SPCs are summarized in detail in Figure 7.52. The influences of different processing conditions or parameters, such as the compaction temperature, pressure, molecular weight, constraining, melt flow indices, polymorphism, and fabric structure, were discussed. Table 7.1 summarized the main processing parameters and their influence on the processing and properties for PP SPCs. Most experimental results were referred from different researchers' papers. All of these results could prove the three main advantages of the PP SPCs: lightweight, eco-friendly recyclability and excellent fire/matrix interfacial adhesion.

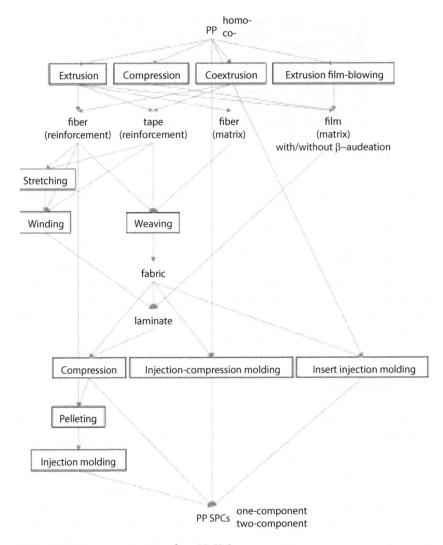

Figure 7.52 Process routes to produce PP SPCs.

In most cases, PP SPCs have been obtained through the so-called "hot compaction" technique. The use of temperature and pressure to control the melting of the outer surface layer of polymeric fibers implies that the processing window for the hot compaction process is quite narrow. Other processing techniques include film stacking and coextrusion-based copolymer/polymorphism/overheating/undercooling as described in this chapter, which are also related to compression molding. The most challenging

Table 7.1 The effects of different parameters regarding the processing of PP SPCs.

Parameter	Effect
Temperature	High temperature results in low viscosity, better flow-ability, and the best mechanical properties until full consolidation occurs; thereafter much higher temperature results in fiber melting, relaxation of oriented reinforcement and low mechanical strength.
Holding time	Short and low to prevent shrinkage in compaction, but should be longer in injection molding.
Pressure	High pressure results in better permeation, increases T_m of the reinforcement and prevents shrinkage.
Stretching	Higher melting point, higher orientation then higher mechanical properties.
Molecular weight	Higher molecular weight can improve ductility.
Melt flow indices	Higher MFI, better tensile properties, lower perforation energy.
Fiber style	Continuous fibers have higher tensile strength.
Reinforcement geometry	Tape presents greater surface for compaction than round fiber, and flat tape gives less crimp when woven.
Reinforcement content	Increases mechanical strength and perforation energy.
Weave style	Balanced weave style is beneficial.
Interleaved film	Benefits interface adhesion and decreases void content.
β-nucleation	Larger for the homo- than for the random copolymer; wider transcrystalline layer in β-rPP matrix compared to the α-rPP version; β-rPP-based SPCs provide higher mechanical properties and perforation energy than the α-rPP- and β-PP-based ones; β-PP-based SPCs have similar mechanical properties to the α-rPP-based ones.

technique is injection molding where the SPC pellets have to work their way through the harsh environment of a conventional injection molding screw and barrel and are subjected to only a small processing window. Injection-compression molding, insert-injection molding, injection molding based on thermoplastic elastomers, and coextrusion followed by injection molding have taken PP SPCs into industrial volume production, where even products with complex geometry can be prepared at short cycle times; however, relatively low mechanical properties limit their application.

The applications of PP SPCs were also introduced. Although various products are already commercially available, the current forms of PP SPCs on the market are only available in sheet or fabric form, limiting the range and types of components that can be manufactured. Future works still need to focus on the development of techniques that will extend the processing window (heating of the matrix only and not the reinforcement polymer fibers) by materials development and process optimization, and the development of techniques sufficiently versatile to be scaled up to an industrial level, also with improved mechanical properties. Control systems are also being improved to facilitate the use of these new materials with molding techniques capable of rapid heating and cooling with fine control. Additionally, numerical simulation techniques can also be used. Therefore, PP SPCs and their processing technologies will remain in the spotlight in the future.

Acknowledgments

The author would like to thank the Research Fund for the Doctoral Program of Higher Education of China (Project No. 20131101120047), the Basic Research Fund of Beijing Institute of Technology (Project No. 20141042002) and the National Natural Science Foundation of China (Project No. 51403019).

References

1. Banik, K.., Abraham, T.N., and Karger-Kocsis, J., Flexural creep behavior of unidirectional and cross-ply all-poly(propylene) (pure) composites. *Macromol. Mater. Eng.* 292(12), 1280–1288, 2007.
2. Capiati, N.J., and Porter, R.S., Concept of one polymer composites modeled with high-density polyethylene. *J. Appl. Polym. Sci.* 10(10), 1671–1677, 1975.
3. Kmetty, A., Bárány, T., and Karger-Kocsis, J., Self-reinforced polymeric materials: A review. *Prog. Polym. Sci.* 35(10), 1288–1310, 2010.

4. Karger-Kocsis, J., and Bárány, T., Single-polymer composites (SPCs): Status and future trends. *Compos. Sci. Technol.* 92(3), 77–94, 2014.

5. Loos, J., and Schimanski, T., Morphological investigation of polypropylene single fiber reinforced polypropylene model composites. *Polymer* 42(8), 3827–3834, 2001.

6. Ward, I.M., and Hine, P.J., The science and technology of hot compaction. *Polymer* 45(5), 1413–1427, 2004.

7. Abo El-Maaty, M.I., Bassett, D.C., Olley, R.H., Hine, P.J., and Ward, I.M., The hot compaction of polypropylene fibres. *J. Mater. Sci.* 31(5), 1157–1163, 1996.

8. Lacroix, F.V., Loos, J., Lu, H., and Schulte, K., Morphological investigations of polyethylene fibre reinforced polyethylene. *Polymer* 40(4), 843–847, 1999.

9. Hine, P.J., Ward, I.M., and Teckoe, J., (1998). The hot compaction of woven polypropylene tapes. *J. Mater. Sci.* 33(11), 2725–2733, 1998.

10. Mead, W.T., and Porter, R.S., Annealing characteristics of ultraoriented high-density polyethylene. *J. Appl. Phys.* 47(10), 4278–4288, 1976.

11. Barkoula, N.M., Peijs, T., Schimanski, T., and Loos, J., Processing of single polymer composites using the concept of constrained fibers. *Polym. Compos.* 26(1), 114–120, 2005.

12. Copolymer, in: *Wikipedia*, https://en.wikipedia.org/wiki/Copolymer. (Retrieved 2016)

13. Alcock, B., Cabrera, N.O., Barkoula, N.-M., Loos, J., and Peijs, T., The mechanical properties of unidirectional all-polypropylene composites. *Compos. Part A: Appl. Sci. Manuf.* 37(5), 716–726, 2006.

14. Varga, J., Beta-modification of isotactic polypropylene: Preparation, structure, processing, properties, and application. *J. Macromol. Sci. Part B: Phys.* 41(4), 1121–1171, 2002.

15. Bárány, T., Izer, A., and Czigány, T., On consolidation of self-reinforced polypropylene composites. *Plast. Rubber and Compos.* 35(9), 375–379, 2006.

16. Dai, P., Zhang, W., Pan, Y., Chen, J., Wang, Y., and Yao, D., Processing of single polymer composites with undercooled polymer melt. *Compos. Part B: Eng.* 42(5), 1144–1150, 2011.

17. Kmetty, Á., Bárány, T., and Karger-Kocsis, J., Injection moulded all-polypropylene composites composed of polypropylene fibre and polypropylene based thermoplastic elastomer. *Compos. Sci. Technol.* 73(73), 72–80, 2012.

18. Hine, P.J., and Ward, I.M., Novel composites by hot compaction of fibers. *Polym. Eng. Sci.* 37(11), 1809–1814, 1997.

19. Hine, P.J., Ward, I.M., Jordan, N.D., Olley, R., and Bassett, D.C., The hot compaction behavior of woven oriented polypropylene fibres and tapes. I. Mechanical properties. *Polymer* 44(4), 1117–1131, 2003.

20. Le Bozec, Y., Kaang S., Hine, P.J., and Ward, I.M., The thermal-expansion behaviour of hot-compacted polypropylene and polyethylene composites. *Compos. Sci. Technol.* 60, 333–344, 2000.

21. Hine, P.J., Ward, I.M., Jordan, N.D., Olley, R.H., and Bassett, D.C., The hot compaction behaviour of woven oriented polypropylene fibres and tapes. II.

Morphology of cloths before and after compaction. *Polymer* 44, 1133–1143, 2003.

22. Hine, P.J., Olley, R.H., and Ward, I.M., The use of interleaved films for optimising the production and properties of hot compacted, self reinforced polymer composites. *Compos. Sci. Technol.* 68(6), 1413–1421, 2008.

23. Izer, A., and Bárány, T., Hot consolidated all-PP composites from textile fabrics composed of isotactic PP filaments with different degrees of orientation. *Express Polym. Lett.* 1(12), 790–796, 2007.

24. Kitayama, T., Utsumi, S., Hamada, H., Nishino, T., Kikutani, T., and Ito, H., Interfacial properties of PP/PP composites. *J. Appl. Polym. Sci.* 88(13), 2875–2883, 2003.

25. Bárány, T., Karger-Kocsis, J., and Czigány, T., Development and characterization of self-reinforced poly(propylene) composites: Carded mat reinforcement. *Polym. Adv. Technol.* 17(9–10), 818–824, 2006.

26. Abraham, T.N., and Karger-Kocsis, J., Dynamic mechanical thermal analysis of all-PP composites based on β and α polymorphic forms. *J. Mater. Sci.* 43(10), 3697–3703, 2008.

27. Bárány, T., Izer, A., and Karger-Kocsis, J., Impact resistance of all-polypropylene composites composed of alpha and beta modifications. *Polym. Test.* 28(2), 176–182, 2009.

28. Izer, A., Bárány, T., and Varga, J., Development of woven fabric reinforced all-polypropylene composites with beta nucleated homo- and copolymer matrices. *Compos. Sci. Technol.* 69(13), 2185–2192, 2009.

29. Abraham, T.N., Wanjale, S.D., Bárány, T., and Karger-Kocsis, J., Tensile mechanical and perforation impact behavior of all-PP composites containing random PP copolymer as matrix and stretched PP homopolymer as reinforcement: Effect of beta nucleation of the matrix. *Compos. Part A: Appl. Sci. Manuf.* 40(5), 662–668, 2009.

30. Wang, J., Chen, J., Dai, P., Wang, S., and Chen, D., Properties of polypropylene single-polymer composites produced by the undercooling melt film stacking method. *Compos. Sci. Technol.* 107, 82–88, 2015.

31. Wang, J., Chen, J., and Dai, P., Polyethylene naphthalate single-polymer-composites produced by the undercooling melt film stacking method. *Compos. Sci. Technol.* 91, 50–54, 2014.

32. Alcock, B., Cabrera, N.O., Barkoula, N.M., and Peijs, T., Low velocity impact performance of recyclable all-polypropylene composites. *Compos. Sci. Technol.* 66(11–12), 1724–1737, 2006.

33. Alcock, B., Cabrera, N.O., Barkoula, N.-M., Spoelstra, A.B., Loos, J., and Peijs, T., The mechanical properties of woven tape all-polypropylene composites. *Compos. Part A: Appl. Sci. Manuf.* 37(1), 147–161, 2007.

34. Alcock, B., Cabrera, N.O., Barkoula, N.-M., Wang, Z., and Peijs, T., The effect of temperature and strain rate on the impact performance of recyclable all-polypropylene composites. *Compos. Part B: Eng.* 39(3), 537–547, 2008.

35. Cabrera, N.O., Alcock, B., and Peijs, T., Design and manufacture of all-PP sandwich panels based on co-extruded polypropylene tapes. *Compos. Part B: Eng.* 39(7–8), 1183–1195, 2008.

36. Alcock, B., Cabrera, N.O., Barkoula, N.-M., Loos, J., and Peijs, T., Interfacial properties of highly oriented coextruded polypropylene tapes for the creation of recyclable all-polypropylene composites. *J. Appl. Polym. Sci.* 104(1), 118–129, 2007.

37. Khondker, O.A., Yang, X., Usui, N., and Hamada, H., Mechanical properties of textile-inserted PP/PP knitted composites using injection–compression molding. *Compos. Part A: Appl. Sci. Manuf.* 37(12), 2285–2299, 2006.

38. Mao, Q., Hong, Y., Wyatt, T.P., Chen, J., Wang, Y., Wang, J., and Yao, D., Insert injection molding of polypropylene single-polymer composites. *Compos. Sci, Technol.* 106, 47–54, 2014.

39. Wang, J., Mao, Q., and Chen, J., Preparation of polypropylene single-polymer composites by injection molding. *J. Appl. Polym. Sci.* 130(130), 2176–2183, 2013.

40. Wang, J., Wang, S., and Chen, D., Development and characterization of insert injection moulded polypropylene single-polymer composites with sandwiched woven fabric. *Compos. Sci. Technol.* 117, 18–25, 2015.

41. Andrzejewski, J., Szostak, M., Barczewski, M., Krasucki, J., and Sterzynski, T., Fabrication of the self-reinforced composites using co-extrusion technique. *J. Appl. Polym. Sci.* 131(23), 205–212, 2014.

42. Peijs, T., Composites for recyclability. *Mater. Today* 6(4), 30–35, 2003.

43. Fakirov, S., Nano- and microfibrillar single-polyymer composites: A review. *Maromol. Mater. Eng.* 298, 9–32, 2012.

44. Plastemart, CURV can be used as an enabling material for new composites and applications, http://www.plastemart.com/upload/Literature/CURV_self_reinforced_PP.asp, 2003.

45. PURE, http://www.ditweaving.com/ (Retrieved August 7, 2017).

46. Armordon, Advanced, high performance composite materials, http://www.donlow.com/Armordon/, 2016.

47. Para-Lite® Vonroll, http://www.vonroll.com, 2016.

48. Compomeer, Innovative, fibre reinforced thermoplastic composite sheets, sandwich panels and consolidated laminates, http://compomeer.eu/, 2014.

8

Polypropylene/Plant-Based Fiber Biocomposites and Bionanocomposites

Amir Ghasemi[1,2], Ehsan Pesaran Haji Abbas[1,2], Leila Farhang[1] and Reza Bagheri[1,2,*]

[1]Polymeric Materials Research Group, Department of Materials Science and Engineering, Sharif University of Technology, Tehran, Iran
[2]R&D Department, Parsa Polymer Sharif Co., Tehran, Iran

Abstract

Natural fiber reinforced polypropylene (PP) biocomposites and nanobiocomposites are gaining more attraction in research and industrial applications because of their advantages, including low cost, low density, high specific mechanical properties, and less abrasion for processing machineries. These biocomposites can be processed by various conventional plastic processing methods like extrusion, thermoforming, and injection molding. Because of incompatibility between polypropylene and natural fibers, various physical and chemical methods are applied on natural fibers for improving interfacial adhesion between polymer and fiber, and, consequently, mechanical properties of final product. At present, natural fiber reinforced PP biocomposites are used in automobile, packaging, and construction sectors.

Keywords: Biocomposites, natural fibers, surface treatment, processing, mechanical properties

8.1 Introduction

Petroleum resources shortage and increasing awareness about environmental and sustainability issues have prompted many researchers to think of utilizing bioresources as a substitution for petroleum-based materials [1, 2]. Natural fibers obtained from plants are one of the sources that have

Corresponding author: rezabagh@sharif.edu

Visakh. P. M. and Matheus Poletto. (eds.) Polypropylene-Based Biocomposites and Bionanocomposites, (247–286) 2018 © Scrivener Publishing LLC

gained a lot of attention during the last decades. Application of natural fibers (biofibers) as reinforcement in polymers is very attractive because of their advantages, including flexibility during processing, high mechanical properties, biodegradability, and low cost [3–6]. These biocomposites are gaining more and more acceptance in various industries and structural applications. Natural fiber reinforced polypropylene (PP) biocomposites are a good example of this class of materials which have been utilized in sectors like automobile and packaging. In this chapter, different types of natural fibers and their chemical structure are studied. Moreover, processing, mechanical properties, and applications of natural fiber reinforced PP biocomposites are reviewed.

8.2   Types of Natural Fibers

There are different classifications for natural fibers in the literature based on their origin, properties, applications, etc. Considering the origin of natural fibers, they can be classified as [7]:

1. Plant-based fibers (PBFs) like flax and cotton.
2. Animal-based fibers like silk and wool.
3. Mineral-based fibers like asbestos and basalt.

8.2.1   Plant-Based Fibers

PBFs can be extracted from different sections of the plants. Accordingly, they can be classified as [8]:

- Wood
- Stem/bast: Flax, Hemp
- Seed/fruit: Cotton, Coir
- Leaf: Abaca, Sisal
- Straw: Wheat/Rice straw

Although the origin of these fibers is different, their chemical composition and structure are almost similar. The major constituents of these fibers are cellulose, hemicellulose and lignin while minor ingredients include pectin, waxes, etc. [9]. Cellulose is a homopolymer with linear macromolecules consisting of repeating D-anhydroglucose ($C_6H_{11}O_5$) units joined by β-1,4-glycosidic linkages with a degree of polymerization of about 10000 [10]. The chemical structure of cellulose is shown in Figure 8.1.

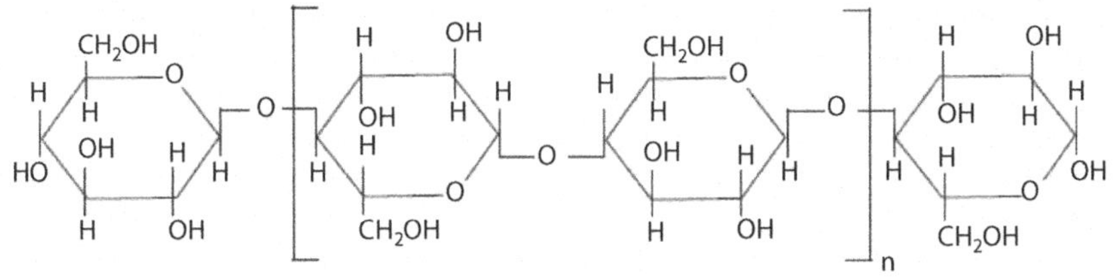

Figure. 8.1 Chemical structure of cellulose. (Reproduced with permission from [12]; Copyright © 2011 Elsevier)

As seen in this figure, each repeating unit of cellulose contains 3 hydroxyl groups which contribute to two characteristics of this constituent in PBFs: the hydrophilic nature of cellulose and strong hydrogen bonds between cellulose molecules play the main role in the formation of cellulose crystals. Cellulose forms a semicrystalline structure containing highly crystalline units and amorphous regions [11].

In contrast to cellulose, hemicellulose is a copolymer which contains different monomers, including glucose, xylose, mannose, galactose, rhamnose, and arabinose. Hemicellulose has a branched structure that hinders its crystallization and makes it an amorphous polymer. Its degree of polymerization is between 500 and 3000 [11]. Hemicellulose is also a hydrophilic polymer and is soluble in alkaline solutions.

Lignin is the most complex constituent of the PBFs. Its structure has not been completely understood until now because there is no method to extract it from fiber without any change in its structure. However, its functional groups are studied and identified. Lignin is a phenolic compound which contains hydroxyl and methoxy functional groups [8]. High ratio of carbon to hydrogen atoms in lignin indicates that it is an unsaturated (aromatic) polymer. Hydroxyl groups in lignin can be linked with hydroxyl groups in cellulose and carboxyl groups in hemicellulose, which make lignin compatible with two other components. The most probable structure of the lignin is shown in Figure 8.2.

PBFs are natural composites in which the reinforcement constituent is distributed within a flexible matrix. Each fiber at the macroscopic level is a bundle of microfibers with the diameter in the range of 20 to 40 μm [11]. Each microfiber contains a lumen in the center that is surrounded by 4 layers: outer layer, or primary layer, and the secondary layer (S), which itself consists of three distinct layers (S1, S2, and S3). In all of these layers, cellulose molecules with their semicrystalline structure pack together and form microfibrils with the diameter of 10–30 nm [11]. These microfibrils are embedded within a matrix mainly composed of lignin and hemicellulose.

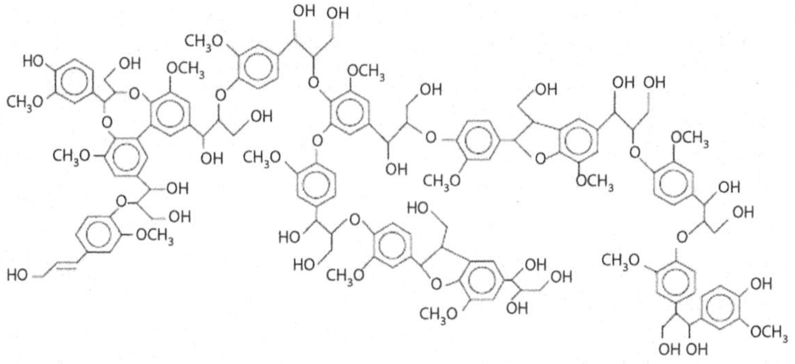

Figure 8.2 Chemical structure of lignin. (Reproduced from [13]; Copyright © 2015 Elsevier)

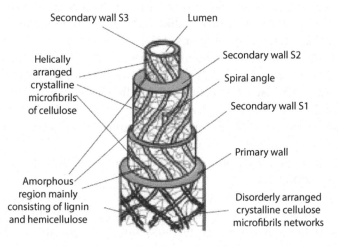

Figure 8.3 Structure of a single microfiber. (Reproduced with permission from [14]; Copyright © 2001 Elsevier)

The angle between these microfibrils and fiber axis is called microfibrillar angle. This parameter is an important structural characteristic of the fibers. The proportion of these constituents, as well as microfibrillar angle, vary from layer to layer. In the primary layer, cellulose content is small, and lignin is the dominant component. In the secondary layers, cellulose content is higher in comparison to primary layer. Figure 8.3 presents a general view of a microfiber.

Various parameters can affect the mechanical properties of PBFs such as the origin of plant of the fiber, fiber extraction method, cellulose content of fiber, and cellulose microfibrillar angle. Generally, increase in

Table 8.1 Mechanical properties of some PBFs.

Fiber	Tensile strength (MPa)	Elastic modulus (GPa)	Elongation at break (%)
Jute	393–800	13–26.5	1.16–1.5
Cotton	287–800	5.5–12.6	7–8
Flax	345–1500	27.6	2.7–3.2
Sisal	468–700	9.4–22	3–7
Hemp	690	70	1.6

cellulose content improves tensile strength (TS) and elastic modulus of fibers. Moreover, the arrangement of cellulose microfibrils in the direction of fiber axis means smaller fibrillar angle, which results in increasing fiber tensile strength. Table 8.1 shows the mechanical properties of some common PBFs.

8.2.2   Animal-Based Fibers

Fibers obtained from animals have been utilized by mankind for thousands of years. Chemical structure of these fibers is based on specific proteins like keratin, fibroin and collagen. These fibers can be obtained from:

- Animal hair: like wool and goat hair
- Silk
- Avian fiber: fibers from birds like feathers

Animal-based fibers are usually used for clothing and they are not commonly utilized as reinforcement for polymer matrices.

8.2.3   Mineral-Based Fibers

Mineral fibers are obtained from mineral sources and used in their natural form or after some modifications. They can be divided into three groups:

- Asbestos: a naturally occurring mineral fiber
- Ceramic fibers: like glass, silicon and basalt fiber
- Metal fibers: Aluminum fibers

Asbestos is a well-known naturally occurring mineral fiber with desirable properties such as fire resistance, acceptable tensile strength, heat, sound and electrical insulation. Asbestos was once extensively used in different applications like electrical and sound insulation. However, nowadays its use is limited in many countries because of its health hazards.

8.3 Processing of PP/Plant-Based Fiber Biocomposites and Bionanocomposites

Similar to man-made fiber/PP composites, there are several processing methods for plant-based fiber/PP biocomposites. The selection of a proper processing method depends on several important factors, including the shape and geometry of the final product, ease of manufacturing, cost, and desired properties of the final product [15]. The most applicable processing methods for this class of PP composites are compounding extrusion, injection molding and thermoforming.

8.3.1 Compounding Extrusion

The goal of the compounding process is to uniformly distribute PBFs in the PP matrix by using thermal and mechanical energy. The product of this process is in the form of plastic granules which can be used as the feedstock for other processes like injection molding. Typically, twin-screw extruders, co-rotating or counter-rotating, with modular configuration are utilized for compounding of plant-based fiber PP composites. In this process, ingredients are fed into the extruder and flow along the heated barrels by rotating screws. During this movement, the heat energy generated by extruder heaters and friction forces between the materials and barrel wall melt PP, while high shear forces disperse and distribute fibers uniformly within the polymer matrix. After composite melt exits the extruder die, it is cooled by air or water and granulated by proper devices. Pellets made by this technique can be formed by second processes such as injection molding. Alternatively, the compounding device can be completed by a simultaneous sheet forming line to produce fiber reinforced PP sheets. One of the challenges in this process is the friability of natural fibers, which can make the feeding step challenging. Another point is the high moisture content of natural fibers, which is addressed in different ways. Fibers can be dried before the process or, if required, venting is performed during the extrusion, which results in slower extrusion rates.

8.3.2 Thermoforming

The main products of the thermoforming process are composites made from natural fiber and thermoplastic mats. The input of this process are the pre-cut layers or mats of natural fiber and thermoplastic polymer fibers. As these layers are placed into a heated mold, heat is transferred from the mold to the polymer fibers and melts them. Parallel to the heating phase, a proper pressure is applied by the mold, which causes the thermoplastic melt to penetrate into the natural fiber mat. After the heating phase is completed, cold press starts, which consolidates the hybrid material into the final composite.

8.3.3 Injection Molding

Injection molding is one of the most versatile processing methods for manufacturing of natural fiber PP composites. This process is usually used for producing interior parts of automobiles with the thickness of 2–5 mm like instrument panel components, consoles, door handles, and load floors. In addition to the automobile industry, other sectors also use natural fiber PP composite injected-molded products like home and office furniture, lawn and garden products, toys, housewares, power tools, sporting goods, storage, and containers. The main advantages of this method are short processing cycle time (which is important for natural fibers sensitive to thermal energy), high volume productivity, few post-processing operations, capability to produce components with intricate shapes, and excellent dimensional tolerance.

The feedstock of this process is the pellet of the composite, which means that a pre-compounding process for distributing natural fibers within the PP matrix is necessary. Granules are fed from the hopper into the heated barrel. A rotating screw plunger moves the granules slowly within the heated barrel, where the plastic is melted. Then, by advancing the plunger, the plastic melt is forced through a nozzle that rests against a mold and enters the mold cavity through a gate and running system. After the melt enters into the mold cavity the cooling phase starts, during which plastic solidifies inside the mold and forms the final shape.

8.3.4 Compatibilization of Polypropylene and Plant-Based Fibers

In fiber reinforced PP composites, the mechanical performance of the material depends strongly on the fiber-matrix interaction quality. In fact, the degree of adhesion between these two phases determines the stress transfer from the matrix to the fiber. A strong adhesion at the interface is necessary for an effective stress transfer and load distribution throughout the interface [8].

Natural fibers mainly consist of cellulose which contains three hydroxyl groups per monomer in its structure. The persistence of these polar groups makes natural fibers hydrophilic. On the other hand, PP is hydrophobic with its non-polar macromolecules. This difference in the polarity of composite constituents can deteriorate the mechanical properties of the final product. So, a proper method should be applied to compatibilize natural fibers and PP and improve the adhesion between them. For this purpose, several surface treatments of natural fibers have been developed. During these treatments, a structural property of the fiber at the surface is changed in a way that makes it compatible with PP. This change can be in the physical or chemical structure of the natural fiber [3]. In the following subsections, physical and chemical surface treatment approaches are discussed.

8.3.4.1   Physical Treatment

In physical treatments, the interphase adhesion is enhanced by increasing the mechanical bonding and entanglements between fiber and matrix. The most common physical treatments are plasma and corona treatments [3, 8, 16]. Corona treatment is actually a surface oxidation activation technique that changes the surface energy of the natural fiber. During this treatment, the number of polar components of free surface energy increases, which results in better compatibilization with PP [3]. In plasma treatment, under atmospheric pressure and in the presence of an inert gas, some elements like ions, electrons, and free radicals are produced on the fiber surface. As a result, surface characteristics of the natural fibers are changed, which positively affects adhesion of fibers to the PP matrix. The main advantage of plasma treatment is that various modifications can be introduced to the fiber surface by utilizing different types of gases in the process [3].

8.3.4.2   Chemical Modifications

The existence of hydroxyl groups in natural fibers makes them hydrophilic and incompatible with hydrophobic PP. A reasonable way to compatilize these ingredients is introducing a third bi-functional material which can react simultaneously with hydrophilic natural fiber and hydrophobic PP. In fact, this coupling agent acts like a bridge between natural fiber and PP and improves interfacial adhesion, resulting in better mechanical properties. The aim of chemical treatments is to substitute some of the hydroxyl

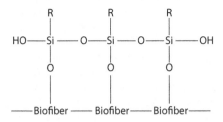

Figure 8.4 Covalent bond between natural fiber and silane coupling agent. (Reproduced with permission from [17]; Copyright © 2011)

groups in natural fiber surface with these bi-functional materials. Based on the coupling agent introduced to the fibers, there are various chemical treatments such as silane treatment, acetylation, and so on [3, 8].

Silane is a multifunctional chemical compound with the general formula X_3SiR. In this formula, R is a chemical group which is capable of reacting with the nonpolar molecules of PP. The most common R groups are amins, vinyls, and methacreloxypropene [8]. X group is capable of hydrolyzing in an aqueous solution to form a silanol group. Silanol groups can react with hydroxyl groups on the surface of natural fiber. The most common X groups include methoxy, chloro, and ethoxy [8]. Figure 8.4 illustrates covalent bonding at the fiber surface between silane coupling agent and cellulose.

Acetyl with the formula CH_3CO is another multifunctional chemical group that is utilized as a coupling agent in natural fiber PP biocomposites. In this method, acetyl group reacts with the hydroxyl groups of the natural fiber and substitutes their hydrogen atoms [18]. The substitution of acetyl group decreases the hydrophilicity of natural fibers and improves the interaction of these fibers with PP, resulting in better mechanical properties. Usually, acetic anhydride obtained from acetic acid is utilized for this treatment.

Maleic anhydride is the most common coupling agent used in natural fiber PP biocomposites. The main difference of maleic anhydride with other coupling agents is that it is mostly used to modify PP instead of natural fibers [3]. Nowadays, different grades of maleic anhydride grafted PP (MAgPP) are in the market which are utilized to compatilize PP with not only natural fibers, but also other reinforcements like glass fibers. The chemical structure of maleic anhydride grafted PP is illustrated in Figure 8.5. MAgPP acts like a coupling agent by bonding with PP matrix through its PP backbone and to the polar molecules at the surface of the natural fibers through its polar maleic anhydride group [19].

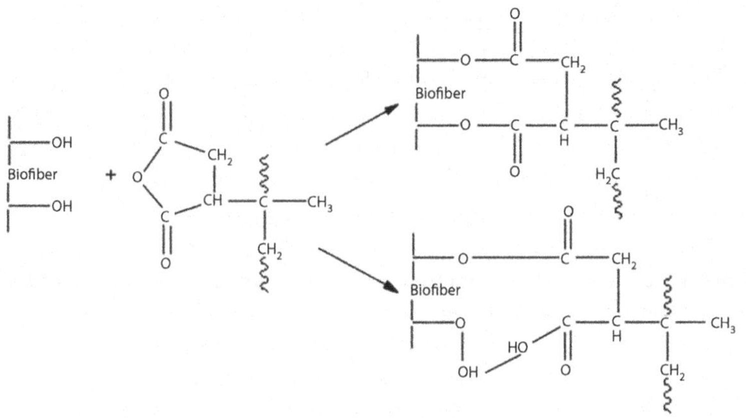

Figure 8.5 Chemical structure of maleic anhydride grafted PP. (Reproduced with permission from [19]; Copyright © 2007 Express Polymer Letters)

Mercerization or alkali treatment is another chemical treatment commonly applied to natural fibers. This method includes treatment of the natural fibers in an alkaline solution. Parameters like solution concentration, treatment time, and temperature determine the efficiency of the treatment. This treatment causes some material removal from the fiber's surface and increases its surface roughness [20]. The rough surface of the fiber then strengthens mechanical entanglement of the polymer chains to the fiber's surface and improves interfacial adhesion.

8.4   Characterization and Properties of Plant-Based Fiber Reinforced Polypropylene Biocomposites and Bionanocomposites

Thermoplastics provide several advantages over thermoset polymers for use with PBFs like ease of molding complex parts, good dimensional stability, low processing cost, and design flexibility. Among thermoplastics, PP and PE are suitable matrix materials for natural fibers. This is mainly thanks to the fact that their processing temperature does not exceed 230 °C. Because natural fibers have low thermal stability they need to be processed at temperatures below 200–230 °C to avoid thermal degradation. Natural fiber reinforced PP biocomposite characteristics are significantly influenced by a great number of variables, including fiber's source and type, aspect ratio, processing methods, content, and interface modification. In this part, the

recent studies on various aspects of PBF reinforced PP composites characterization are reviewed. Moreover, this section addresses some of the basic
issues in the chemical, mechanical, and physical properties of such composites and nanocomposites.

8.4.1   PP/Wood Fiber Biocomposites and Bionanocomposites

In recent years, wood fibers have gained significant interest as natural
reinforcing materials for commercial thermoplastics like PP mainly due to
their high specific properties. Wood fibers could potentially replace synthetic fibers and inorganic fillers in various applications. However, they
are highly limited by their high moisture absorption and weak interfacial
adhesion between polar-hydrophilic fibers and nonpolar-hydrophobic
petroleum-based plastics. The recent studies indicate an improvement
in wood fiber– PP matrix bonding in the presence of compatibilizers or
coupling agents [21–23]. For example, maleated polypropylene (MAPP),
the most common coupling agent, enhances the interfacial adhesion
between wood fibers and PP, and improves the mechanical properties
accordingly [21, 23].

A comprehensive study on the effect of interface modifiers on the properties of PP/wood fiber composites has been carried out by Sain *et al.* [24].
Their invaluable results demonstrated that the properties of composites
were very poor in the absence of interface modifiers. In contrast, it was
noted that the use of maleated PP, itaconic anhydride, and bismaleimide-
modified PP resulted in a stable surface and thus improved tensile strength.
In similar studies, Maldas and Kokta [25] and Karnani *et al.* [26] show an
improvement in the mechanical strength of wood fiber-filled PP composites with the use of silane and isocyanate surface modifiers.

In addition, different chemical treatments on fibers aimed at improving the adhesion between the fiber surface and matrix may not only
increase the mechanical properties but also modify the surface of the
fibers. Moreover, results showed that the water absorption of natural fiber
reinforced biocomposites is decreased significantly due to such chemical
modifications on natural fibers as acetylation, alkali, acrylation, silane,
benzoylation, permanganate, and isocyanates [27–29]. Another equally
important factor which improves the physical and mechanical properties
of wood fibers/PP biocomposites and bionanocomposites is the dispersion
of natural fibers in matrix. In fact, the better dispersion of wood fibers in
matrix guarantees an effective wetting of fibers, as well as a better adhesion
between the two phases [30].

Czvikovszky [31] has developed a reactive extrusion procedure for preparing wood fiber/PP composites. The results demonstrated that wood fiber and PP were bound together through applying a small amount of reactive additive. In fact, the additive compatibilizes both natural fiber and the synthetic components, and then electron beam treatment creates active sites on PP and wood fibers. This method has not only improved thermal stability over the conventional wood fiber/PP composites, and over the PP itself, but also has significantly enhanced the modulus of elasticity, flexural strength, and tensile strength.

To simultaneously investigate the effects of chemical treatments on fibers and coupling agent in composite, Coutinho *et al.* [32] have investigated the mechanical properties of silane-treated wood fibers/PP biocomposite in the presence or absence of MAPP at three different temperatures: 170 °C, 180 °C, and 190 °C. The tensile and three-point bending tests showed that 180 °C is the best mixing temperature, while the use of vinyltris (2-methoxy ethoxy) silane produced improved adhesion with or without MAPP coating.

8.4.2   PP/Jute Fiber Biocomposites and Bionanocomposites

Jute is known as the "Golden Fiber," which is produced from plants of the genus *Corchorus*. Jute fibers containing 82–85% of holocellulose are classified as bast fiber (like flax and hemp) [33]. Although in recent decades the production of jute declined from 3.7 million tons to 0.8 million tons, the production and global consumption of jute is still second only to cotton. Jute is an easy plant to grow and has a high yield per acre, so that it has little need for pesticides and fertilizers. India, Bangladesh, and China provide the ideal conditions for the growth of jute [34].

In recent years, jute fibers have attracted the attention of researchers because of their excellent specific properties like low density and high strength. However, a notable weakness of jute fibers is their polarity and hydrophilicity, which makes them unsuitable for the hydrophobic PP. This feature is mainly because of the presence of the hydroxyl groups that can form hydrogen bonds instead of interacting with the resin matrix. The presence of surface impurities, such as pectin and waxy substances, hinders the hydroxyl groups from reacting with the nonpolar components and interlocks adhesion with such matrixes. With this in mind, several studies have been conducted on physical and/or chemical treatments and modification of the surface of jute fibers for improving the mechanical and physical properties of the biocomposite materials [35, 36]. For instance, a systematic study of the effect of surface treatments on the properties of

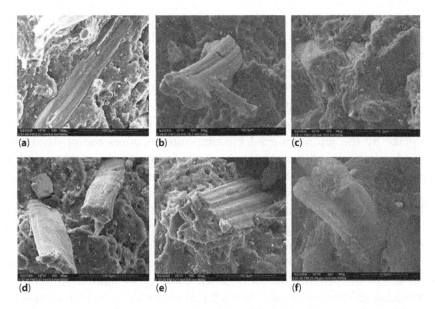

Figure 8.6 Scanning electron microscopy images of the impact fracture surfaces of the PP composites reinforced by (a) untreated jute fibers, (b) alkali treated fibers, (c) alkali and silane treated jute fibers, (d) alkali and MAPP treated jute fibers, (e) ASMT (3600), and (f) ASMT (35000). (Reproduced with permission from [37]; Copyright © 2010 Express Polymer Letters)

PP/jute fiber has been carried out by Wang *et al.* [37]. They investigated the effects of four treatments, including alkali treatment, alkali and silane treatment, alkali and MAPP treatment, and alkali, silane, and MAPP (ASMT) treatment on the physical and mechanical properties of jute fibers/PP biocomposites. The results convincingly indicated that all of the mentioned treatments effectively enhanced interfacial compatibility and the mechanical properties accordingly. Among these methods, ASMT provided the best interfacial compatibility and properties. The microscopic studies on the fractured surfaces of the composites displayed a good adhesion between PP and jute fibers. The tensile strength and impact toughness were also increased by 46 and 37 wt%, respectively, in the presence of 15 wt% of ASMT treated fibers (Figure 8.6).

Other studies have indicated similar improvement in mechanical and physical properties by various fiber surface treatments such as monomer grafting under UV radiation [38–40], silane [41, 42], alkali [43–45], maleic anhydride grafted PP [46, 47], and others [48]. In a similar study, the influences of post-treated jute fibers on physicomechanical properties of PP composites were evaluated by oxidizing the raw jute fibers and post-treatment of them with urea. The results demonstrated that the better compatibility of post-treated jute fiber with the PP matrix improved tensile, water absorption, flexural, impact, and hardness properties of composites compared to the composites which were reinforced with oxidized or raw ones [49].

Conversely, other researchers have tried to determine how the water uptake by the jute reinforced biocomposite enhances the physical and mechanical properties of the specimens. Karmaker *et al.* have showed that swelling of an embedded jute fiber in PP, caused by water absorption, is able to fill the gaps between fibers and matrix which were mainly formed because of the thermal shrinkage of PP melt. The fill up of such gaps results in a higher shear strength between jute fibers and PP, and increases the mechanical properties accordingly [50].

Tensile modulus (TM), tensile strength (TS), percentage elongation at break (Eb%), bending modulus (BM), bending strength (BS), bending modulus (BM), and impact strength (IS) of the PP/jute fiber composite (50 wt%/50 wt%) were found to be 2305 MPa, 45 MPa, 12%, 4267 MPa, 52 MPa, and 18 kj/m^2, respectively. In addition, they evaluated the effect of jute fiber on mechanical properties of PP/non-vulcanized natural rubber (NR) blends. It was reported that TM, BS, TS, BM and hardness of jute reinforced NR blended (10–50 wt%) PP matrix composites were significantly decreased with the increase of NR in PP. Impact strength (IS) and elongation at break (Eb%), however, were enhanced compared to the control biocomposite [51]. The influence of jute fibers on the properties of jute/ thermoset plastic biocomposites were comprehensively studied, including the crystallinity, biodegradability, fiber orientation on friction and wear behavior, chemical and physical modification, water absorption, thermal stability, durability, and eco-design of automotive components [35, 52–56].

8.4.3   PP/Flax Fiber Biocomposites and Bionanocomposites

Flax fibers, a cellulose-based polymer obtained from the stems of the plant *Linum usitatissimum*, have been used to make linen since prehistoric times. This plant grows best in temperate latitudes of the Northern Hemisphere, where moist and moderately warm summers yield fine flax. Statistics show that the world's leading flax producing countries are France (64,000 tons annually), China (31,000 tons), Belgium (15,300 tons) and Holland (4,600 tons). In comparison with jute and cotton fibers, the structure of flax fiber is more crystalline, making it stiffer, denser, stronger, and crisper. Over the recent decade, flax fibers have been attracting constantly growing attention for many higher value-added textile markets. Nowadays, flax fiber reinforced PP is widely used in the field of biocomposites and nanocomposites.

Flax fibers are available in lengths up to 90 cm and an average diameter of 12 to 16 microns. The mechanical and physical properties of flax fibers depend on the diameter, length, and the location of fibers in the stems. In this regard, the greatest challenge in working with flax fibers is their variation in properties and characteristics, which is mainly linked to the variation of the fiber size along its longitudinal axis [56]. Other variables in flax fiber/PP biocomposites have been extensively studied. For example, the effect of fiber chemical treatments (alkalization) and matrix modification (maleic anhydride, vinyltrimethoxy silane, maleic anhydride-PP) on the mechanical properties of flax fiber/PP biocomposites was investigated by Arbelaiz et al. [57]. Results suggested that matrix modification results in better mechanical performance than fiber surface chemical treatments. Another equally important result is that MA or silanes grafted onto PP lead to even better mechanical properties than MAPP modification of matrix in these biocomposites.

In a similar study, the effect of zein coupling agent, a protein extracted from corn, on the dynamic and static properties of flax fiber/PP biocomposite was investigated by John et al. [58]. Biocomposites containing chemically treated flax fiber were found to possess enhanced tensile, flexural and impact properties. Moreover, the storage modulus of modified flax fiber reinforced biocomposites was significantly increased thanks to the improved interfacial adhesion. Other studies have similarly investigated the influence of different parameters, and various aspects of flax fiber/PP biocompoistes, including the physical structure of flax fibers [59], the effects of processing methods [60], processing parameters [61], aspect ratio [60], fiber morphology stability [62], crystalline and thermal degradation [63], sensitivity to crack propagation [64], environmental assessment [65, 66], water absorption [67], fracture behavior of biocomposites [68], and the feasibility of using flax fiber/PP biocomposites in the automotive and building industries [69–71].

8.4.4 PP/Sisal Fiber Biocomposites and Bionanocomposites

Sisal with the botanical name *Agave sisalana*, is a coarse and strong fiber commercially produced in Brazil, Mexico, and East Africa. Sisal has low tolerance to moist or saline soil conditions, and can be harvested in most soil types. Although the global demand for sisal fibers during the decade 2000–2010 declined by an annual rate of 2.3%, it is being increasingly used in biocomposite materials for a wide range of applications ranging from cars to packaging to buildings in the current decade. Joseph *et al.* [72] comprehensively delved into the effects of processing parameters on the mechanical properties of sisal fiber reinforced PP biocomposites. Such biocomposites were prepared by both melt-mixing and solution-mixing methods. They reported that under optimum mixing conditions (mixing time = 10 min, rotor speed = 50 rpm, and mixing temperature = 170 °C), melt-mixed composites showed better tensile properties than those of solution-mixed biocomposites.

They have also studied the influence of sisal fiber loading, length of fiber, and fiber orientation on the mechanical properties of biocomposites. Generally, at low fiber content, sisal fibers did not restrain PP and the localized strains that occurred caused the bond between PP and fiber to break. However, at higher fiber loading, stress can be distributed more evenly and the strength of biocomposite increases. Sisal fiber length of 2 mm provided the best balance of properties in the case of melt-mixed composites. In addition, biocomposites containing longitudinally oriented fibers presented better mechanical properties than those with random and transverse orientations. The influence of chemical treatment on the tensile properties of sisal/PP biocomposites was carried out by such treatments as sodium hydroxide, urethane derivative of PPG, maleic anhydride, and permanganate [72, 73]. Chemically treated sisal fibers achieved better interfacial bonding with PP, and presented enhancement in mechanical properties compared to untreated fibers accordingly [73, 74].

The burning rate of sisal/PP biocomposites has been reduced by adding different flame retardants (FRs) like magnesium hydroxide ($Mg(OH)_2$) and zinc borate. The results showed that no synergistic effect was observed when flame retardants were incorporated into the sisal/PP biocomposites. Moreover, their test results showed that FR agents enhanced the thermal stability without sacrificing the mechanical properties. They have also demonstrated that the FR additives had no effect on the processability of the composites [75, 76]. Sisal fiber reinforced PP was studied regarding the effect of fiber on biodegradability of the biocomposites [77], water uptake behavior [78], the influences of aging on mechanical properties [79], and fatigue behavior [80].

8.4.5 PP/Hemp Fiber Biocomposites and Bionanocomposites

Historically, the cultivation of hemp has taken place in the United States from the 17th to mid-20th centuries. It belongs to the *Cannabis* family and grows in temperate climates. Two principal types of fibers are derived from the hemp plant's stalk—long (bast) fibers and inner short fibers (hurds).

The typical length of long bast fiber is similar to soft wood fibers and is very low in lignin content. However, the short core fibers are more similar to hard wood fibers. Generally, hemp fiber is extremely versatile and has many qualities, including strength and durability, that makes it a very desirable reinforcement for countless products such as automotive interior trim products, accessories, packaging, shoes, furniture, and buildings.

Remarkable improvements in tensile strength, Young's modulus, fiber separation, crystallinity index, lignin reduction and thermal stability of hemp fiber reinforced PP were reported by chemical treatments of hemp fibers [81, 82]. For example, compared to the untreated fibers, the fibers treated with a solution of 5 wt% NaOH/2 wt% Na_2SO_3 have higher interfacial adhesion as a consequence of relatively strong chemical bonding between the hemp fibers and the PP. The optimum hemp fiber reinforced PP biocomposite, consisting of 40 wt% $NaOH/Na_2SO_3$ treated hemp fibers and 4 wt% MAPP, had a Young's modulus of 5.31 GPa and tensile strength of 50.5 MPa [83].

In similar studies, the influence of white rot fungi, chelator, and enzyme treatments on the separation of hemp fibers from bundles, and the interfacial bonding of the fibers with the PP matrix were investigated [84–86]. The results showed that the separation of treated fiber composites was better than that for untreated fiber composites. Biocomposites consisting of hemp fibers treated with chelator presented higher interfacial shear strength and tensile strength than the unmodified composites. Several studies have focused on the spherulitic morphology and isothermal crystallization rate of PP. Their results showed that the increasing content of modified hemp fibers provides more nucleation sites for facilitating crystal growth [82]. Thermal properties of hemp fiber reinforced 1-pentene/PP were investigated by Khoathane et al. [87]. They have illustrated that the thermal stability of the composites was enhanced compared to that of the fibers or the matrix as individual entities.

The effects of fiber content, and anisotropy of fibers on the water sorption and mechanical properties of nonwoven PP/hemp mats which were prepared by hot press, was studied. A strong decrease in three-point bending strength was reported by immersing the biocomposite specimen in distilled water for 19 days; however, impact strength of samples was noticeably increased. Optimal mechanical properties were found in samples with 40–50 wt% hemp fiber cut from the biocomposite sheets parallel to the carding direction [65]. The recycling of hemp fiber reinforced PP has been studied extensively for many years [88]. As expected, the length of fibers, molecular weight, and Newtonian viscosity have been decreased by reprocessing of composite. Despite the number of reprocessing cycles, the mechanical properties of recycled hemp fiber/PP biocomposites remain well preserved. Thermal conductivity for hemp fiber reinforced PP [89], mechanical performance of hemp fiber reinforced PP biocomposites at different operating temperatures [90], and rheological properties of hemp fiber/PP composites [91] were also evaluated.

8.4.6   PP/Kenaf Fiber Biocomposites and Bionanocomposites

Like many other PBFs indigenous to East-Central Africa, Kenaf (*Hibiscus cannabinus* L.) is easy to grow and high yielding, and has a unique combination of long bast and short core fibers (35 wt%:65 wt%) which makes it suitable for biocomposite industries. It has been introduced since the Second World War in South Africa, Mexico, Egypt, China, Thailand, and Cuba. A single kenaf fiber can have a modulus and tensile strength as high as 60 GPa and 11.9 GPa, respectively [92]. Various chemical treatments have been successfully examined on kenaf fibers in order to improve the mechanical performance of the kenaf fiber reinforced PP. Kenaf fibers were also treated by means of melt grafting reactions with glycidyl methacrylate (GMA) to enhance their suitability for use as reinforcements in composite materials. Studies show that the tensile properties of the treated kenaf fibers are significantly enhanced compared to untreated kenaf fibers [93–96]. The influence of compatibilizer on the structure-property relationships of kenaf fiber/PP biocomposites have been studied by Feng *et al.* [97]. They used MAPP to enhance the compatibility between fibers and matrix. Their results indicated that the biocomposites with coupling agent have better impact strength, high temperature moduli, and lower creep compliance than the uncoupled systems. They have also showed that the crystallization and melting behavior of these blends were affected by compatibilizers. Owing to the improved adhesion between the polymer molecules and kenaf fibers, the coupled samples had more restricted molecules than the uncoupled ones. Accordingly, a sharp decrease in the crystallization rate and crystallization temperature (T_c) of the coupled high molecular weight blends was observed in the presence of compatibilizers. However, in the lower molecular weight systems, the coupling agent increases the crystallization temperature (T_c) of polymer matrix [97].

In a similar study, the effect of fiber content and coupling agent on the impact strength of kenaf fiber/PP and wood flour/PP biocomposites were compared. At room temperature and in reversed notched tests, PP/kenaf composites showed significantly lower energy to maximum load (EML) values than neat PP samples. But in notched tests, addition of kenaf fibers to PP did not change the EML values notably. In addition, MAPP improved all impact test values. In notched impact tests and over the studied temperature range, kenaf fiber/PP biocomposites containing MAPP consistently yielded higher EML values than both PP specimens and wood flour reinforced PP composites [98].

Han *et al.* [99] have compared the influence of kenaf fiber treatment by electron beam irradiation (EBI) and alkalization on dynamic mechanical and thermal properties of kenaf fiber/PP biocomposites. The results showed that the modification of kenaf fiber surfaces at 200 kGy EBI and treatment with 5 wt% NaOH was most effective for enhancing the mechanical performance of kenaf/PP biocomposites. The effect of processing conditions on the morphology and properties of kenaf fiber/PP [98, 100], rheological properties of kenaf fiber/PP composites [101], and the hybrid effect on the composite properties [102–104] were also investigated.

8.4.7 PP/Other Plant-Based Fiber Biocomposites and Bionanocomposites

Abaca fiber (banana fiber) is a bast fiber obtained from the pseudo-stem of banana plant and grown in tropical countries such as the Philippines, Ecuador, and Costa Rica [105]. Nowadays, abaca fiber reinforced PP is gaining interest due to its strength. Abaca is the strongest of the natural fibers and is commercially used in innovative applications such as underfloor protection for passenger cars [106]. The mechanical properties, odor emission, and structure properties of abaca fiber reinforced PP with different fiber loadings (20, 30, 40, 50 wt% and in some cases 35 and 45 wt%) were investigated. TM, BM, and IS were found to increase gradually as fiber loading increased up to 40 wt%, and then decreased. However, with the addition of coupling agent (MAPP-MAH-PP), the mechanical properties were found to increase at different fiber loadings (30 to 80%). The odor emission of the biocomposites was compared with 30 wt% flax and jute fiber reinforced PP. Results showed that the odor emission of flax fiber biocomposites were smaller than jute and abaca fiber composites [107].

Mechanical properties of 10 wt% PBF reinforced PP composites are shown in Table 8.2. It is generally accepted that the strength of a natural fiber reinforced PP biocomposite is mainly dependent on the fiber's strength. Among all PBFs listed in Table 8.1, flax fibers have the highest TS. Accordingly, in the same reinforcement content, high strength is preferred for flax fiber/PP composite over other plant-based fiber composites, which is reflected in the results shown in Table 8.2.

Table 8.2 Mechanical properties of different PBFs reinforced PP.

Composite Fiber(10)/PP(90)	Tensile Strength (MPa)	Tensile Modulus (MPa)	Bending Strength (MPa)	Bending Modulus (MPa)	Impact Strength (kJ/m²)	Refs.
Flax fiber/PP	49	4100	–	–	1.8	[58, 108, 109]
Sisal fiber/PP	26	700	–	–	20	[74]
Hemp fiber/PP	27	–	42	–	50	[110]
Bagasse fiber/PP	18	330	23	1050	–	[111]
Wood fiber/PP	32	600	60	–	35	[112]
Bamboo fiber/PP	36	800	37	1900	–	[113, 114]
Abaca fiber/PP	25	1800	48	1850	–	[115]
Coir fiber/PP	26	600	34	650	6	[115]
Palm fiber/PP	25	1050	50	1780	–	[116]
Calcium alginate fiber/PP	24	1000	38	1750	16.5	[117]
Jute fiber/PP	43	900	48	2000	11	[118–120]
Phosphate glass fiber/PP	39	900	50	2100	7.5	
E-glass fiber (50)/PP (50)-	85	7100	88	12000	35	[33]

8.5   Applications of Plant-Based Fiber Reinforced Polypropylene Biocomposites and Bionanocomposites

Natural fiber reinforced polymer composites have many advantages such as biodegradability, low density, low cost, better finishing of molded components, high stiffness-to-weight ratio, and high degree of flexibility which causes a natural fiber to bend rather than fracture, thus giving them suitability for different applications (building and construction, automotive industry, aerospace, and sports) [121–127]. It is also worth mentioning that the environmental impact assessment of natural fiber reinforced PP biocomposites by sustainability assessment methodologies is one of the most influential stages of the development of new materials for all products. In this regard, life cycle analysis (LCA) has been developed to assess environmental impacts associated with the entire product life cycle from cradle to grave (i.e., from raw materials, energy consumption, ultimate disposal, and recycling characteristics) [128]. This analysis provides a reliable evaluation to avoid a narrow outlook on environmental issues by the following points [129]:

- Compiling an inventory of relevant inputs and environmental outputs;
- Evaluating the potential environmental impacts associated with identified inputs and outputs;
- Interpreting the results of the inventory to make a more informed decision

In this section, an overview of the recent achievements of natural fiber reinforced composites usage in different industries that are in full compliance with the LCA's instructions is provided.

8.5.1   Automotive Industries

The use of natural fiber reinforced biocomposites in automotive industries continues to grow rapidly [130, 131]. The fundamental driving forces behind the efforts for replacement of synthetic fibers with natural fibers are price, weight reduction, and marketing attractiveness rather than technical demands. In fact, several automobile companies in Europe are replacing synthetic fibers with natural fibers not only to comply with increasing environmental considerations but also to satisfy the public demand for green vehicle [6].

In 1941, hemp and flax fibers were used in PP for the bodywork of a Henry Ford car. Although these composites were able to withstand ten times the impact of an equivalent metal panel, the automotive part suppliers did not put it into general production due to economic limitations at the time [132]. Thermoformable natural fiber reinforced PP can be extruded in sheet form for use in various interior components, such as door panels, seatbacks, speaker trays, and console box liner, reducing the consumption of traditional plastics by less than 50% [133]. In an invaluable study, Bledzki *et al.* investigated the influence of manufacturing processes on different aspects of products. They reported that the compression molding process is favorable for the automotive sector mainly because of the relatively lower odor concentrations compared to injection molding. It seems that the decomposition of materials in the injection molding process is more than in the compression molding process which causes unfavorable odor [134]. A recent annual report describes that the use of natural fiber biocomposites in automotive industries is increasing and is gaining preference over glass fiber (GF) and carbon fiber (CF). The report also mentioned that the original equipment manufacturer (OEM) and major automotive part suppliers interest in "green" technologies is mainly because of the fact that such materials have converged with tough new European regulations on fogging, odor, and volatile organic compound (VOC) emissions to increase the pressure on the manufacturers of automotive interior parts [135, 136]. For example, what is known as Section 01350 standard testing forces the manufacturers to consider strict regulations on material emissions.

Although the range of products is restricted to nonstructural components, several researchers have investigated the possibility of using natural fiber reinforced biocomposites in structural automotive parts, considering the social, environmental, economical, and technical advantages. For instance, Figure 8.7 compares the performance of jute and glass fiber based composites in the frontal bonnet of an off-road vehicle. In this study, the frontal bonnet was assigned as a functional unit which covers the engine and requires structural and mechanical performance. It is clearly observed that jute fiber reinforced PP composites present a better performance than GF/PP in all aspects except the technical aspect [55].

In operational/commercial phase, automotive companies like Mercedes Benz and BMW are planning to completely substitute the structural and nonstructural parts with such biocomposites. For example, Mercedes Benz used such European renewable fibers as flax, ramie, and hemp to produce natural reinforced composites for door panels and car roofs in the K-Series, where the "K" stands for "kraut" and "compost" [131]. The use of flax and sisal fiber reinforced PP biocomposites to increase the strength and impact

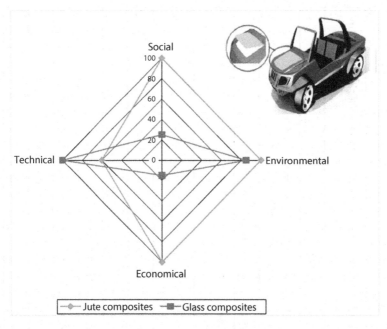

Figure 8.7 Comparison of the performance of jute and glass fiber-based composite bonnet from different aspects. (Reproduced with permission from [55]; Copyright © 2010 Elsevier)

resistance of interior door linings and panels of BMW 3, 5 and 7 series is another example of the technical applications of natural fibers [137]. Table 8.3 summarizes the various components made of natural fiber reinforced PP in major vehicle manufacturers around the world [127, 137–139].

8.5.2   Building and Construction

Green building technologies and innovations are essential for creation of ecologically sustainable, healthy, and suitable places for people to live and work. There is historical evidence of the existence of primitive natural fiber reinforced products as far back as 10,000 BC. In addition, the indigenous people of America also used plant-based fiber reinforcement to strengthen the structures of shelters [142]. The selection of advanced engineering materials can save energy, reduce costs, and increase inhabitant's satisfaction. With this in mind, lightweight, high stiffness, and high strength biocomposites and nanocomposites produced from biodegradable reinforcements are very promising alternatives for applications in green building industries. Moreover, these new multifunctional

Table 8.3 The components made of natural fiber reinforced PP in major automotive manufacturers [15, 140, 141].

Model	Manufacturer	Components
A2, A3, A4, A4 Avant, A6, A8, Roadstar, Coupe	Audi	Seatback, side and back door panel, boot lining, hat rack, spare tire lining
C5	Citroen	Interior door paneling
3, 5, 7 series	BMW	Door panels, headliner panel, boot lining, seatback, noise insulation panels, molded foot well linings
Eco Elise	Lotus	Body panels, spoiler, seats, interior carpets
Punto, Brava, Marea, Alfa Romeo 146, 156	Fiat	Door panel
Astra, Vectra, Zafira	Opel	Instrumental panel, headliner panel, door panels, pillar cover panel
406	Peugeot	Front and rear door panels
2000 and others	Rover	Insulation, rear storage shelf/panel
Raum, Brevis, Harrier, Celsior	Toyota	Door panels, seatbacks, floor mats, spare tire cover
Golf A4, Passat Variant, Bora	Volkswagen	Door panel, seatback, boot-lid finish panel, boot liner
Space Star, Colt	Mitsubishi	Cargo area floor, door panels, instrumental panels
Clio, Twingo	Renault	Rear parcel shelf

Manufacturer	Model	Applications
Daimler-Benz	Mercedes A, C, E, S class, Trucks, EvoBus (exterior)	Door panels, windshield/dashboard, business table, pillar cover panel, glove box, instrumental panel support, insulation, molding rod/apertures, seat backrest panel, trunk panel, seat surface/backrest, internal engine cover, engine insulation, sun visor, bumper, wheel box, roof cover
Honda	Pilot	Cargo area
Volvo	C70, V70	Seat padding, natural foams, cargo floor tray
General Motors	Cadillac DeVille, Chevrolet Trailblazer	Seatbacks, cargo area floor
Saturn	L3000	Package trays and door panel
Ford	Mondeo CD 162, Focus, Freestar	Floor trays, door panels, B-pillar, boot liner

materials aim to respond to customer demand for greater comfort, convenience and luxury, and also reduce the dependence on petroleum-based plastics.

The demand for natural fiber reinforced thermoplastics is forecast to rise 15–20 percent per year with a growth rate of 50% in building applications. Among the various matrices, special attention has been devoted to PP. Plant-based fiber reinforced PP biocomposites have opened up tremendous opportunities for creative solutions based on reinforced plastics in two principle products [143]:

i. Structural products, including pedestrian bridges, floors, walls, multipurpose panels, and roof structures.
ii. Nonstructural products, including exterior construction, water tanks, decking, windows, as well as door frames and panels.

For example, Van de Weyenberg *et al.* [144] have characterized the mechanical properties of biocomposites reinforced with flax fibers in structural and nonstructural products such as shutters, door panels, door frames, roofing sheets, etc. They have shown that the good properties (strength and modulus) of biocomposites made from 40 wt% flax fiber reinforced PP, give them a wide area of applications. Similar studies reported that other PBFs like jute, wood, bamboo, sisal, and coir have great potential to replace conventional synthetic fibers in building industries [136, 145–151].

8.5.3 Packaging

With a growing awareness for environmental considerations in the manufacturing life cycle, packaging industries are demanding a shift of their design from traditional polymers to natural biocompoistes, focusing on the biodegradability or recyclability of green products [141].

As mentioned previously, natural fibers are often considered for nonstructural applications that require low costs, low energy for processing, biodegradability, and light weight. For example, non-load-bearing indoor components in civil engineering are successfully produced with natural fiber reinforced composites because of their vulnerability to environmental attack [121, 152]. In addition to mentioned industries, natural fiber reinforced PP biocomposites have contributed to various markets. Table 8.4 presents just a few applications of natural fiber reinforced PP that are already helping molders and their customers find solutions for the evolving needs in different industries.

Table 8.4 Applications of PBF reinforced PP biocomposites and bionanocomposites [153, 154].

Category of Fibers	Fiber	Applications
Bast/Skin	Jute	Building panels, roofing sheets, door frames, machinery, door shutters, transport, packaging, geotextiles, and chip boards
	Hemp	Construction products, roofing covers, office product textiles, cordage, geotextiles, paper and packaging, furniture, electrical components, ceilings, boats, manufacture of bank notes, and manufacture of pipes
	Flax	Window frame, machinery, panels, decking, railing systems, fencing, roofing covers, tennis racket, bicycle frame, fork, seat post, snowboards, and laptop cases
	Kenaf	Packing material, partition boards, mobile cases, bags, insulations, clothing-grade cloth, soilless potting mixes, animal bedding, and material that absorbs oil and liquids
	Ramie	Roofing covers, industrial sewing thread, packing materials, partition boards, fishing nets, and filter cloths. Fabrics for household furnishings (upholstery, canvas) and clothing, paper manufacturing
Fruit	Coir	Building panels, flush door shutters, roofing sheets, storage tank, packing material, helmets and postboxes, mirror casing, paper weights, projector cover, voltage stabilizer cover, filling material for seat upholstery, brushes and brooms, ropes and yarns for nets, bags, and mats, as well as padding for mattresses, seat cushions
Seed	Cotton	Furniture industry, roofing covers, textile and yarn, machinery, ropes, goods, and cordage
	Rice husk	Building materials such as building panels, bricks, window frame, partition boards, panels, decking, railing systems, and fencing

(Continued)

Table 8.4 Cont.

Category of Fibers	Fiber	Applications
Leaf	Sisal	In construction industry for panels, doors, shutting plate, and roofing sheets; also, manufacturing of paper and pulp
Stalk	Bagasse	Building panel, window frame, office products, furniture panels, decking, railing systems, bricks, fencing, drains and pipelines construction
Wood	Wood	Partition boards, window frame, panels, door shutters, decking, office products, railing systems, and fencing

8.6   Future Perspectives and the Global Market

As mentioned in previous sections of this chapter, there have been major achievements in the field of natural fiber reinforced biocomposites. Such invaluable achievements promise a very bright future for PBF/PP biocomposites, and are a big motivation for scientists and industrial leaders to strive for and focus on innovation and product development for the next scientific breakthrough. However, it seems that the environmental benefits of PBF reinforced biocomposites compared to synthetic fiber composites are still questionable because of their relatively excessive processing requirements. So, the main mission of future research should be the conduction of a step-by-step life cycle sustainability assessment of biocomposites in order to retain the main superior features of the process when developing high-performance biocomposites.

Although the prospects for PBF/PP composites in commercial markets are substantial for value-added products in various industries, they need to outperform synthesized fibers in terms of performance, reliability, durability, and serviceability in order to be a viable alternative. Therefore, further studies are necessary to overcome the existing obstacles. For example, future research should focus more on the enhancement of adhesion between polar natural fibers and nonpolar PP with various chemical modifications on natural fibers to get the best mechanical properties in PBF/PP composites.

Future studies also need to address drawbacks such as water absorption, variability, nonlinear mechanical behavior, low thermal resistance, and insufficient fracture toughness. Many of these weaknesses can be overcome with the development of more advanced processing techniques for natural fibers and their biocomposites. Another equally important challenge for these kinds of composites is their long-term stability for structural applications. For instance, the physical and mechanical properties of PBF/PP biocomposites are affected by humidity, UV radiation, and temperature in outdoor applications. The major detrimental effects of hygrothermal and UV exposure are discoloration, brittleness, deformation, surface cracks, and deterioration in mechanical properties. Significant research is underway to develop PBF/PP biocomposites with improved characteristics through overcoming the aforementioned obstacles, and expanding their use for load-bearing applications.

According to a January 2015 Markets and Markets report (www.marketsandmarkets.com), the natural fiber composites market is expected to reach $5.83 billion by 2019, at a compound annual growth rate (CAGR) of 12.31% between 2014 and 2019. North America dominated the global market for natural fiber reinforced biocomposites in 2013. The trend is projected to continue until 2019, with a CAGR of 11.16% between 2014 and 2019.

This report illustrated that compression molding held the major share in the natural fiber biocomposites market in 2013, and is expected to remain the major market by 2019. Injection molding process held the second biggest market for such composites in 2013, and it is projected that this market will grow at a high rate of 12.58% between 2014 and 2019 thanks to the continuously increasing application areas for the natural fiber-based reinforced composites.

8.7   Conclusion

Plant-based fibers, such as wood, hemp, kenaf, flax, cotton, jute, sisal, bamboo, banana, etc., have been considered as environmentally friendly reinforcements for composites from time immemorial. The interest in PBF reinforced PP biocomposites is rapidly growing both in terms of basic research and industrial applications. This interest is warranted due to several advantages of natural reinforcements compared to man-made fibers, including low relative density, low environmental impact, availability, low cost, high specific strength, and their potential across a wide range of applications. These characteristics make the PBF reinforced PP composites predestined to find more and more applications like transportation (automobiles, railway coaches, aerospace), building and construction (roofing covers, door frames), sporting goods (tennis rackets, bicycle frames, snowboards), packaging, etc. As far as the mechanical properties of such composites are concerned, if the adhesion between polar PBFs and nonpolar polyolefin is not strong enough to sustain high loads, the benefits of the high tensile strength of these reinforcements will be lost. Many studies have been carried out to enhance the adhesion between natural fibers and PP, which is the most critical challenge. Also, different coupling agents like MAPP and chemical modifications of fiber surface (alkalization/mercerization, diazotization, and oxidation) have been used to improve the interfacial strength.

References

1. Satyanarayana, K.G., Arizaga, G.G., and Wypych, F., Biodegradable composites based on lignocellulosic fibers—An overview. *Prog. Polym. Sci.* 34(9), 982–1021, 2009.
2. Bogoeva-Gaceva, G., Avella, M., Malinconico, M., Buzarovska, A., Grozdanov, A., Gentile, G., *et al.*, Natural fiber eco-composites. *Polym. Compos.* 28(1), 98–107, 2007.
3. Faruk, O., Bledzki, A.K., Fink, H.-P., and Sain, M., Biocomposites reinforced with natural fibers: 2000–2010. *Prog. Polym. Sci.* 37(11), 1552–1596, 2012.
4. Lee, B.-H., Kim, H.-J., and Yu, W.-R., Fabrication of long and discontinuous natural fiber reinforced polypropylene biocomposites and their mechanical properties. *Fiber. Polym.* 10(1), 83–90, 2009.
5. Faruk, O., Bledzki, A.K., Fink, H.P., and Sain, M., Progress report on natural fiber reinforced composites. *Macromol. Mater. Eng.* 299(1), 9–26, 2014.
6. Monteiro, S.N., Lopes, F.P.D., and Ferreira, A.S., and Nascimento, D.C.O., Natural-fiber polymer-matrix composites: cheaper, tougher, and environmentally friendly. *JOM* 61(1), 17–22, 2009.
7. Cheung, H.-Y., Ho, M.-P., Lau, K.-T., Cardona, F., and Hui, D., Natural fibre-reinforced composites for bioengineering and environmental engineering applications. Compos. Part B: Eng. 40(7), 655–663, 2009.
8. Mohanty, A.K., Misra, M., and Drzal, L.T. (Eds.), *Natural Fibers, Biopolymers, and Biocomposites*, CRC Press, 2005.
9. Biagiotti, J., Puglia, D., and Kenny, J.M., A review on natural fibre-based composites–Part I: Structure, processing and properties of vegetable fibres. *J. Nat. Fibers* 1(2), 37–68, 2004.
10. John, M.J., and Thomas, S., Biofibres and biocomposites. *Carbohydr. Polym.* 271(3), 343–364, 2008.
11. Gibson, L.J., The hierarchical structure and mechanics of plant materials. *J. R. Soc. Interface* 9(76), 2749–2766, 2012.
12. Akil, H., Omar, M., Mazuki, A., Safiee, S., Ishak, Z.M., and Bakar, A.A., Kenaf fiber reinforced composites: A review. *Mater. Des.* 32(8), 4107–4121, 2011.

13. Sadeek, S.A,, Negm, N.A., Hefni, H.H., and Wahab, M.M.A., Metal adsorption by agricultural biosorbents: Adsorption isotherm, kinetic and biosorbents chemical structures. *Int. J. Biol. Macromol.* 81, 400–409, 2015.

14. Rong, M.Z., Zhang, M.Q., Liu, Y., Yang, G.C., and Zeng, H.M., The effect of fiber treatment on the mechanical properties of unidirectional sisal-reinforced epoxy composites. *Compos. Sci. Technol.* 61(10), 1437–1447, 2001.

15. Holbery, J., and Houston, D., Natural-fiber-reinforced polymer composites in automotive applications. *JOM* 58(11), 80–86, 2006.

16. Mukhopadhyay, S., and Fangueiro, R., Physical modification of natural fibers and thermoplastic films for composites—A review. *J. Thermoplast. Compos. Mater.* 22(2), 135–162, 2009.

17. Kumar, R., Obrai, S., and Sharma, A., Chemical modifications of natural fiber for composite material. *Der Chemica Sinica* 2(4), 219–228, 2011.

18. Bledzki, A., and Gassan, J., Composites reinforced with cellulose based fibres. *Prog. Polym. Sci.* 24(2), 221–274, 1999.

19. Liu, X., and Dai, G., Surface modification and micromechanical properties of jute fiber mat reinforced polypropylene composites. *Express Polym. Lett.* 1(5), 299–307, 2007.

20. Kalia, S., Kaith, B., and Kaur, I., Pretreatments of natural fibers and their application as reinforcing material in polymer composites—A review. *Polym. Eng. Sci.* 49(7), 1253–1272, 2009.

21. Schloesser, T., and Knothe, J., Vehicle parts reinforced with natural fibres. *Kunststoffe Plast Europe* 87, 25–26, 1997.

22. Colberg, M., and Sauerbier, M., Injection moulding of natural fibre-reinforced plastics. *Kunststoffe Plast Europe* 87(12), 1780–1782, 1997.

23. Schneider, J., Myers, G., Clemons, C., and English, B., Biofibres as reinforcing fillers in thermoplastic composites. *Eng. Plast. (UK)* 8(3), 207–222, 1995.

24. Sain, M., Kokta, B., and Imbert, C., Structure-property relationships of wood fiber-filled polypropylene composite. *Polym. Plast. Technol. Eng.* 33(1), 89–104, 1994.

25. Maiti, S., and Singh, K., Influence of wood flour on the mechanical properties of polyethylene. *J. Appl. Polym. Sci.* 32(3), 4285–4289, 1986.

26. Karnani, R., Krishnan, M., and Narayan, R., Biofiber-reinforced polypropylene composites. *Polym. Eng. Sci.* 37(2), 476–483, 1997.

27. Lu, J.Z., Qinglin, W., and McNabb Jr., H.S., Chemical coupling in wood fiber and polymer composites: A review of coupling agents and treatments. *Wood Fiber Sci.* 32(1), 88–104, 2000.

28. Kokta, B., Maldas, D., Daneault, C., and Beland, P., Composites of poly (vinyl chloride) and wood fibers. Part II: Effect of chemical treatment. *Polym. Compos.* 11(2), 84–89, 1990.

29. Quiroga, A., Marzocchi, V., and Rintoul, I., Influence of wood treatments on mechanical properties of wood–cement composites and of Populus Euroamericana wood fibers. *Compos. Part B: Eng.* 84, 25–32, 2016.

30. Kokta, B., Raj, R., and Daneault, C., Use of wood flour as filler in polypropylene: Studies on mechanical properties. *Polym. Plast. Technol. Eng.* 28(3), 247–259, 1989.
31. Czvikovszky, T., Electron-beam processing of wood fiber reinforced polypropylene. *Radiat. Phys. Chem.* 47(3), 425–430, 1996.
32. Coutinho, F., Costa, T.H., and Carvalho, D.L., Polypropylene–wood fiber composites: Effect of treatment and mixing conditions on mechanical properties. *J. Appl. Polym. Sci.* 65(6), 1227–1235, 1997.
33. Khan, R.A., Khan, M.A., Zaman, H.U., Pervin, S., Khan, N., Sultana, S., *et al.*, Comparative studies of mechanical and interfacial properties between jute and E-glass fiber-reinforced polypropylene composites. *J. Reinf. Plast. Compos.* 29(7), 1078–1088, 2010.
34. Faruk, O., Bledzki, A.K., Fink, H.-P., and Sain, M., Biocomposites reinforced with natural fibers: 2000–2010. *Prog. Polym. Sci.* 37(11), 1552–1596, 2012.
35. Seki, Y., Innovative multifunctional siloxane treatment of jute fiber surface and its effect on the mechanical properties of jute/thermoset composites. *Mater. Sci. Eng.: A* 508(1), 247–52, 2009
36. Mwaikambo, L.Y., and Ansell, M.P., Chemical modification of hemp, sisal, jute, and kapok fibers by alkalization. *J. Appl. Polym. Sci.* 84(12), 2222–2234, 2002.
37. Wang, X., Cui, Y., Xu, Q., Xie, B., and Li, W., Effects of alkali and silane treatment on the mechanical properties of jute-fiber-reinforced recycled polypropylene composites. *J. Vinyl Addit. Technol.* 16(3), 183–188, 2010.
38. Masudul Hassan, M., Islam, M.R., and Khan, M.A., Improvement of physicomechanical properties of jute yarn by photografting with 3-(trimethoxysilyl) propylmethacrylate. *J. Adhes. Sci. Technol.* 17(5), 737–750, 2003.
39. Gassan, J., and Gutowski, V.S., Effects of corona discharge and UV treatment on the properties of jute-fibre epoxy composites. *Compos. Sci. Technol.* 60(15), 2857–2863, 2000.
40. Khan, M.A., Rahman, M.M., and Akhunzada, K., Grafting of different monomers onto jute yarn by *in situ* UV-radiation method: Effect of additives. *Polym. Plast. Technol. Eng.* 41(4), 677–689, 2002.
41. Khan, M.A., Mina, F., and Drzal, L., Influence of silane coupling agents of different functionalities on the performance of jute-polycarbonate composite, paper 5, pp. 1–8, in: *3rd International Wood and Natural Fibre Composite Symposium*, 2000.
42. Gassan, J., and Bledzki, A.K., Effect of moisture content on the properties of silanized jute-epoxy composites. *Polym. Compos.* 18(2), 179–184, 1997.
43. Mwaikambo, L., and Ansell, M., Hemp fibre reinforced cashew nut shell liquid composites. *Compos. Sci. Technol.* 63(9), 1297–1305, 2003.
44. Ray, D., Sarkar, B., and Rana, A., Fracture behavior of vinylester resin matrix composites reinforced with alkali-treated jute fibers. *J. Appl. Polym. Sci.* 85(12), 2588–2593, 2002.

45. Ray, D., Sarkar, B., Das, S., and Rana, A., Dynamic mechanical and thermal analysis of vinylester-resin-matrix composites reinforced with untreated and alkali-treated jute fibres. *Compos. Sci. Technol.* 62(7), 911–917, 2002.

46. Gassan, J., and Bledzki, A.K., Influence of fiber surface treatment on the creep behavior of jute fiber-reinforced polypropylene. *J. Thermoplast. Compos. Mater.* 12(5), 388–398, 1999.

47. Gassan, J., and Bledzki, A.K., The influence of fiber-surface treatment on the mechanical properties of jute-polypropylene composites. *Compos. Part A: Appl. Sci. Manuf.* 28(12), 1001–1005, 1997.

48. Mitra, B., Basak, R., and Sarkar, M., Studies on jute-reinforced composites, its limitations, and some solutions through chemical modifications of fibers. *J. Appl. Polym. Sci.* 67(6), 1093–1100, 1998.

49. Siddika, S., Mansura, F., and Hasan, M., Physico-mechanical properties of jute-coir fiber reinforced hybrid polypropylene composites. *World Acad. Sci. Eng. Technol.* 73, 1145–1149, 2013.

50. Karmaker, A., Hoffmann, A., and Hinrichsen, G., Influence of water uptake on the mechanical properties of jute fiber-reinforced polypropylene. *J. Appl. Polym. Sci.* 54(12), 1803–1807, 1994.

51. Zaman, H.U., Khan, R.A., Haque, M., Khan, M.A., Khan, A., Huq, T., *et al.*, Preparation and mechanical characterization of jute reinforced polypropylene/ natural rubber composite. *J. Reinf. Plast. Compos.* 29(20), 3064–3065, 2010.

52. Sarikanat, M., The influence of oligomeric siloxane concentration on the mechanical behaviors of alkalized jute/modified epoxy composites. *J. Reinf. Plast. Compos.* 29(6), 807–817, 2009.

53. Vimal, R., Subramanian, K.H.H., Aswin, C., Logeswaran, V., and Ramesh, M., Comparisonal study of succinylation and phthalicylation of jute fibres: Study of mechanical properties of modified fibre reinforced epoxy composites. *Mater. Today: Proc.* 2(4), 2918–2927, 2015.

54. Pawar, M., Patnaik, A., and Nagar, R., Investigation on mechanical and thermo-mechanical properties of granite powder filled treated jute fiber reinforced epoxy composite. *Polym. Compos.* 38(4), 736–748, 2017.

55. Abdallah, M., Zitoune, R., Collombet, F., and Bezzazi, B., Study of mechanical and thermomechanical properties of jute/epoxy composite laminate. *J. Reinf. Plast. Compos.* 29(11), 1669–1680, 2010.

56. Alves, C., Ferrão, P., Silva, A., Reis, L., Freitas, M., Rodrigues, L., *et al.*, Ecodesign of automotive components making use of natural jute fiber composites. *J. Clean. Prod.* 18(4), 313–327, 2010.

57. Arbelaiz, A., Fernandez, B., Cantero, G., Llano-Ponte, R., Valea, A., and Mondragon, I., Mechanical properties of flax fibre/polypropylene composites. Influence of fibre/matrix modification and glass fibre hybridization. *Compos. Part A: Appl. Sci. Manuf.* 36(12), 1637–1644, 2005.

58. John, M.J., and Anandjiwala, R.D., Chemical modification of flax reinforced polypropylene composites. *Compos. Part A: Appl. Sci. Manuf.* 40(4), 442–448, 2009.

59. Van den Oever, M., Bos, H., and Van Kemenade, M., Influence of the physical structure of flax fibres on the mechanical properties of flax fibre reinforced polypropylene composites. *Appl. Compos. Mater.* 7(5–6), 387–402, 2000.

60. Bos, H.L., Müssig, J., van den Oever, M.J., Mechanical properties of short-flax-fibre reinforced compounds. *Compos. Part A: Appl. Sci. Manuf.* 37(10), 1591–1604, 2006.

61. Van de Velde, K., and Kiekens, P., Effect of material and process parameters on the mechanical properties of unidirectional and multidirectional flax/polypropylene composites. *Compos. Struct.* 62(3), 443–448, 2003.

62. Ausias, G., Bourmaud, A., Coroller, G., and Baley, C., Study of the fibre morphology stability in polypropylene-flax composites. *Polym. Degrad. Stab.* 98(6), 1216–1224, 2013.

63. Arbelaiz, A., Fernandez, B., Ramos, J., and Mondragon, I., Thermal and crystallization studies of short flax fibre reinforced polypropylene matrix composites: Effect of treatments. *Thermochim. Acta* 440(2), 111–121, 2006.

64. Czigany, T., An acoustic emission study of flax fiber-reinforced polypropylene composites. *J. Compos. Mater.* 38(9), 769–778, 2004.

65. Hargitai, H., Rácz, I., and Anandjiwala, R.D., Development of HEMP fiber reinforced polypropylene composites. *J. Thermoplast. Compos. Mater.* 21(2), 165–174, 2008.

66. Patel, M., Bastioli, C., Marini, L., and Würdinger, E., Life-cycle assessment of bio-based polymers and natural fiber composites. *Biopolymers Online*, 2005.

67. Stamboulis, A., Baillie, C., Garkhail, S., Van Melick, H., and Peijs, T., Environmental durability of flax fibres and their composites based on polypropylene matrix. *Appl. Compos. Mater.* 7(5-6), 273–294, 2000.

68. Romhany, G., Karger-Kocsis, J., and Czigany, T., Tensile fracture and failure behavior of technical flax fibers. *J. Appl. Polym. Sci.* 90(13), 3638–3645, 2003.

69. Malkapuram, R., Kumar, V., and Negi, Y.S., Recent development in natural fiber reinforced polypropylene composites. *J. Reinf. Plast. Compos.* 28(10), 1169–1189, 2008.

70. Pacheco-Torgal, F., and Jalali, S., Cementitious building materials reinforced with vegetable fibres: A review. *Constr. Build. Mater.* 25(2), 575–581, 2011.

71. Humphreys, M.F., The use of polymer composites in construction, in: *SASBE 2003 – Smart and Sustainable Built Environment*, 2003.

72. Joseph, P., Joseph, K., and Thomas, S., Effect of processing variables on the mechanical properties of sisal-fiber-reinforced polypropylene composites. *Compos. Sci. Technol.* 59(11), 1625–1640, 1999.

73. Dwivedi, U., and Chand, N., Influence of MA-g-PP on abrasive wear behaviour of chopped sisal fibre reinforced polypropylene composites. *J. Mater. Process. Technol.* 209(12), 5371–5375, 2009.

74. Kaewkuk, S., Sutapun, W., and Jarukumjorn, K., Effects of interfacial modification and fiber content on physical properties of sisal fiber/polypropylene composites. *Compos. Part B: Eng.* 45(1), 544–549, 2013.

75. Suppakarn, N., and Jarukumjorn, K., Mechanical properties and flammability of sisal/PP composites: Effect of flame retardant type and content. *Compos. Part B: Eng.* 40(7), 613–618, 2009.

76. Jeencham, R., Suppakarn, N., and Jarukumjorn, K., Effect of flame retardants on flame retardant, mechanical, and thermal properties of sisal fiber/polypropylene composites. *Compos. Part B: Eng.* 56, 249–253, 2014.

77. Zhang, M.Q., Rong, M.Z., and Lu, X., Fully biodegradable natural fiber composites from renewable resources: All-plant fiber composites. *Compos. Sci. Technol.* 65(15), 2514–2525, 2005.

78. Bismarck, A., Mohanty, A.K., Aranberri-Askargorta, I., Czapla, S., Misra, M., Hinrichsen, G., *et al.*, Surface characterization of natural fibers; surface properties and the water up-take behavior of modified sisal and coir fibers. *Green Chem.* 3(2), 100–107, 2001.

79. Inácio, W., Lopes, F., and Monteiro, S., Charpy toughness behavior of continuous sisal fiber reinforced polyester matrix composites, in: *EPD Congress*, pp. 151–158, Minerals, Metals and Materials Society/AIME: Warrendale PA, USA, 2010.

80. Belaadi, A., Bezazi, A., Bourchak, M., and Scarpa, F., Tensile static and fatigue behaviour of sisal fibres. *Mater. Des.* 46, 76–83, 2013.

81. Pracella, M., Haque, M.M.-U., and Alvarez, V., Functionalization, compatibilization and properties of polyolefin composites with natural fibers. *Polymers* 2(4), 554–574, 2010.

82. Pracella, M., Chionna, D., Anguillesi, I., Kulinski, Z., and Piorkowska, E., Functionalization, compatibilization and properties of polypropylene composites with hemp fibres. *Compos. Sci. Technol.* 66(13), 2218–2230, 2006.

83. Beckermann, G., and Pickering, K.L., Engineering and evaluation of hemp fibre reinforced polypropylene composites: Fibre treatment and matrix modification. *Compos. Part A: Appl. Sci. Manuf.* 39(6), 979–988, 2008.

84. Li, Y., Pickering, K., and Farrell. R.. Determination of interfacial shear strength of white rot fungi treated hemp fibre reinforced polypropylene. *Compos. Sci. Technol.* 69(7), 1165–1171, 2009.

85. Li, Y., and Pickering, K.L., Hemp fibre reinforced composites using chelator and enzyme treatments. *Compos. Sci. Technol.* 68(15), 3293–3298, 2008.

86. Li, Y., and Pickering, K., The effect of chelator and white rot fungi treatments on long hemp fibre-reinforced composites. *Compos. Sci. Technol.* 69(7), 1265–1270, 2009.

87. Khoathane, M., Vorster, O., and Sadiku, E., Hemp fiber-reinforced 1-pentene/polypropylene copolymer: The effect of fiber loading on the mechanical and thermal characteristics of the composites. *J. Reinf. Plast. Compos.* 27(14), 1533–1544, 2008.

88. Bourmaud, A., and Baley, C., Rigidity analysis of polypropylene/vegetal fibre composites after recycling. *Polym. Degrad. Stab.* 94(3), 297–305, 2009.

89. Behzad, T., and Sain, M., Measurement and prediction of thermal conductivity for hemp fiber reinforced composites. *Polym. Eng. Sci.* 47(7), 977–983, 2007.

90. Tajvidi, M., Motie, N., Rassam, G., Falk, R.H., and Felton, C., Mechanical performance of hemp fiber polypropylene composites at different operating temperatures. *J. Reinf. Plast. Compos.* 29(5), 664–674, 2010.

91. Twite-Kabamba, E., Mechraoui, A., and Rodrigue, D., Rheological properties of polypropylene/hemp fiber composites. *Polym. Compos.* 30(10), 1401–1407, 2009.

92. Karnani, R., Krishnan, M., and Narayan, R., Biofiber-reinforced polypropylene composites. *Polym. Eng. Sci.* 37(2), 476–483, 1997.

93. Edeerozey, A.M., Akil, H.M., Azhar, A., and Ariffin, M.Z., Chemical modification of kenaf fibers. *Mater. Lett.* 61(10), 2023–2025, 2007.

94. Meon, M.S., Othman, M.F., Husain, H., Remeli, M.F., and Syawal, M.S.M., Improving tensile properties of kenaf fibers treated with sodium hydroxide. *Procedia Eng.* 41, 1587–1592, 2012.

95. Nirmal, U., Lau, S.T., and Hashim, J., Interfacial adhesion characteristics of kenaf fibres subjected to different polymer matrices and fibre treatments. *Journal of Composites* 2014, Article ID 350737, 2014.

96. Asumani, O., Reid, R., and Paskaramoorthy, R., The effects of alkali–silane treatment on the tensile and flexural properties of short fibre non-woven kenaf reinforced polypropylene composites. *Compos. Part A: Appl. Sci. Manuf.* 43(9), 1431–1440, 2012.

97. Feng, D., Caulfield, D., and Sanadi, A., Effect of compatibilizer on the structure-property relationships of kenaf-fiber/polypropylene composites. *Polym. Compos.* 22(4), 506–517, 2001.

98. Clemons, C., and Sanadi, A.R., Instrumented impact testing of kenaf fiber reinforced polypropylene composites: Effects of temperature and composition. *J. Reinf. Plast. Compos.* 26(15), 1587–1602, 2007.

99. Han, Y.H., Han, S.O., Cho, D., and Kim, H.-I., Kenaf/polypropylene biocomposites: Effects of electron beam irradiation and alkali treatment on kenaf natural fibers. *Composite Interfaces* 14(5–6), 559–578, 2007.

100. Aji, I., Sapuan, S., Zainudin, E., and Abdan, K., Kenaf fibres as reinforcement for polymeric composites: A review. *Int. J. Mech. Mater. Eng.* 4(3), 239–248, 2009.

101. Schemenauer, J.J., Osswald, T.A., Sanadi, A.R., and Caulfield, D.F., Melt rheological properties of natural fiber-reinforced polypropylene, in: *ANTEC 2000 Society of Plastic Engineers Conference*, pp. 2206–2210, 2000.

102. Alavudeen, A., Rajini, N., Karthikeyan, S., Thiruchitrambalam, M., and Venkateshwaren, N., Mechanical properties of banana/kenaf fiber-reinforced hybrid polyester composites: Effect of woven fabric and random orientation. *Mater. Des.* 66, 246–257, 2015.

103. Ghani, M., Salleh, Z., Hyie, K.M., Berhan, M., Taib, Y., and Bakri, M., Mechanical properties of kenaf/fiberglass polyester hybrid composite. *Procedia Eng.* 41, 1654–1659, 2012.

104. Diharjo, K., Hastuti, S., Triyasmoko, A., Sumarsono, A.G., Putera, D.P., Riyadi, F., *et al.,* The application of kenaf fiber reinforced polypropylene

composite with clay particles for the interior panel of electrical vehicle, in: *Rural Information & Communication Technology and Electric-Vehicle Technology (rICT & ICeV-T), 2013 Joint International Conference*, IEEE, 2013.

105. Joseph, S., Sreekala, M., Oommen, Z., Koshy, P., and Thomas, S., A comparison of the mechanical properties of phenol formaldehyde composites reinforced with banana fibres and glass fibres. *Compos. Sci. Technol.* 62(14), 1857–1868, 2002.

106. Bledzki, A.K., Faruk, O., and Sperber, V.E., Cars from bio-fibres. *Macromol. Mater. Eng.* 291(5), 449–457, 2006.

107. Bledzki, A., Mamun, A., and Faruk, O., Abaca fibre reinforced PP composites and comparison with jute and flax fibre PP composites. *Express Polym. Lett.* 1(11), 755–762, 2007.

108. Le Moigne, N., van Den Oever, M., and Budtova, T., A statistical analysis of fibre size and shape distribution after compounding in composites reinforced by natural fibres. *Compos. Part A: Appl. Sci. Manuf.* 42(10), 1542–1550, 2011.

109. Le Duc, A., Vergnes, B., and Budtova, T., Polypropylene/natural fibres composites: Analysis of fibre dimensions after compounding and observations of fibre rupture by rheo-optics. *Compos. Part A: Appl. Sci. Manuf.* 42(11), 1727–1737, 2011.

110. Yan, Z.L., Wang, H., Lau, K., Pather, S., Zhang, J., Lin, G., et al., Reinforcement of polypropylene with hemp fibres. *Compos. Part B: Eng.* 46, 221–226, 2013.

111. Luz, S., Gonçalves, A., and Del'Arco, A., Mechanical behavior and microstructural analysis of sugarcane bagasse fibers reinforced polypropylene composites. *Compos. Part A: Appl. Sci. Manuf.* 38(6), 1455–1461, 2007.

112. Karmarkar, A., Chauhan, S., Modak, J.M., and Chanda, M., Mechanical properties of wood–fiber reinforced polypropylene composites: Effect of a novel compatibilizer with isocyanate functional group. Compos. Part A: Appl. Sci. Manuf. 38(2), 227–233, 2007.

113. Chattopadhyay, S.K., Khandal, R., Uppaluri, R., and Ghoshal, A.K., Bamboo fiber reinforced polypropylene composites and their mechanical, thermal, and morphological properties. *J. Appl. Polym. Sci.* 119(3), 1619–1626, 2011.

114. Chen, X., Guo, Q., and Mi, Y., Bamboo fiber-reinforced polypropylene composites: A study of the mechanical properties. *J. Appl. Polym. Sci.* 69(10), 1891–1899, 1998.

115. Haque, M.M., Rahman, R., Islam, M.N., Huque, M.M., and Hasan, M., Mechanical properties of polypropylene composites reinforced with chemically treated coir and abaca fiber. *J. Reinf. Plast. Compos.* 29(15), 2253–2261, 2009.

116. Haque, M.M., Hasan, M., Islam, M.S., and Ali, M.E., Physico-mechanical properties of chemically treated palm and coir fiber reinforced polypropylene composites. *Bioresour. Technol.* 100(20), 4903–4906, 2009.

117. Khan, A., Huq, T., Saha, M., Khan, R.A., Khan, M.A., and Gafur, M., Effect of silane treatment on the mechanical and interfacial properties of calcium alginate fiber reinforced polypropylene composite. *J. Compos. Mater.* 44(24), 2875–2886, 2010.

118. George, G., Jose, E.T., Jayanarayanan, K., Nagarajan, E., Skrifvars, M., and Joseph, K., Novel bio-commingled composites based on jute/polypropylene yarns: Effect of chemical treatments on the mechanical properties. *Compos. Part A: Appl. Sci. Manuf.* 43(1), 219–230, 2012.
119. Shubhra, Q.T., Alam, A., Gafur, M., Shamsuddin, S.M., Khan, M.A., Saha, M., *et al.*, Characterization of plant and animal based natural fibers reinforced polypropylene composites and their comparative study. *Fiber. Polym.* 11(5), 725–731, 2010.
120. Acha, B.A., Reboredo, M.M., and Marcovich, N.E., Creep and dynamic mechanical behavior of PP–jute composites: Effect of the interfacial adhesion. *Compos. Part A: Appl. Sci. Manuf.* 38(6), 1507–1516, 2007.
121. Kalia, S., Thakur, K., Celli, A., Kiechel, M.A., and Schauer, C.L., Surface modification of plant fibers using environment friendly methods for their application in polymer composites, textile industry and antimicrobial activities: A review. *J. Environ. Chem. Eng.* 1(3), 97–112, 2013.
122. Koronis, G., Silva, A., and Fontul, M., Green composites: A review of adequate materials for automotive applications. *Compos. Part B: Eng.* 44(1), 120–127, 2013.
123. Mwaikambo, L., Review of the history, properties and application of plant fibres. *Afr. J. Sci. Technol.* 7(2), 121, 2006.
124. Bongarde, U., and Shinde, V., Review on natural fiber reinforcement polymer composites. *IJSRIT* 3(2), 431–436, 2014.
125. Sen, T., and Reddy, H.J., Various industrial applications of hemp, kinaf, flax and ramie natural fibres. *IJIMT* 2, 192–198, 2011.
126. Kakroodi, A.R., Kazemi, Y., and Rodrigue, D., Mechanical, rheological, morphological and water absorption properties of maleated polyethylene/hemp composites: Effect of ground tire rubber addition. *Compos. Part B: Eng.* 51, 337–344, 2013.
127. Mohammed, L., Ansari, M.N., Pua, G., Jawaid, M., and Islam, M.S., A review on natural fiber reinforced polymer composite and its applications. *Int. J. Polym. Sci.* 2015, Article ID 243947, 2015.
128. Akil, H., Omar, M., Mazuki, A., Safiee, S., Ishak, Z., and Bakar, A.A., Kenaf fiber reinforced composites: A review. *Mater. Des.* 32(8), 4107–4121, 2011.
129. Williams, A.S., *Life Cycle Analysis: A Step by Step Approach*, Illinois Sustainable Technology Center: Champaign, IL, 2009.
130. Zah, R., Hischier, R., Leão, A., and Braun, I., Curauá fibers in the automobile industry—A sustainability assessment. *J. Clean. Prod.* 15(11), 1032–1040, 2007.
131. Marsh, G., Next step for automotive materials. *Mater. Today* 6(4), 36–43, 2003.
132. Shahzad, A., Hemp fiber and its composites—A review. *J. Compos. Mater.* 46(8), 973–986, 2012.
133. Leao, A.L., Rowell, R., and Tavares, N., Applications of natural fibers in automotive industry in Brazil—Thermoforming process. in: *Science and Technology of Polymers and Advanced Materials*, Prasad, P.N., Mark, J.E., Kandil, S.H., and Kafafi, Z.H. (Eds.), pp. 755–761, Springer, 1998.

134. Bledzki, A.K., Faruk, O., and Mamun, A.A.. Influence of compounding processes and fibre length on the mechanical properties of abaca fibre-polypropylene composites. *Polimery* 53(2), 120–125, 2008.

135. Monteiro, S., Satyanarayana, K., Ferreira, A., Nascimento, D., Lopes, F., Silva, I., *et al.,* Selection of high strength natural fibers. *Matéria (Rio de Janeiro)* 15(4), 488–505, 2010.

136. Dittenber, D.B., and Gangarao, H.V., Critical review of recent publications on use of natural composites in infrastructure. *Compos. Part A: Appl. Sci. Manuf.* 43(8),1419–1429, 2012.

137. Suddell, B., Industrial fibres: Recent and current developments. in: *Proceedings of the Symposium on Natural Fibres*, FAO and CFC, Rome, 2008.

138. Pickering, K. (Ed.), *Properties and Performance of Natural-Fibre Composites*, Elsevier, 2008.

139. Bos, H.L., The potential of flax fibres as reinforcement for composite materials, PhD thesis, Technische Universiteit Eindhoven, 2004.

140. Al-Oqla, F.M., and Sapuan, S., Natural fiber reinforced polymer composites in industrial applications: Feasibility of date palm fibers for sustainable automotive industry. *J. Clean. Prod.* 66, 347–354, 2014.

141. Saba, N., Tahir, P.M., and Jawaid, M., A review on potentiality of nano filler/natural fiber filled polymer hybrid composites. *Polymers* 6(8), 2247–2273, 2014.

142. Zsiros, J.A., Natural fibers and fiberglass: A technical and economic comparison, MSci thesis, Brigham Young University, 2010.

143. Uddin, N. (Ed.), *Developments in Fiber-Reinforced Polymer (FRP) Composites for Civil Engineering*, Elsevier, 2013.

144. Van de Weyenberg, I., Ivens, J., De Coster, A., Kino, B., Baetens, E., and Verpoest, I., Influence of processing and chemical treatment of flax fibres on their composites. *Compos. Sci. Technol.* 63(9), 1241–1246, 2003.

145. Singh, B., and Gupta, M., Natural fiber composites for building applications, in: *Natural Fibres, Biopolymers and Biocomposites*, Mohanty, A.K., Misra, M., and Lawrence T. Drzal, L.T. (Eds.), p. 37, CRC Press: Boca Raton. 2005.

146. Kalia, S., Kaith, B., and Kaur, I. (Eds.), *Cellulose Fibers: Bio- and Nano-Polymer Composites: Green Chemistry and Technology*, Springer Science & Business Media, 2011.

147. Christian, S., and Billington, S., Sustainable biocomposites for construction, in: *Composites & Polycon 2009*, pp. 15–17, American Composites Manufacturers Association, 2009.

148. Burgueno, R., Quagliata, M.J,, Mehta, G.M., Mohanty, A.K., Misra, M., and Drzal, L.T., Sustainable cellular biocomposites from natural fibers and unsaturated polyester resin for housing panel applications. *J. Polym. Environ.* 13(2), 139–149, 2005.

149. Burgueno, R., Quagliata, M.J., Mohanty, A.K., Mehta, G., Drzal, L.T., and Misra, M., Hybrid biofiber-based composites for structural cellular plates. *Compos. Part A: Appl. Sci. Manuf.* 36(5), 581–593, 2005.

150. Burgueño, R., Quagliata, M.J., Mohanty, A.K., Mehta, G., Drzal, L.T., and Misra, M., Hierarchical cellular designs for load-bearing biocomposite beams and plates. *Mater. Sci. Eng. A* 390(1), 178–187, 2005.

151. Burgueño, R., Quagliata, M.J., Mohanty, A.K., Mehta, G., Drzal, L.T., and Misra, M., Load-bearing natural fiber composite cellular beams and panels. *Compos. Part A: Appl. Sci. Manuf.* 35(6), 645–656, 2004.

152. Azwa, Z., Yousif, B., Manalo, A., and Karunasena, W., A review on the degradability of polymeric composites based on natural fibres. *Mater. Des.* 47, 424–442, 2013.

153. Riedel, U., and Nickel, J., Applications of natural fiber composites for constructive parts in aerospace, automobiles, and other areas. *Biopolymers Online*, 2005.

154. Stevens, C., and Müssig, J. (Eds.), *Industrial Applications of Natural Fibres: Structure, Properties and Technical Applications*, John Wiley & Sons, 2010.

9

Polypropylene Composite with Oil Palm Fibers: Method Development, Properties and Applications

Muhammad Shahid Nazir[1], Mohd Azmuddin Abdullah[2],*
and Muhammad Rafi Raza[3]

[1] *Department of Chemistry, COMSATS Institute of Information Technology, Lahore, Punjab, Pakistan*
[2]*Institute of Marine Biotechnology, Universiti Malaysia Terengganu, Kuala Nerus, Terengganu, Malaysia*
[3]*Department of Mechanical Engineering, COMSATS Institute of Information Technology Sahiwal, Pakistan*

Abstract

Reinforcing materials are pertinent to the improvement of the mechanical and chemical properties of composite material. Glass and carbon fibers are normally used as inorganic filler, while aromatic fibers are used as organic filler. Cellulose from oil palm fibers has a big potential to be developed as composites with synthetic polymers such as polypropylene, polyethylene or polyvinylchloride. These have varied applications as high strength materials in the automotive, aerospace, manufacturing and construction sectors or in general environmental remediation. The selection of pretreatment method of the lignocellulosic materials is of paramount importance to provide a specific, suitable and more exposed structure of filler. The techniques may be physical, chemical, thermal or a combination of these. Composite characterization using techniques such as FTIR, XRD, SEM and TEM investigate the interface, stress point, structure of matrix and fiber surface, and the sites of interaction. The properties of composite material can be studied by controlling the conditions of fabrication, processing time and temperature, cooling rate and annealing conditions. In this review, the method development of eco-friendly cellulose extraction from oil palm empty fruit bunches, the development of cellulose-polypropylene composite materials, and the characterization of composite properties and their potential applications are discussed.

Corresponding author: azmuddin@umt.edu.my; joule1602@gmail.com

Visakh. P. M. and Matheus Poletto. (eds.) Polypropylene-Based Biocomposites and Bionanocomposites, (287–314) 2018 © Scrivener Publishing LLC

Keywords: Oil palm fibers, green composite, nanofibers, fabrication, composite characterization, eco-friendly polymer, polypropylene

9.1   Introduction

Oil palm (*Elaeis guineensis* Jacq.) is a monoecious plant, originally from Guinea in West Africa, and brought to Peninsular Malaysia as an ornamental plant in 1875, and has become a cash crop since 1917. The standard population density of oil palm in Malaysia is 148 palm trees per hectare, or 5 million ha total oil palm cultivated land, that could produce 11 million tonnes of trunks. Each plant can produce approximately 150 kg of fresh fruit bunches (FFB) per year [1]. The weight of the FFB may vary from 10–40 kg, depending on the number of compactly bound fruitlets in the bunch (Figure 9.1a). The oil extracted from the reddish brown mesocarp (Figure 9.1b) is called crude palm oil (CPO), used for edible purposes, whilst the oil extracted from the white seed or palm kernel is used for oleochemicals [2]. After the fruit is detached from the FFB, the remainder is called an empty fruit bunch (EFB) (Figure 9.1c). Palm oil has become the major contributor of the total oil and fat production in the world, increasing from 13% in 1990 to 28% in 2011 [3]. A total of 17 million tonnes of palm oil is generated from 416 mills which produce 77.2 million tonnes of biomass. Out of these, 19.8 million tonnes on a wet basis or 6.93 million tonnes on a dry basis are EFBs [4].

During mill processing, EFB constitutes 90 million tonnes of waste remaining after oil extraction. The agro-based industries, with oil palm becoming the leading crop, have now increasingly been giving emphasis to turning the biomass wastes into high value-added products such as fertilizer, biopolymers, oleochemicals, soap, cheap sugar source for fermentation, biodiesel or ethanol [2]. EFB can be used in the manufacturing of

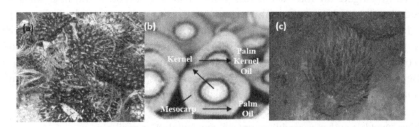

Figure 9.1 (a) Oil palm fresh fruit bunch, (b) fruit cross section [3], and (c) empty fruit bunch.

pulp and paper, paneling composite, medium density fiberboard, compost and other value-added biomaterials such as organic acid, biocompost, activated carbon, bioplastics, fertilizer and animal feed [5, 6].

Cellulose from EFB can be explored for potential applications in the field of composite preparation with synthetic polymer such as polyprolylene, polyethylene or polyvinylchloride [7, 8]. These may find applications as high strength materials in car or airplane manufacturing or construction materials. Another potential application is in the development of biosorbent for metal ion sorption or general environmental remediation [9, 10]. For these, cellulose must be extracted from EFB via non-green or green methods. The non-green method for the delignification of lignocellulosic biomass has been extensively reported using acidified sodium chlorite ($NaClO_2 + H_3O^+$) as a standard reagent for delignification and extraction of cellulose from wood materials [11]. Chlorite (ClO_2^-) may produce chlorine radical, $Cl^{\bullet}$, which reacts and fragments the lignocellulosic material into highly toxic organochlorine compounds [9].

In this review, we explore the method development of eco-friendly cellulose extraction from EFB, the development of cellulose-polypropylene composite materials, the characterization of composite properties and their potential applications.

9.2   Method Development

9.2.1   Pretreatment Methods

The importance of pretreatment method is to provide the specific, suitable and more exposed structure of filler to the composite. The techniques may be physical, chemical, thermal or a combination of these. Oil palm is a lignocellulosic material that is chemically made up of parallel cellulose rod-like structure, filled with hemicelluloses, lignin and other materials such as pectin, wax and inorganic elements. Naturally, oil palm solid mass chemical composition may vary in different parts of the same plant and also from other plants, depending on the abiotic and biotic factors. This compact and solid lignocellulose may necessitate the use of vigorous methods to open up the cellulose parallel chains to become exposed structure, thereby loosening up the cellulose crystallinity exhibited due to the presence of H-bonding [12]. The different pretreatment methods include milling, irradiation, ozonolysis, organosolv, acidic or basic hydolysis, steam explosion, wet oxidation, freeze explosion, hydrothermal and biological [7–8, 12].

Pretreatment of oil palm fibers begins with the washing of fibers with detergent to remove the organic and inorganic impurities, including oil, grease, sand and dust particles. These fibers are flushed with plenty of water to remove dark brown stains of raw fiber. Fibers are cut off into different sizes depending on the type of further treatment. The soluble extractive portion is isolated by using different combinations of organic and aqueous phases. Outermost wax layer and oil contents are removed using a Soxhlet apparatus. Raw fibers are placed in a cellulose thimble (cup) and a round-bottom flask is filled with solvents (ethanol, n-hexane, chloroform, water or a combination of these), placed on the heating mantle at 85 °C and refluxed for 6 hours. Oil is separated and fibers are washed with deionized water to completely remove the solvent [12].

9.2.1.1   Physical Method

Physical pretreatment is a green technique, completely without the assistance of any chemicals. This technique includes treatment with mechanical, electromagnetic wave exposure or cutting machines to loosen the compact bound fibers and provides high surface area with increased amorphous composition. This makes the fibers easily accessible for any further treatment [13]. Fibers can be cut down into smaller sizes using a ball mill or hammer. In ball milling, steel balls of different sizes can grind the fibers into the required sizes depending on the number, ball sizes and also duration of treatment [14]. Though simple and completely done without any chemicals, it is highly energy intensive [15]. Ultrasound or ultrasonication (US) uses high frequency sound waves where fibers are suspended in selected liquid and exposed to it. The wave travels through the liquid (as medium) and starts penetrating the suspended solid. From the principle of wave propagation, the push and pull penetration mechanism is observed where the sound wave travels in the form of "compression and expansion." The liquid molecules partly push the fibers and the other part of liquid molecules pull, and in this way, the grooves or cavitation are generated at the hot spot of liquid. The high pressure (2000 atm) and temperature (500 °C) [16] causes the biomass to explode at the point of contact with the breaking down of cellulosic crystalline structure to an amorphous one.

Electromagnetic radiation has the characteristic energy to break up the cellulose linkages and open up the compact lignocelluloses [17]. Lignocelluloses contain adsorbed and bound water which meets the primary requirement for microwave application, and is especially effective at 300 MHz–300GHz frequency [18]. Microwave transfers the energy to the biomass in the form of rotational motion though the moisture content present in the biomass and starts penetrating from there. A large amount of heat is generated instantaneously in the process. Gamma irradiation can also be utilized for depolymerization of complex lignocellulosic structure. Naturally, the characteristic radiation is at a high frequency of approximately 10^{18} hertz, and artificially, the γ-rays can be produced from the radioactive decay of cobalt 60 or others elements. The SI unit used to measure the extent of radiation dose is gray (Gy) where gamma radiation must be 1 Gy when a body of mass 1 kg absorbs 1 joule energy. Depolymerization of soft wood and hard wood observed has been reported at 70 kGy and 90 kGy, respectively [7, 13]. Mechanical, thermal properties and crystallinity will be affected with the exposure to γ-radiations for enzyme or chemical accessibility [7].

9.2.1.2 Chemical Method

Chemical pretreatment involves the dissolution of the lignin, hemicelluloses and crystalline part of oil palm lignocellulose with the help of mineral or organic compounds. The depolymerization is achieved by hydrolysis brought about by acid or alkali. Acid hydrolysis is normally associated with the use of phosphoric, sulphuric, formic, peracetic or hydrochloric acid [19], and is carried out with singular acid or with a combination of different compositions, depending upon the type and the chemical nature of biomass. Concentrated acid may cause serious dehydration of biomass rather than hydrolysis. Hence, dilute acid is recommended as being preferable for hydrolysis.

Alkali treatment or mercerization of cellulosic fibers using calcium, sodium or potassium hydroxide [20], individually or in combination, provides hydroxyl ions (⁻OH) to facilitate the hydrolysis, depending on the type of biomass [21–23]. This could lead to high quality and purity of fibers with higher fiber-matrix adhesion, and improve the aspect ratio with the reduction in diameter and higher fiber roughness. The higher fiber surface roughness and aspect ratio help to enhance the mechanical properties of a composite due to the improvement in adhesion. The following reaction takes place as a result of alkali treatment: Alkali typically dissolves lignin, the filler component of lignocellulosic composite, to provide the porous structure which is best for developing interlinkages between polar cellulose and nonpolar polypropylene.

9.2.1.3 Thermal Method

Autoclave (AUTO) pretreatment at 121 °C and 15 psi with different chemicals and for different durations disassemble the lignocelluloses [8, 12, 24–25]. The effective autoclaving treatment not only disinfects, but also opens up lignocelluloses for penetration of reagents and for hydrolysis to take place. Steam produced at low temperature and pressure provides surface preparation of substrate for reagent attack, and it is effective for further synthesis. Autoclaving treatment with reactive chemicals may produce significant effects in the sterilization and partial and complete depolymerization of lignocelluloses to monomer constituents. Treatment duration and reagents concentration could determine the end products.

Steam explosion is a green method with high efficiency to process biomass and could be performed on a large scale [26]. In the process, steam is forced into fibrous tissues and cells of biomass after pressurization, and rapid release of the pressure results in an explosive decompression. There are mainly two steam explosion modes—the valve blow mode and the catapult explosion mode. The catapult mode completes the explosion within 0.0875 s and the valve blow mode requires at least 0.5 s [27]. Steam flash-explosion, which adopts the catapult mode, is a sustainable and practical [28] process where the high explosion speed provides enough force to rupture the biomass compact structure whilst avoiding violent treatment under high temperature or pressure for longer duration.

9.2.2   Cellulose Extraction

Delignification and cellulose extraction from oil palm fibers (EFB, mesocarps, frond and trunk) is challenging because of its inert nature and compact structure. Very strong acid and alkali may have to be used to treat the fibers and often it takes a long time to achieve the desired result. To speed up the process, multi-pronged approaches can be adopted such as the combination of different solvents, reactive species and assistance of electromagnetic radiations or ultrasonic and mechanical treatments. Table 9.1 shows the different methods that can be utilized for delignification and cellulose extraction from oil palm as purely extracted cellulose (PEC) with microcrystalline (MCC) or nanocrystalline (NCC) structures. The use of green cellulose extraction by sonication and autoclave for oil palm empty fruit bunch fibers using less hazardous or toxic chemicals at low temperature and pressure could pave the way for a more economical and environmentally friendly process route [8, 12].

Table 9.2 shows the comparison of physical properties of purely extracted cellulose (PEC) from EFB and Table 9.3 the elemental analysis of raw EFB and PEC from ultrasonication (SONO-CHEM) and autoclave (AUTO-CHEM) treatment [8]. While some pretreatment techniques may be successful in the laboratory environment, especially if the aim is to achieve cellulose of high-crystallinity, the cost benefit analysis will ultimately determine their suitability, especially for large-scale application, the intended final product use and whether or not it meets industrial standards [8].

Table 9.1 Different treatments for cellulose extraction from oil palm fibers.

Materials	Methods	Ref.
EFB	The 2% alkali treatment at 95 °C for 60 min RSM simulated results give 98% cellulose and α-cellulose 97.4%	[49]
EFB	Acid hydrolysis (2% H_2SO_4) with ultrasonic assistance at 20 kHz and 2 kW power with different amplitudes, such as 15, 60 and 90%, for durations of 15, 45 and 60 min; ultrasonically pretreated fibers are acid hydrolyzed at 140 °C and 2 bar	[50]
EFB	The extraction of MCC involves pretreatment with soda anthraquinone method and bleaching with ozone and peroxide; later hydrolysis with 2.5 M HCl at 105 °C for 15 min	[51]
EFB	Use of 2.5 N HCl at 105 °C for 30 min and ammonium hydroxide; neutralized with 5% NH_4OH and dried at 105 °C.	[52]
EFB	Autoclave extraction with a mixture of hydrogen peroxide and formic acid at 120 °C and further bleached with hydrogen peroxide at 80 °C; yields 64% of cellulose with α-cellulose content of 93.7% and crystallinity of 70%	[12]
EFB	Ultrasonic extraction with hydrogen peroxide at 40 kHz and room temperature; yields 49% cellulose with α-cellulose content of 91.3% and crystallinity of 68.7%	[8]
EFB	TEMPO-oxidation with sonication assistance; yields 93% NCC	[53]
Pulp	Soda pulping of oil palm and evaluating pulping variable with 60% cellulose yield	[54]
EFB Pulp	58% sulphuric acid and sonication treatment of oil palm to NCC	[55]
Trunk	Soda process, ozone bleach of oil palm trunk to NCC	[56]
Frond	Alkaline H_2O_2 translation of oil palm frond to MCC	[57]

Table 9.2 Comparison of physical properties of purely extracted cellulose (PEC) from EFB after ultrasonication (SONO-CHEM) and autoclave (AUTO-CHEM) treatment [8].

Properties	SONO-CHEM-PEC	AUTO-CHEM-PEC
Moisture (%)	9.9	9.89
α-Cellulose (%)	91.3	93.7
Particle size (μm)	21.48	21.72
Density (g cm^{-3})	1.58	1.59
Molecular weight (g mol^{-1})	1.49×10^5	1.87×10^5
Degree of polymerization	919	1154

Table 9.3 Elemental analysis of raw EFB and PEC from ultrasonication (SONO-CHEM) and autoclave (AUTO-CHEM) treatment [8].

Element	Raw FEB		SONO-CHEM-PEC		AUTO-CHEM-PEC	
	Weight %	Atomic %	Weight %	Atomic %	Weight %	Atomic %
C	44.79	54.31	66.34	72.42	63.16	69.54
O	44.41	40.43	33.66	27.58	36.84	30.46
Al	0.49	0.27	---	---	---	---
Ca	1.50	0.54	---	---	---	---
K	1.86	0.69	---	---	---	---
Mg	0.82	0.49	---	---	---	---
Na	1.00	0.64	---	---	---	---
S	0.52	0.24	---	---	---	---
Si	4.61	2.39	---	---	---	---
Total	100.00	100.00	100.00	100.00	100.00	100.00

9.2.3   Composite Development

Lignocellulosic fibers blended with different polymer matrices have been developed to form composites such as with polypropylene (PP), polylactic acid (PLA), polyhydroxyl alkanoates (PHA) and polyhydroxyl butyrates (PHB) and polyolefin [7–8, 29]. There has especially been extensive reporting done on PP matrix with lignocellulosic filler material such as fibers of jute, kenaf, wheat straw, bamboo, sugar cane bagasse, coir, rice husks, pineapple leaf and oil palm. The three types of lignocellulose fibers used in the blending are long, short and particulate fibers. The composite fabrication is normally achieved by compression molding and injection molding, as shown in Table 9.4 [30].

Table 9.4 Different fabrication techniques for blending lignocellulosic fibers to form composites [30].

Materials	Fabrication methods
Jute yarn Flax and sisal Lyocell kapok Unidirectional/multidirectional flax	pultrusion and compression molding
cellulose filament yarn jute yarn flax yarn	Single/twin-screw extruder and injection molding
cellulose fiber sisal fiber kenaf fibers	hyduralic/compression molding
flax fibers jute fibers vetiver grass bamboo fibers	twin-screw extruder/internal mixture and injection molding

(Continued)

Table 9.4 Cont.

Materials	Fabrication methods
cellulose pulp, sisal, coir, luff sponge cellulose whiskers soft wood, avicel fiber, alfa pine saw dust wood fiber hemp fiber saw dust luffa fiber paper slug	Screw extruder/internal mixture and compression molding
rice husk, wood powder lignocel fiber hard wood dust liquefied wood mill hemp pine soft wood fiber wheat straw	screw extruder/internal mixing and injection molding

Table 9.5 shows the materials and methods for oil palm-based PP composite fabrication. In the vertical injection molding machine, the fibers and PP blend are filled into a hopper, which leads to the thermocouple barrel, and is melted at 170 °C. Some treated fibers become very fragile and thermally unstable at such high temperature and the PP also melts. The selection of cellulose extraction techniques for fabrication is therefore a very important step. When cellulose is segregated from the natural composite structure of lignocellulose, lignin and hemicelluloses are depolymerized and leached out from the parent composite structure, leaving behind straight cellulose chains. If the treatment time is not optimized and the cellulose chain is not filtered out and instead allowed to stay in the mixture, the virgin cellulose may be further depolymerized into glucose or

Table 9.5 Materials and methods for oil palm-based PP composite fabrication.

Composite materials	Methods	Ref.
EFB/PP EFB Cellulose/PP	Injection molding	[8]
EFB/PP	Injection molding	[58]
Frond/PP	Injection molding	[59]
EFB-PP	Twin-screw extruder, Injection molding	[60]
Oil palm fiber/PP	Twin-screw extruder, Injection molding	[61]
Chemically modified palm fiber/PP		[42]
EFB/PP		[62]
EFB/PP		[63]
Oil palm fiber/high density PP		[64]

carbon. At 170 °C, it must be ensured that the PP gets melted before a small amount of fibers are periodically added to avoid the lumping of fibers in the PP matrix. Mixing can be achieved by an internal mixer to get evenly distributed and homogeneous fiber-PP blend before it is sent to an injection molding machine or is hot compacted to draw its desired shape. In the molding machine, the composite may stick to the inner vessel surface and this can be controlled by using a releasing agent.

9.2.4   Characterization

Characterization of the composite structure can be carried out using a Fourier transmission infrared spectrometer (FTIR), X-ray diffractometer (XRD) or thermogravimetric analyzer (TGA) to evaluate the chemical characteristics of the composite. A scanning electron microscope (SEM) is useful in assessing the composite morphology and the ASTM D638-03 method is used for tensile testing of the samples, with the tensile properties of the fabricated composite being determined by a universal tensile testing machine.

Figure 9.2 shows that the raw EFB exhibits a tightly packed, compact structure while the EtOH/EFB and NaOH/EFB show the rough or partially torn structure with cavitations which expose the cellulose strand

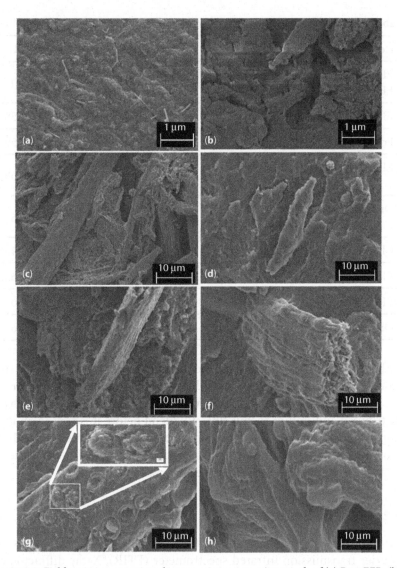

Figure 9.2 Field emission scanning electron microscope micrograph of (a) Raw EFB, (b) EtOH-EFB, (c) NaOH-EFB, (d) Raw/PP, (e) EtOH/PP, (f) NaOH/PP, (g) PEC-US/PP, and (h) PEC-AUTO/ PP at 10% PEC loading [8].

due to the partial removal of hemicelluloses, lignin and waxy layer. At 10% loading, the raw/PP and EtOH/PP suggest a well-mixed composite (Fig. 9.2d,e), while the partial opening and roughness of the NaOH/PP provide better substrate for the intervening PP (Fig. 9.2f). Both PEC-US/ PP and PEC-AUTO/PP show better filling and intermixing of the PEC

Table 9.6 The FTIR peaks and the corresponding functional groups of cellulose and PP.

Peaks (cm^{-1})	Functional group	Ref.
1432 1375 1326 895	-CH$_2$ C-H, C-O of cellulose Glucose-Glucose linkage	[12]
2947, 2869 2921, 2839	CH$_3$ CH$_2$ of polypropylene	[65]

with the PP and may have developed joints with the molten PP to give greater reinforcement [8].

Fourier transmission infrared spectrometry elucidates the functional groups retained in the composite structure and formation of new cross-linkages between the filler and the matrix. For sample preparation, the powdered polymer composite is made into a pellet with KBr, an inorganic salt which does not have any interaction with the composite molecule. The selection of KBr salt for pellet formation is due to its transparency in the IR region, ranging from 400 to 4000 cm^{-1}. The composite powder is mixed with KBr and cast into pellet using a hyduralic press, with the pellet placed in the sample holder through which IR light passes and data recorded. The FTIR spectrum contains the peaks which are the characteristics of the corresponding functional group shown in Table 9.6 for the characteristic peaks of cellulose and PP.

The physical technique of XRD principally decides whether the compound is amorphous or crystalline and also its significant purity. The sample powder is compactly packed in the glass holder and placed in the path of an X-ray source where diffraction occurs when the light strikes the surface and generates the information about the nature of the material in the form of a diffractogram. The diffractogram is a profile between 2Θ and counts per second which is shown either as low intensity or high intensity peak. The characteristic peaks are matched and compared with the known crystal structure standard cards of the Joint Committee on Powder Diffraction Standards (JCPDS) for cellulose and polypropylene, which are the standard cards published by the International Center for Diffraction Data (ICDD). Figure 9.3 shows the high intensity peak for cellulose, which matches with JCPDS card number 03-0289 [12], while the PP standard card number (JCPDS 00-050-2397) confirms the purity of PP in the composite.

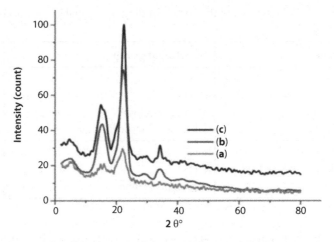

Figure 9.3 X-ray diffractogram of (a) raw fiber, (b) extracted EFB cellulose, and (c) commercial cellulose [12].

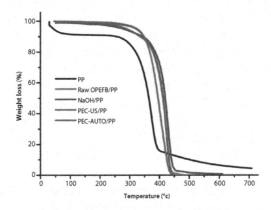

Figure 9.4 Thermogravimetric analysis of PP, Raw EFB/PP, EtOH/PP, NaOH/PP, PEC-US/PP, and PEC-AUTO/PP [8].

The thermal stability of composite structure is described by TGA and more than one degradation curve suggests the presence of more than one component in the composite structure. A known amount of sample is placed in the pre-weighed crucible and the program set for a 10 °C/min temperature rise from 30 to 700 °C with an inert gas flow of 10 mL/min. It is the non-oxidative degradation of sample which is recorded in the form of weight losses. Figure 9.4 illustrates the degradation pattern of neat PP, raw fibers and differently treated oil palm fibers with PP composite [8]. The residue left after degradation at 700 °C consists of charcoal, ash and aromatic residue [12]. The degradation temperature of wood, pulp, PP and PP-based composite was found by studying the thermal and structural properties of PP composite with wood pulp as filler, and FTIR analysis helped to understand the linkage between the wood pulp and PP [31].

9.3   Composite Properties

Natural fiber reinforced polymer composites (NFPCs) exhibit optimal performance depending upon the nature of fiber and its modification. Table 9.7 shows the chemical constituents and physical properties of oil palm fibers [32]. Comparing the different compositions present in natural fibers (Table 9.8) [33] with the different mechanical properties exhibited by the natural fibers (Table 9.9 and 9.10) [33–34], the natural plant fibers with higher cellulose content appear to potentially show better mechanical properties [34]. Cellulose whiskers (CWs) extracted from grass used as filler to fabricate polymer composite show improved composite mechanical properties and thermal stability after alkali treatment on the whiskers. Better mechanical properties are observed at 5% filler but any further loading reduces thermal stability and also the elongation due to the phase separation [35].

Table 9.7 The chemical constituents and physical properties of oil palm fibers [32].

Chemical constituents (%)	
Cellulose	65
Hemicellulose	---
Lignin	19
Ash Content	2
Physical properties of oil palm fiber	
Diameter (μm)	150–500
Density (g/cc)	0.7–1.55
Tensile strength (MPa)	248
Young's modulus (MPa)	6700
Elongation at break (%)	14
Microfibrilllar angle (°)	46

Table 9.8 Comparison of different compositions present in natural fibers [33].

Fiber	Cellulose (wt%)	Hemicellulose (wt%)	Lignin (wt%)	Waxes (wt%)
Bagasse	55.2	16.8	25.3	---
Bamboo	26–43	30	21–31	---
Flax	71	18.6–20.6	2.2	1.5
Kenaf	72	20.3	9	---
Jute	61–71	14–20	12–13	0.5
Hemp	68	15	10	0.8
Ramie	68.6–76.2	13–16	0.6–0.7	0.3
Abaca	56–63	20–25	7–9	3
Sisal	65	12	9.9	2
Coir	32–43	0.15–0.25	40–45	---
Oil palm	65	---	29	---
Pineapple	81	---	12.7	---
Curaua	73.6	9.9	7.5	---
Wheat straw	38–45	15–31	12–20	---
Rice husk	35–45	19–25	20	14–17
Rice straw	41–57	33	8–19	8–38

Table 9.9 Comparison of different mechanical properties existing in natural fibers [34].

Fiber	Tensile Strength (Mpa)	Elastic Modulus (Gpa)	Specific Modulus (E/1000 ρ)	Elongation at break (%)	Density (g cm³)
Kenaf	240–260	14–38	12–32	1.6	1.45
Flax	500–1500	58	38	2.7–3.2	1.5
Jute	393–773	60	39	1.5–1.8	1.3
Hemp	690	70	46	2.0–4.0	1.47
Cotton	400	5.5–12.6	---	7.0–8.0	1.5–1.6
Carbon	4000	230–240	---	1.4–1.8	1.4
E-glass	3400	71	28	3.4	2.55

Table 9.10 Comparison of mechanical properties of natural fibers [33].

Fiber	Tensile Strength (Mpa)	Young's (Gpa)	Elongation at break (%)
Abaca	400	12	3–10
Bagasse	290	17	---
Bamboo	140–230	11–17	---
Flax	345–1035	27.6	2.7–3.2
Hemp	690	70	1.6
Jute	393–773	26.5	1.5-1.8
Kenaf	930	53	1.6
Sisal	511-635	9.4-22	2.0–2.5
Ramie	560	24.5	2.5
Oil palm	248	3.2	25
Pineapple	400–627	1.44	14.5
Coir	175	4–6	30
Curaua	500–1150	11.8	3.7–4.3

Significant improvement in mechanical properties has also been reported with 20% alkali treatment, followed by grafting with 50% ethylenedimethylacrylate (EMA), in the fabrication of low-cost *Cocos nucifera* (coir) fiber reinforced polymer composite as compared to the controlled formulation of coir fiber reinforced polymethacrylate composite [21]. The fabrication of cellulose nanowhiskers (CNW)/PLA through solution casting shows stable TGA temperature range at 25–220 °C, while the dynamic mechanical thermal analysis (DMTA) shows improved storage modulus at higher temperatures [36]. The DMA, tensile tests, differential scanning calorimetry (DSC), XRD and FTIR analysis on CW fabricated in the low density polyethylene (LDPE) nanocomposite through hot compression molding show no variation in crystalline structure but enhancement in the mechanical properties is due to the uniform dispersion of filler and the bonding with the matrix [37].

However, another study with CW/PLA nanocomposites suggests no significant change in mechanical properties with increasing filler loading but the CW shows more sensitivity to degradation at higher temperatures as compared to MCC, attributable to the existence of treating elements [38]. In fact, a study on the effects of kenaf loading on tensile and morphological properties of kenaf reinforced LDPE/poly(vinyl alcohol) (LDPE/PVA/ KNF) through compression molding has even suggested that an increase in kenaf loading actually reduces the tensile strength and elongations by 2.7% at 10% loading and 15.8% at 40% loading. The morphological analysis shows that there is an adhesion between kenaf and LDPE/PVA matrix which may have led to the reduced mechanical properties [39].

Fabrication of composites for wood fibers and cellulose with PP has been reported [30]. Some, such as wheat husk and rye husk, exhibit better mechanical, thermal and surface properties of PP composite than the soft wood filler [40]. The fabricated coir fiber reinforced PP composite through compression molding has shown improved adhesive property of the fiber after alkali treatment on the fiber, while higher alkali concentration (<10%) damaged the fiber [41]. The palm and coir fiber reinforced PP composites fabricated through injection molding show improved strength with increasing fiber loading until the optimal value of 30%, beyond which there will be a reduction in mechanical properties. This is suggested as being due to the non-uniform dispersion or weak bonding between the fibers and the matrix. The coir fiber reinforced PP composite, however, shows better mechanical properties than the palm reinforced PP composite [42]. Rod-like CWs have been used to reinforce the isotactic PP composite, and the effects of aggregated CWs without surface modification (AGWH), aggregated CWs grafted with maleated PP (GRWH) and surfactant modified SWs (SUWH) on the mechanical properties of the PP composite have been investigated. The 6% GRWH and SUWH show improved tensile strength and elongation at break as compared to neat PP [43]. The mechanical properties of the composite for the three types of CWs in the preparation of atactic PP composite depend upon the amount of filler and the interaction of filler with the matrix [44].

Oil palm fibers have been utilized as biocomposites with other fibers or to reinforce other bio/synthetic polymers. The woven hybrid biocomposites using EFB and jute fiber have reportedly exhibited improved mechanical properties. The stability of the EFB composite is increased with the addition of jute fibers, attributable to the better stability of jute fiber than the EFB [45]. However, the use of sisal/oil palm hybrid fiber to reinforce natural rubber composite shows reduced strength, while the modulus increases with increasing fiber content. The increased modulus is assumed to be due to the interaction between the fiber and the matrix [32].

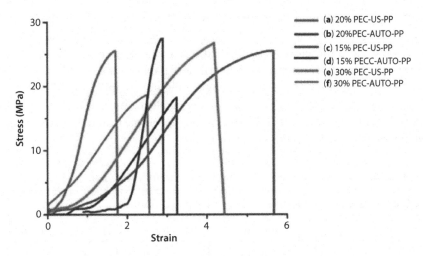

Figure 9.5 Tensile strength of (a) 20% PEC-US-PP, (b) 20% PEC-AUTO-PP, (c) 25% PECUS-PP, (d) 25% PEC-AUTO-PP, (e) 30% PEC-US-PP, and (f) 30% PEC-AUTO-PP [8].

The oil palm MCC in PLA composites, fabricated by solution casting, also shows improved thermal stability with 30% increment in Young's modulus, although the tensile strength and elongation are reduced due to increased MCC content and improper dispersion of MCC within the matrix [46]. The purely extracted cellulose from EFB based on ultrasonication (US) and autoclave (AUTO) treatment, used in the preparation of PP composite through injection molding, has shown optimal 25% cellulose loading, resulting in high tensile strength of 27 MPa (Figure 9.5). This is comparable to the reported tensile strength of neat PP, but achieved with reduced PP content and without any addition of coupling agents. The fabricated cellulose reinforced PP composite also shows thermal stability of 150 °C higher than the individual cellulose and PP. This has addressed both the needs of reduced use of petroleum-based products, whilst solving the environmental problems associated with abundant unutilized agro-wastes [8].

9.4   Applications

Biomass wastes can be harnessed for value-added products such as bioenergy, biochemical, biopolymer or biomaterials. Hybrid biocomposite especially has strategic applications in structural engineering of buildings in earthquake-prone areas, structural automotive parts, and high-performance aerospace components [7–8, 15, 45]. The NFPCs have advantages over the conventional polymer composites, such as low cost, light weight, high strength, corrosion resistance, better fatigue properties and biodegradability, and therefore are suitable for miscellaneous applications especially in automotive, construction, textile and paper industries (Tables 9.11 and 9.12) [23], bioengineering and environmental engineering [47]. Apart from these, other sectors include consumer electronics, sporting equipment and medical devices. Table 9.13 shows a comparison of the attributes of biocomposites and their applications [48].

Table 9.11 Utilization of NFPCs in the automotive industry [23].

Manufacturer	Model	Applications
Rover	2000 and others	Rear storage shelf/panel, insulations
Opel	Vectra, Astra, Zafira	Door panels, pillar cover panel, headliner panel, instrumental panel
Volkswagen	Passat, Variant, Golf, A4, Bora	Seatback, door panel, boot-lid finish panel, boot liner
Audi	A2, A3, A4, Avant, A6, A8, Roadster, Coupe	Boot liner, spare tire lining, side and back door panel, seatback, hat rack
Daimler-Chrysler	A, C, E, and S class, EvoBus (exterior)	Pillar cover panel, door panels, car windshield/car dashboard, business table
BMW	3, 5 and 7 series other pilot	Seatback, headliner panel, boot liner, door panels, noise insulation panels, molded footwell linings
Peugeot	406	Front and rear door panels, seatbacks, parcel shelf
Fiat	Punto, Brava Marea, Alfa Romeo 146,156, 159	Door panel
General Motors	Cadillac DeVille, Chevrolet TrailBlazer	Seatbacks, cargo area floor mat

(Continued)

Table 9.11 Cont.

Manufacturer	Model	Applications
Toyota	ES3	Pillar garnish, other interior parts
Saturn	L300	Package trays, door panel
Volvo	V70, C70	Seat padding, natural foams, cargo floor tray
Ford	Mondeo CD 162, Focus	Floor trays, door inserts, door panels, B-pillar, boot liner
Saab	9S	Door panels
Renault	Clio, Twingo	Rear parcel shelf
Toyota	Raum, Brevis, Harrier, Celsior	Floor mats, spare tire cover, door panels, seat backs
Mitsubishi		Cargo area floor, door panels, instrument panel
Mercedes-Benz	C, S, E and A classes Trucks	Door panels (flax/sisal/wood fibers with epoxy resin/UP matrix), glove box (cotton fibers/wood mold, flax/sisal), instrument panel support insulation (cotton fiber), molding rod/apertures, seat backrest panel (cotton fiber), trunk panel (cotton with PP/PET fibers), seat surface/backrest (coconut fiber/natural rubber) Internal engine cover, engine insulation, sun visor, interior insulation, bumper, wheel box, roof cover
Citroen	C5	Interior door paneling
Lotus	Eco Elise (July 2008)	Body panels, spoiler, seats, interior carpets
Rover	2000 and other	Insulation, rear storage shelf/panel
Vauxhall	Corsa, Astra, Vectra, Zafira	Headliner panel, interior door panels, pilar cover panels, instrument panel

Table 9.12 Utilization of NFPCs in various industries [23].

Fiber	Application in building, construction, and others
Hemp fiber	Construction products, textiles, cordage, geotextiles, paper and packaging, furniture, electrical, manufacture of banknotes, manufacture of pipes
Oil palm fiber	Building materials such as windows, door frames, structural insulated panel building systems, siding, fencing, roofing, decking, other building materials
Wod fiber	Window frame, panels, door shutters, decking, railing systems, fencing
Flax fiber	Window frame, panels, decking, railing systems, fencing, tennis rackets, bicycles frames, fork, seat post, snowboards, laptop cases
Rice husk fiber	Building materials such as panels, bricks, window frames, decking, railing systems, fencing
Bagasse fiber	Building materials such as panels, decking, railing systems, fencing
Sisal fiber	In construction industry such as panels, doors, shutting plates, and roofing sheets; also manucaturing of paper and pulp
Stalk fiber	Building panel, furniture panels, bricks and constructing drains and pipelines
Kenaf fiber	Packing material, mobile cases, bags, insulations, clothing-grade cloth, soilless potting mixes, animal bedding and material that absorbs oil/liquids
Cotton fiber	Furniture industry, textile/yarn, goods, cordage
Coir fiber	Building panels, flush door shutters, roofing sheets, storage tank, packing materials, helmets, postboxes, mirror casing, paper weights, projector cover, voltage stabilizer cover, filling materials for seat upholstery, brushes/brooms, ropes/yarns for nets, bags and mats, as well as padding for mattresses, seat cushions.
Ramie fiber	Use in products as industrial sewing thread, packing materials, fishing nets and filter cloths; it is also made into fabrics for household furnishing (upholstery, canvas) and clothing, paper manufacture
Jute fiber	Building panels, roofing sheets, door frames, door shutters, transport, packaging, geotextiles, chip boards

Table 9.13 Comparison of attributes between biocomposites and their applications [48].

Material attributes	Application attribute
Excellent weight, specific stiffness; Good weight, specific strength	Weight critical vehicles/products (transport, mobile, electronics, sport equipment)
Variable fiber properties	Non-safety critical/low required reliability applications
Renewable resource; Low embodied energy; Biodegradable	Short life-span products (disposible and high obsolescence rate products)
Nontoxic	Children's toys, consumer handled items, hobbyist built items
Biocompatible	Medical devices and implants
Low cost	Competitive consumer products
High water absorption	Dry use products
Poor durability	Short life-span products, limited exposure to harsh environments

9.5 The Way Forward

Oil palm fibers have great potential to be utilized as composites with other natural fibers and bio/synthetic polymers. The eco-friendly cellulose extraction can be improved using environmentally benign and clean solvents, while purification of other biomolecules and biochemicals, such as hemicelluloses, lignin and vanillin, is put in place for more integrated bioprocesses. The mechanical and thermal properties of cellulose-PP composite could be improved with the addition of coupling agent such as the cheap silica from EFB-ash modified materials or by the addition of new natural fibers or other polymer matrices. For environmental application, such as biosorbents for diesel or oil sorption or wastewater treatment, the fabrication technique can be optimized to manufacture composites with porous structure. The biocomposites are environmentally friendly and have great potential for offsite installation for easy installation along with fine finishing.

References

1. Yusoff, S., Renewable energy from palm oil—Innovation on effective utilization of waste. *J. Clean. Prod.* 14, 87–93, 2006.
2. Malaysian Palm Oil Board, Oil palm and the environment, http://www.mpob. gov.my/palm-info/environment/520-achievements#MillJun, 2012.
3. Malaysian Palm Oil Council, Malaysian Palm Oil: Assuring sustainable supply of oil & fats into the future, http://www.ceopalmoil.com, 2012.
4. Foo, Y.N., Foong, K.Y., Yousof, B., and Kalyana, S., A renewable future driven with Malaysian palm oil-based green technology. *JOPEH* 2, 1–7, 2011.
5. Rosnah, M.S., Astimar, A.A., Hasamudin, W., Hassan, W., and Gapor, M.T., Solid-state characteristics of microcrystalline cellulose from oil palm empty fruit bunch fiber. *J. Oil Palm Res.* 21, 613–620, 2009.
6. Abdullah, M.A., Nazir, M.S., and Wahjoedi, B.A., Development of value-added biomaterials from oil palm agro-wastes, in: *2nd International Conference on Biotechnology and Food Sciences (ICBFS)*, Bali, Indonesia, 2011.
7. Abdullah, M.A., Nazir, M.S., Ajab, H., Daneshfozoun, S., and Almustapha, S., Advances in eco-friendly pre-treatment methods and utilization of agro-based lignocelluloses, in: *Industrial Biotechnology: Sustainable Production and Bioresource Utilization*, Thangadurai, D., and Sangeetha, J. (Eds.), pp. 371–419, Apple Academic Press: New Jersey, USA/CRC Press, Florida, USA, 2016.
8. Abdullah, M.A., Nazir, M.S., Raza, M.R., Wahjoedi, B.A., and Yussof, A.W., Autoclave and ultra-sonication treatments of oil palm empty fruit bunch fibers for cellulose extraction and its polypropylene composite properties. *J. Clean. Prod.* 126, 686–697, 2016.
9. Nazir, M.S., Eco-friendly extraction, characterization and modification of microcrystalline cellulose from oil palm empty fruit bunches, PhD Thesis, Universiti Teknologi PETRONAS, Malaysia, 2013.
10. Daneshfozoun, S., Nazir, M.S., Abdullah, B., and Abdullah, M.A., Surface modification of celluloses extracted from oil palm empty fruit bunches for heavy metal sorption. *Chem. Eng. Trans.* 37, 679–684, 2014.
11. Wise, L.E., Murphy, M., and Addieco, A.A.D., Chlorite hollocellulose, its fractionation and bearing on summative wood analysis and on studies on the hemicelluloses. *Paper Trade* 122, 35–43, 1946.
12. Nazir, M.S., Wahjoedi, B.A., Yussof, A.W., Abdullah, M.A., Eco-friendly extraction and characterization of cellulose from oil palm empty fruit bunches. *Bioresources* 8, 2161–2172, 2013.
13. Singh, R., Krishna, B.B., Kumar, J., and Bhaskar, T., Opportunities for utilization of non-conventional energy sources for biomass pretreatment. *Bioresour. Technol.* 199, 398–407, 2016.
14. Crofton, A.R., Hudson, S.M., Howard, K., Pender, T., Abdelgawad, A., Wolski, D., and Kirsch, W.M., Formulation and characterization of a plasma sterilized, pharmaceutical grade chitosan powder. *Carbohydr. Polym.* 146, 420–426, 2016.

15. Aziz, M., Kurniawan, T., Oda, T., and Kashiwagi, T., Advanced power generation using biomass wastes from palm oil mills. *Appl. Therm. Eng.* 114, 1378–1386, 2017.
16. Priego Capote, F.P., and Luque de Castro, M.D., *Analytical Applications of Ultrasound*, 1st ed., Elsevier: Netherlands, 2007.
17. Li, H., Qu, Y., Yang, Y., Chang, S., and Xu, J., Microwave irradiation—A green and efficient way to pretreat biomass. *Bioresour. Technol.* 199, 34–41, 2016.
18. Zheng, Y., Zhao, J., Xu, F., and Li, Y., Pretreatment of lignocellulosic biomass for enhanced biogas production. *Prog. Energy Combust. Sci,* 42, 35–53, 2014.
19. Coral Medina, J.D., Woiciechowski, A.L., Zandona Filho, A., Bissoqui, L., Noseda, M.D., de Souza Vandenberghe, L.P., Zawadzki, S.F., and Soccol, C.R., Biological activities and thermal behavior of lignin from oil palm empty fruit bunches as potential source of chemicals of added value. *Ind. Crops Prod.* 94, 630–637, 2016.
20. Singh, J., Suhag, M., and Dhaka, A., Augmented digestion of lignocellulose by steam explosion, acid and alkaline pretreatment methods: A review. *Carbohydr. Polym.* 117, 624–631, 2015.
21. Rahman, M.M., and Khan, M.A., Surface treatment of coir (*Cocos nucifera*) fibers and its influence on the fibers' physico-mechanical properties. *Compos. Sci. Technol.* 67, 2369–2376, 2007.
22. Kalia, S., Kaith, B., and Kaur, I., Pretreatments of natural fibers and their application as reinforcing material in polymer composites—A review. *Polym. Eng. Sci.* 49, 1253–1272, 2009.
23. Mohammed, L., Ansari, M.N.M., Pua, G., Jawaid, M., and Saiful Islam, M., A review on natural fiber reinforced polymer composite and its applications. *International J. Polym. Sci.* 2015, Article ID 243947, 2015.
24. Foston, M., and Ragauskas, A.J., Changes in lignocellulosic supramolecular and ultrastructure during dilute acid pretreatment of *Populus* and switchgrass. *Biomass Bioenergy* 34, 1885–1895, 2010.
25. Gabriele, B., Cerchiara, T., Salerno, G., Chidichimo, G., Vetere, M.V., Alampi, C., Gallucci, M.C., Conidi, C., and Cassano, A., A new physical–chemical process for the efficient production of cellulose fibers from Spanish broom (*Spartium junceum L.*), *Bioresour. Technol.* 101, 724–729, 2010.
26. Oliveira, F.M.V., Pinheiro, I.O., Souto-Maior, A.M., Martin, C., Goncalves, A.R., and Rocha, G.J.M., Industrial-scale steam explosion pretreatment of sugarcane straw for enzymatic hydrolysis of cellulose for production of second generation ethanol and value-added products. *Bioresour. Technol.* 130, 168–173, 2013.
27. Yu, Z., Zhang, B., Yu, F., Xu, G., and Song, A., A real explosion: The requirement of steam explosion pretreatment. *Bioresour. Technol.* 121, 335–341, 2012.
28. Zhao, W., Yang, R., Zhang, Y., and Wu, L., Sustainable and practical utilization of feather keratin by an innovative physicochemical pretreatment: High density steam flash-explosion. *Green Chem.* 14, 3352–3360, 2012.

29. Yu, L. (Ed.), *Biodegradable Polymer Blends and Composites from Renewable Resources*, John Wiley and Sons: Hoboken, New Jersey, USA, 2009.
30. Malkapuram, R., Kumar, V., and Negi, Y.S., Recent development in natural fiber reinforced polypropylene composites. *J. Reinf. Plast. Compos.* 28, 1169–1189, 2009.
31. Awal, A., Ghosh, S., and Sain, M., Thermal properties and spectral characterization of wood pulp reinforced bio-composite fibers. *J. Therm. Anal. Calorim.* 99, 695–701, 2009.
32. Jacob, M., Thomas, S., and Varughese, K.T., Mechanical properties of sisal/oil palm hybrid fiber reinforced natural rubber composites. *Compos. Sci. Technol.* 64, 955–965, 2004.
33. Faruk, O., Bledzki, A.K., Fink, H.-P., and Sain, M., Biocomposites reinforced with natural fibers: 2000-2010. *Prog. Polym. Sci.* 37, 1552–1596, 2012.
34. Akhtar, M.N., Sulong, A.B., Nazir, M.S., Majeed, K., Radzi, M.K.F., Ismail, N.F., and Raza, M.R., Kenaf-biocomposites: Manufacturing, characterization, and applications, in: *Green Biocomposites: Manufacturing and Properties*, Jawaid, M., Sapuan, S.M., and Alothman, O.Y. (Eds.), pp. 225–254, Springer International Publishing, Cham, 2017.
35. Pandey, J.K., Chu, W.S., Kim, C.S., Lee, C.S., and Ahn, S.H., Bio-nano reinforcement of environmentally degradable polymer matrix by cellulose whiskers from grass. *Compos. Part B: Eng.* 40, 676–680, 2009.
36. Petersson, L., Kvien, I., and Oksman, K., Structure and thermal properties of poly(lactic acid)/cellulose whiskers nanocomposite materials. *Compos. Sci. Technol.* 67, 2535–2544, 2007.
37. Junior de Menezes, A., Siqueira, G., Curvelo, A.A.S., and Dufresne, A., Extrusion and characterization of functionalized cellulose whiskers reinforced polyethylene nanocomposites. *Polymer* 50, 4552–4563, 2009.
38. Oksman, K., Mathew, A.P., Bondeson, D., and Kvien, I., Manufacturing process of cellulose whiskers/polylactic acid nanocomposites. *Compos. Sci. Technol.* 66, 2776–2784, 2006.
39. Pang, A.L., Ismail, H., and Abu Bakar, A., Tensile properties and morphological studies of kenaf-filled linear low density polyethylene/poly (vinyl alcohol) (LLDPE/PVA/KNF) composites: The effects of KNF loading, *Adv. Mater. Res.* 1133, 156–160, 2016.
40. Bledzki, A.K., Mamun, A.A., and Volk, J., Physical, chemical and surface properties of wheat husk, rye husk and soft wood and their polypropylene composites. *Compos. Part A: Appl. Sci. Manuf.* 41, 480–488, 2010.
41. Gu, H., Tensile behaviours of the coir fibre and related composites after NaOH treatment. *Mater. Des.* 30, 3931–3934, 2009.
42. Haque, M.M., Hasan, M., Islam, M.S., and Ali, M.E., Physico-mechanical properties of chemically treated palm and coir fiber reinforced polypropylene composites. *Bioresour. Technol.* 100, 4903–4906, 2009.
43. Ljungberg, N., Cavaillé, J.Y., and Heux, L., Nanocomposites of isotactic polypropylene reinforced with rod-like cellulose whiskers. *Polymer* 47, 6285–6292, 2006.

44. Ljungberg, N., Bonini, C., Bortolussi, F., Boisson, C., Heux, L., and Cavaillé, J.Y., New nanocomposite materials reinforced with cellulose whiskers in atactic polypropylene: Effect of surface and dispersion characteristics. *Biomacromolecules* 6, 2732–2739, 2005.

45. Jawaid, M., Abdul Khalil, H.P.S., and Alattas, O.S., Woven hybrid biocomposites: Dynamic mechanical and thermal properties. *Compos. Part A: Appl. Sci. Manuf.* 43, 288–293, 2012.

46. Haafiz, M.K.M., Hassan, A., Zakaria, Z., Inuwa, I.M., Islam, M.S., and Jawaid, M., Properties of polylactic acid composites reinforced with oil palm biomass microcrystalline cellulose. *Carbohydr. Polym.* 98, 139–145, 2013.

47. Cheung, H.Y., Ho, M.P., Lau, K.T., Cardona, F., and Hui, D., Natural fibrereinforced composites for bioengineering and environmental engineering applications. *Compos. Part B: Eng.* 40, 655–663, 2009.

48. Dicker, M.P., Duckworth, P.F., Baker, A.B., Francois, G., Hazzard, M.K., and Weaver, P.M., Green composites: A review of material attributes and complementary applications. *Compos. Part A: Appl. Sci. Manuf.* 56, 280–289, 2014.

49. Leh, C.P., Rosli, W.D.W., Zainuddin, Z., and Tanaka, R., Optimisation of oxygen delignification in production of totally chlorine-free cellulose pulps from oil palm empty fruit bunch fibre. *Ind. Crops Prod.* 28, 260–267, 2008.

50. Yunus, R., Salleh, S.F., Abdullah, N., and Biak, D.R.A., Effect of ultrasonic pretreatment on low temperature acid hydrolysis of oil palm empty fruit bunch. *Bioresour. Technol.* 101, 9792–9796, 2010.

51. Wanrosli, W.D., Rohaizu, R., and Ghazali, A., Synthesis and characterization of cellulose phosphate from oil palm empty fruit bunches microcrystalline cellulose. *Carbohydr. Polym.* 84, 262–267, 2011.

52. Mohamad Haafiz, M.K., Eichhorn, S.J., Hassan, A., and Jawaid, M., Isolation and characterization of microcrystalline cellulose from oil palm biomass residue. *Carbohydr. Polym.* 93, 628–634, 2013.

53. Rohaizu, R., and Wanrosli, W.D., Sono-assisted TEMPO oxidation of oil palm lignocellulosic biomass for isolation of nanocrystalline cellulose. *Ultrason. Sonochem.* 34, 631–639, 2017.

54. Wanrosli, W.D., Zainuddin, Z., and Lee, L.K., Influence of pulping variables on the properties of *Elaeis guineensis* soda pulp as evaluated by response surface methodology. *Wood Sci. Technol.* 38, 191–205, 2004.

55. Al-Dulaimi, A.A., and Wanrosli, W.D., Isolation and characterization of nanocrystalline cellulose from totally chlorine free oil palm empty fruit bunch pulp. *J. Polym. Environ.* 25, 192–202, 2016.

56. Lamaming, J., Hashim, R., Leh, C.P., and Sulaiman, O., Properties of cellulose nanocrystals from oil palm trunk isolated by total chlorine free method. *Carbohydr. Polym.* 156, 409–416, 2017.

57. Owolabi, A.F., Haafiz, M.K.M., Hossain, M.S., Hussin, M.H., and Fazita, M.R.N., Influence of alkaline hydrogen peroxide pre-hydrolysis on the isolation of microcrystalline cellulose from oil palm fronds. *Int. J. Biol. Macromol.* 95, 1228–1234, 2017.

58. Karuppuchamy, S., Andou, Y., Jang, S.S., Nishida, H., Hassan, M.A., and Shirai, Y., Eco-friendly superheated steam treated oil palm empty fruit bunch fibers and their application in polymer composites. *Adv. Sci. Eng. Med.* 8, 131–134, 2016.
59. Karuppuchamy, S., Andou, Y., Nishida, H., Nordin, N.I.A.A., Ariffin, H., Hassan, M.A., and Shirai, Y., Superheated steam treated oil palm frond fibers and their application in plastic composites. *Adv. Sci. Eng. Med.* 7, 120–125, 2015.
60. Islam, M.R., Gupta, A., Rivai, M., Beg, M.D.H., and Mina, M.F., Effects of fiber-surface treatment on the properties of hybrid composites prepared from oil palm empty fruit bunch fibers, glass fibers, and recycled polypropylene. *J. Appl. Polym. Sci.* 133(11), 43049, 2016.
61. Suradi, S.S., Yunus, R.M., and Beg, M.D.H., Oil palm bio-fiber-reinforced polypropylene composites: Effects of alkali fiber treatment and coupling agents. *J. Compos. Mater.* 45, 1853–1861, 2011.
62. Karuppuchamy, S., Andou, Y., Baharuddin, A.S., Sulaiman, A., Hassan, M.A., Nishida, H., Shirai, Y., Thermo-mechanical properties of palm fiber plastic (PFP) composites. *Adv. Sci. Eng. Med.* 7, 844–848, 2015.
63. Chee, C.Y., Yong, G.K., Abdullah, L.C., and Nadarajah, K., Effect of nanosilica and titania on thermal stability of polypropylene/oil palm empty fruit fibre composite. *J. Biobased Mater. Bio.* 7, 169–174, 2013.
64. Essabir, H., Boujmal, R., Bensalah, M.O., Rodrigue, D., Bouhfid, R., and Qaiss, A.E.k., Mechanical and thermal properties of hybrid composites: Oil-palm fiber/clay reinforced high density polyethylene. *Mech. Mater.* 98, 36–43, 2016.
65. Mitchell, G., France, F., Nordon, A., Tang, P.L., and Gibson, L.T., Assessment of historical polymers using attenuated total reflectance-Fourier transform infra-red spectroscopy with principal component analysis. *Heritage Science* 1, 28, 2013.

10

Interfacial Modification of Polypropylene-Based Biocomposites and Bionanocomposites

Yekta Karaduman* and Nesrin Sahbaz Karaduman

Akdagmadeni Vocational High School, Bozok University, Akdağmadeni, Yozgat, Turkey

Abstract

Natural fiber reinforced composites, also referred to as "biocomposites" or "green composites," are increasingly being used in structural and semistructural engineering applications due to their distinct advantages over traditional glass fiber reinforced composites. Biocomposites are sustainable, environmentally friendly, low cost, low density and easy to process. Their specific mechanical properties are close to those of glass fiber reinforced plastics. There are a number of commercially important natural fibers, such as hemp, flax, jute, coir, and ramie, although the number of possible natural reinforcement materials are almost limitless; there is an extensive list of natural materials that can be used in fiber or particle form as reinforcement. Researchers constantly introduce new fiber types with improved properties in resulting composites. The greatest problem facing biocomposites and bionanocomposites is the low compatibility between cellulose-based hydrophilic natural fibers and hydrophobic polymer resins, which leads to poor fiber-matrix adhesion and therefore inefficient load distribution between fibers and matrix. In addition, a poor interface increases the moisture absorption in humid or wet environments. So far, several interfacial modification methods have been suggested to address such problems and improve the properties of the resulting biocomposites. This chapter focuses on the interfacial modification of polypropylene (PP)-based biocomposites and bionanocomposites. First, the nature of fiber-matrix interface is introduced and then the qualitative and quantitative methods for determining interface strength are outlined. The interface modification of PP-based biocomposites and bionanocomposites are presented in light of the existing literature.

Corresponding author: yektakaraduman@gmail.com

Visakh. P. M. and Matheus Poletto. (eds.) Polypropylene-Based Biocomposites and Bionanocomposites, (315–348) 2018 © Scrivener Publishing LLC

Keywords: Biocomposite, bionanocomposite, polypropylene (PP), fiber-matrix interface modification

10.1   Introduction

Over the last few decades, there has been increasing interest in sustainable and eco-friendly natural fibers and their composites due to increased environmental awareness. Natural fibers are renewable and readily available at low cost. Their density is approximately half that of glass fibers. Specific mechanical properties of biocomposites are comparable to those of glass fiber reinforced composites. Therefore, biocomposites reinforced with natural fibers provide a solution towards more sustainable economies [1, 2]. There are various application areas for biocomposites, including packaging, automotive, recreation and civil engineering [3, 4]. The overall performance of biocomposites depends on the properties of fibers and resin, fiber volume fraction, fiber architecture, and the quality of the interface between fibers and matrix. The most common fiber types used in biocomposites include flax, hemp, jute, kenaf and coir, although a wide range of natural fibers can be used as reinforcement. In fact, researchers are constantly introducing new fiber types which they claim have certain advantages as reinforcement. Thermoset (e.g., epoxy, polyester, vinyl ester) and thermoplastic (PP, PE, PVA) polymers are mostly used as the matrix material in biocomposites. A good bonding between fibers and matrix generally improves the overall mechanical performance of the composite by enhancing the load distribution capability of the material, even though a poor interface is preferable in some situations where the material is subjected to impact loads. A poor interface has been shown to promote fiber pull-out during impact fracture, which results in an enhanced energy absorption through friction. The fiber-matrix bonding in biocomposites is generally poor due to incompatibility between the surface energies of hydrophilic natural fibers and hydrophobic polymer resins. Researchers have generally focused on improving the fiber-matrix bonding by using various chemical and physical modification techniques such as silane treatment, acetylation, utilization of coupling agents and alkali treatment [5–9].

This chapter deals with the interfacial modification of polypropylene (PP)-based biocomposites and bionanocomposites. First, the nature of fiber-matrix interface is introduced. Then qualitative and quantitative methods for the determination of interface strength are outlined. The interface modification of PP-based biocomposites and bionanocomposites are presented.

10.2 Natural Fibers

In order to understand the nature of fiber-matrix interface in biocomposites, the structure of natural fibers must be known. Natural fibers fall into three major groups depending on their origin: plant fibers, animal fibers and mineral fibers (Figure 10.1) [10]. In terms of composite reinforcement, the most important group is plant fibers. Plant fibers are lignocellulosic in nature; they consist mainly of cellulose and lignin together with other components such as hemicellulose, pectin, waxes, ash and other aromatic compounds. The percentages as well as the characteristics of these ingredients can vary depending upon the fiber type, growing region, growing season and the location on the plant from where the fibers are extracted.

Natural fibers are generally multicellular in structure. A single fiber consists of lumen, primary wall and secondary wall. The secondary wall is made up of three sublayers such as S1, S2 and S3. Cellulose microfibrils, especially those in S2 layer, give the fiber its strength and stiffness whereas other polymers, like lignin and hemicellulose, act as glues that hold individual cellulose microfibrils together and unify the structure [11–14]. Physical properties of some common natural fibers are given in Table 10.1 [11, 15, 16]. Table 10.2 shows the chemical composition of plant fibers [17, 18].

Cellulose is a linear polymer of glucose ($C_6H_{12}O_6$). Glucose molecules add on successively through β-1,4 linkages with the elimination of water to form long cellulose chains [19]. The chemical structure of a cellulose macromolecule is shown in Figure 10.2 [20]. Hydrogen bonds form between the pendant –OH or –CH_2OH groups of cellulose macromolecules to create cellulose microfibrils. These microfibrils typically have a diameter of 2–20 nm and a length of 100–40,000 nm [21]. The regions where these

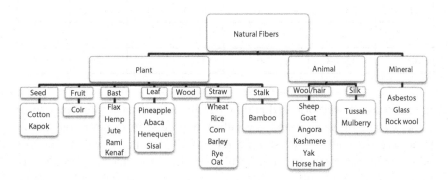

Figure 10.1 Classification of natural fibers [10].

Table 10.1 Physical properties of natural fibers (single fiber) [11, 15, 16].

Fiber	Length l (mm)	Diameter d (µm)	Aspect ratio (l/d)	Microfibril angle (°)	Density (kg/m³)	Moisture uptake (%)
Cotton	20–64	11.5–17	2752	20–30	1550	8.5
Flax	27–36	17.8–21.6	1258	5	1400–1500	12
Hemp	8.3–14	17–23	549	6.2	1400–1500	12
Jute	1.9–3.2	15.9–20.7	157	8.1	1300–1500	12
Kenaf	2–61	17.7–21.9	119	–	1220–1400	17
Ramie	60–250	28.1–35	4639	–	1550	8.5
Sisal	1.8–3.1	18.3–23.7	115	10–22	1300–1500	11
Coir	0.9–1.2	16.2–19.5	64	39–49	1150–1250	13

Table 10.2 Chemical composition of plant fibers [17, 18].

Fiber	Cellulose (% wt)	Hemicellulose (% wt)	Lignin (% wt)	Pectin (% wt)	Moisture (% wt)	Wax/oil (% wt)
Cotton	85–90	5.7	–	0–1	7.85–8.5	0.6
Jute	61–71.5	13.6–20.4	12–13	0.2	12.5–13.7	0.5
Flax	71	18.6–20.6	2.2	2.3	8–12	1.7
Hemp	70–74	17.9–22.4	3.7–5.7	0.9	6.2–12	0.8
Ramie	68.6–76.2	13.1–16.7	0.6–0.7	1.9	7.5–17	0.3
Sisal	66–78	10–14	10–14	10	10-22	2
Pineapple	70–82	–	5–12.7	–	11.8	–
Coir	32–43	0.15–0.25	40–45	3–4	8	–

microfibrils are well oriented and lie parallel to each other forming close-packed structures are called crystalline regions. Most chemicals cannot penetrate through the crystalline regions. Amorphous regions, on the other hand, have a more open structure where the macromolecules are randomly oriented and consequently are easier to manipulate by using chemical agents.

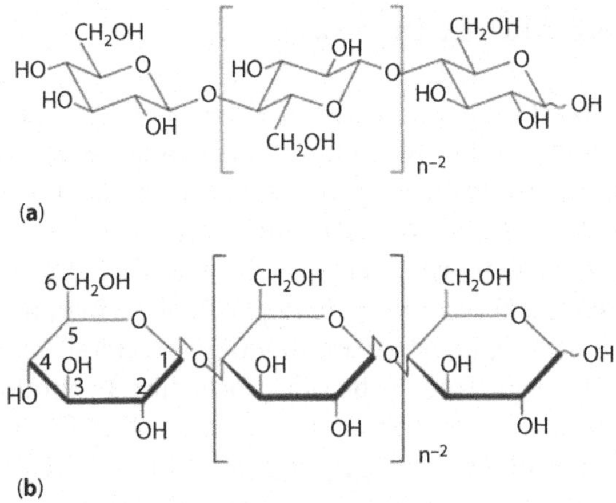

(a)

(b)

Figure 10.2 Chemical structure of cellulose [20].

Hemicellulose is the common name for a group of polysaccharides which act as a binding agent between cellulose macromolecules in plant cell walls [21, 22]. It has an amorphous structure and hence is more vulnerable to chemical action compared with crystalline cellulose. Lignin provides rigidity and strength to plant cell walls and protects them from microbial pathogens [21]. Lignin is made up of oxyphenylpropan units and has a branched structure which results in an amorphous material unlike crystalline cellulose [11, 21, 23]. Therefore, lignin has an open structure permitting the chemical action. Lignin is hydrophobic unlike hemicelluloses and amorphous cellulose [11, 24].

Pectin is a generic name for various complex pectic polysaccharides [25]. Galacturonic acid residues linked through α bonds (poli-α-(1-4)-galacturonic acid) are the major component of most pectins [11, 21, 26]. Pectin acts as a matrix material that binds cellulose and hemicellulose macromolecules together like steel rods in concrete. Pectin is present in low amounts in lignocellulosic plant fibers. Aromatic compounds other than lignin and pectin in plant cell walls include condensed tannins, which are phenolic complexes arising from the condensation of leucoanathocyanidin and catechin [27]; and low molecular weight phenolic acids, e.g., ferulic and p-coumaric acids.

Other ingredients of lignocellulosic plant fibers are low amounts of fats, waxes and lipids: hydrocarbons insoluble in water. Waxes are esters of long-chain alcohols. Lipid compounds include long-chain fatty acids (carboxylic acid), fatty alcohols and wax esters (esters of long-chain fatty alcohols with long-chain fatty acids) [28].

10.3   Fiber-Matrix Interface

Fiber-matrix interface has a profound influence on the mechanical properties and performance of biocomposites. Its main purpose is to transfer the load from matrix to stronger fibers and distribute the load across the main body of the material. If it fails to do so there will be catastrophic consequences. The fiber-matrix interface substantially influences the fracture behavior of composites by undergoing plastic deformation. Therefore, the nature of fiber-matrix interface and bonding mechanisms between fibers and matrices must be known before proceeding to modification methods. It must be borne in mind that the surface composition and energy of natural fibers are highly heterogeneous and can vary in a wide spectrum depending on the fiber type, growing season, growing region and other numerous factors. Consequently, fiber-matrix bonding mechanisms and the type of bonding can vary widely, making it difficult to obtain a standard procedure for surface modification.

10.3.1   Fiber-Matrix Bonding Types

10.3.1.1   *Wetting Phenomena and Intermolecular Interactions*

In broad terms, wetting is the ability of a liquid to make and sustain contact with a solid surface when the two are brought together. Wetting occurs when the adhesive forces between the fiber and resin molecules are greater than the cohesive forces between the resin molecules. Fiber wetting is a result of intermolecular interactions between the fiber and resin molecules. These interactions include ion-ion, ion-dipole, dipole-dipole and dipole-induced dipole interactions, London (dispersion) forces, and hydrogen bonding. Interactions that depend on the inverse sixth power of the separation between the molecules, i.e., the dipole-dipole interactions of rotating polar molecules in the gas phase, all the London interactions, and the dipole-induced dipole interactions are known collectively as van der Waals interactions [29]. Fiber wetting is of crucial importance since the resin must spread over the greatest possible surface area of the fibers before any further physical or chemical interactions, like covalent bonding and mechanical interlocking, can take place.

Wetting is generally determined by surface energies, γ, of the materials (Figure 10.3). Thermodynamic work of adhesion W_a is given by *Dupré equation* [30]:

$$W_a = \gamma_{SV} + \gamma_{LV} - \gamma_{SL} \tag{10.1}$$

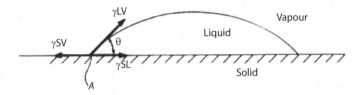

Figure 10.3 Contact angle, θ, and surface energies, γ, for a liquid drop on a solid surface [30].

where subscripts S, L and V stand for solid, liquid and vapor respectively. This equation suggests that the wetting is strongly favored if the surface energies of the two constituents are large and their interfacial surface energy is small, provided that the surface energy of the liquid is not too large to prevent the liquid droplet from spreading over the solid. The state of wetting is measured by contact angle, θ, which is given by Young equation [31]:

$$\gamma_{SV} = \gamma_{SL} + \gamma_{LV} \cdot \cos\theta \qquad (10.2)$$

Complete wetting is said to occur when contact angle θ = 0°. It follows from Equation 10.2 that in order for the complete wetting to take place, the surface energy of the solid must be equal to or greater than the sum of the liquid surface energy and the interface surface energy. Since the interface surface energies are generally small, it could be stated that wetting is favored when the surface energy of the fibers greatly exceeds that of the resins. For instance, glass fibers (γ_{SV} = 560 mJ m^{-2}) are easily wetted by polyester resin (γ_{LV} = 35 mJ m^{-2}) whereas polyethylene fibers (γ_{SV} = 31 mJ m^{-2}) are not [30]. Surface energies and the contact angle offer a quantitative way of evaluating the state of fiber-matrix adhesion in polymer composites. The whole purpose of interfacial modification is to increase the gap between the surface energies of fibers and matrices so that an effective wetting can be obtained.

Surface energies of various fibers and matrices can be deduced from the contact angle by employing surface energy component (SEC) theories. Surface energy measurement is conducted by applying various test liquids with known surface energies on the surface of the substrate (fiber or matrix) and measuring the contact angle. There are two common approaches for estimating the surface energies from the contact angle measurements such as Owens-Wendt and van Oss-Good [32]. Table 10.3 shows the surface energy values of some common natural fibers and matrices.

Table 10.3 Surface energy values of some common natural fibers and matrices [33–35].

	Surface energy (mJ/m²)								
Method	Owens-Wendt			van Oss-Good					
Fiber/ Matrix	γ_s	γ_s^d	γ_s^p	γ_s	γ_s^{LW}	γ^+	γ^-	γ_s^{AB}	Ref.
Coir	40.4	35.1	5.3	37.5	35.5	0.33	3.17	2.0	[33]
Bamboo	44.6	30.9	13.7	38.8	35.4	0.28	10.13	3.4	[34]
Flax	43.7	29.2	14.5	37.0	35.0	0.08	11.7	2.0	[35]
PP	30.7	27.1	3.6	30.9	30.0	0.12	1.87	0.9	[33]
MAPP	28.6	23.6	5.0	28.8	28.3	0.02	3.15	0.5	[33]
PVDF	37.2	30.8	6.4	35.1	31.6	0.88	3.39	3.5	[33]
PLA	40.1	27.4	12.7	34.3	32.7	0.06	10.18	1.6	[35]

The Owens-Wendt method uses dispersive component, γ_s^d, and polar component, γ_s^p, to describe the surface energy, γ_s. The dispersive component accounts for London-dispersion interactions whereas the polar component accounts for polar interactions, including hydrogen bonds and acid-base interactions. In the van Oss-Good approach the surface energy consists of Lifshitz-van der Waals component, γ_s^{LW}, and acid-based component, γ_s^{AB}, which includes electron acceptor, γ^+, and electron donor, γ^-, components. Lifshitz-van der Waals represents London forces, polar-polar and induced interactions [32].

Once the surface energies of the constituent materials are known, these data can be used to determine the four wetting parameters that fully describe the wetting behavior of fibers and can be used to optimize fiber-matrix adhesion in composite materials. These parameters are the work of adhesion (W_a), interfacial energy (γ_{SL}), spreading coefficient (S), and wetting tension (ΔF_i) [32]. The work of adhesion is the energy required to separate a unit area of interface and can be described with *Dupré equation* (Equation 10.1). The interfacial energy is defined as the work necessary to increase the interfacial surface area by unit area [32]. Hence, a small value of interfacial energy indicates a good wetting.

The spreading coefficient is the ability of a liquid to spontaneously spread over a solid. A positive value of the spreading coefficient indicates instantaneous spreading.

$$S = \gamma_{SV} - (\gamma_{SL} + \gamma_{LV})$$ (10.3)

The wetting tension can be defined as the work needed against wetting a porous network by eliminating a unit area of the solid-liquid interface while exposing a unit area of the solid vacuum interface. A positive value of wetting tension is an indication of good wetting [32].

$$\Delta F_i = \gamma_{SV} - \gamma_{SL}$$ (10.4)

Table 10.4 lists the wetting parameters for various biocomposite systems calculated using the surface energies of fiber and matrix [33, 34]. Interfacial modifications, such as alkali treatment and MAPP, generally increase the spreading coefficient, indicating easy wetting. On the other hand, the treatments seem to have little effect on the work of adhesion values of the consolidated composites. In general, when the surface energy of the fiber greatly exceeds that of the matrix the wetting is favored and the

Table 10.4 Wetting parameters (van Oss-Good approach) of various biocomposites calculated using the surface energies of fiber and matrix [33, 34].

Composite	W_a^{ab} (mJ/m²)	γ_{SL} (mJ/m²)	S (mJ/m²)	ΔF (mJ/m²)
Coir/PP	68.1	0.3	6.3	37.2
Coir/MAPP	65.9	0.4	8.3	37.1
Coir/PVDF	72.4	0.2	2.2	37.3
Alkali 5% treated coir/PP	68.6	1.9	6.8	37.7
Alkali 5% treated coir/MAPP	66.0	2.4	8.4	37.2
Alkali 5% treated coir/PVDF	75.0	−0.3	4.8	39.9
Bamboo/PP	68.8	0.9	7.1	37.9
Bamboo/MAPP	66.6	1.3	8.4	37.5
Bamboo/PVDF	74.5	−1.1	5.2	39.9

spreading coefficient increases accordingly. However, this energy margin can have a negative effect on the physical adhesion.

10.3.1.2   Chemical Bonds

Various chemical bonds, such as ionic, covalent and metallic bonds, can form at the interface when two substances come into close contact with one another. The bond strength determines the strength of adhesion. The most common type of chemical bond in biocomposites is the covalent bond that forms between the hydroxyl groups of cellulose and the functional groups in the polymer resin. Functional groups that can make covalent bonds with hydroxyl groups of cellulose can be grafted to polymer resins to form covalent bonds. Covalent bonds are much stronger when compared to intermolecular interactions such as van der Waals forces and hydrogen bonds.

10.3.1.3   Electrostatic Attraction

If the surfaces of the fibers and resin have a net charge of opposite signs, coulombic interactions can occur between them. This effect is used in the deposition of coupling agents on glass fibers [30]. Depending on the net surface charge of glass fibers, anionic or cationic functional groups will be attracted to their surface. Electrostatic forces do not contribute adhesion as much as other kinds of interactions do since the surface charges are not stable and may change depending on factors like humidity, pH and temperature.

10.3.1.4   Mechanical Keying

Fibers with a rough surface can promote a different type of adhesion called mechanical keying or mechanical interlocking. Although this process is insignificant for smooth-surfaced man-made fibers like carbon and glass, it is very important for rough-surfaced natural fibers. For this mechanism to take effect, the resin must penetrate into the cavities of the fiber surface. Some of the fiber modification techniques, such as alkali treatment, aim to increase the surface roughness of natural fibers in order to increase the area of contact and promote mechanical keying between fibers and polymer resins.

10.3.2   Measurement of Interfacial Bond Strength

Measurement of interface bonding strength is carried out using either single-fiber or bulk laminate experiments. In single-fiber experiments the interfacial shear strength (IFSS) is deduced from the shear debonding and sliding behavior of a single fiber embedded in a matrix. Single-fiber tests include pull-out, microbond, and full fragmentation tests. It can be assumed that a high debonding stress in shear mode is an indication of a strong resistance to normal tensile stress. Although this is not always true especially for the mechanical keying-type interface formation, it gives a reasonable approximation to the behavior of the interface under normal tensile loads. Another limitation of single-fiber test methods is that the shear behavior of fibers in real composite materials may differ from that of the single fiber owing to the influence of the neighboring fibers. Bearing in mind these limitations, single-fiber tests still provide an effective way of quantitative determination of the interfacial bond strength. Bulk laminate experiments include transverse tensile and bending tests, short beam shear test and Iosipescu shear test. Complementary techniques for indirect interface assessment can also provide additional information about the chemical and physical nature of fiber-matrix interface. These techniques include XPS, ToF-SIMS, SEM and surface energy analysis. Only single-fiber test methods are discussed here since the full discussion of all the available techniques goes beyond the scope of this chapter.

10.3.2.1   Single-Fiber Pull-Out Test

This is the most common method for polymer matrix composites. It has also been applied to natural fiber composites by several researchers [36, 37]. In this method, a single fiber half embedded within a matrix is pulled out of the matrix under a tensile load. The apparent interfacial shear strength (IFSS, τ_{app}) is calculated from the peak force (F_{max}):

$$\tau_{app} = F_{max}/\pi dL \tag{10.5}$$

in which d is the fiber diameter and L is the embedded fiber length. The apparent IFSS could only provide a rough estimation of good or bad interface without detailed information. A more rigorous treatment of single-fiber pull-out test can be carried out by taking into account the debonding shear stress. In this method, the load-displacement data obtained from the pull-out test is interpreted according to *shear lag theory* [38, 39]. Figure 10.4 shows axial distributions of the normal stress in the fiber and shear stress at the interface. Complete pulling out of the fiber takes place in three stages such as elastic loading up to debonding, propagation of the debonding front, and pull-out by frictional sliding. Basic assumptions of the shear lag model, such as no shear strain in the fiber and no transfer of normal stress across the fiber ends, are adopted in this procedure [30].

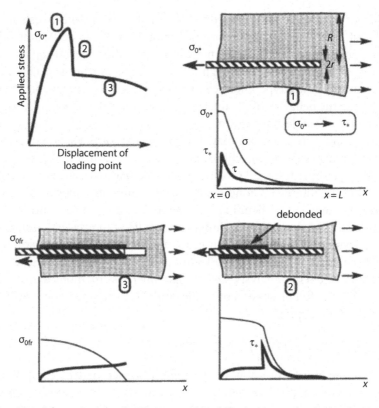

Figure 10.4 Schematic stress distributions and load-displacement plot during the single-fiber pull-out test [30].

The peak in the load-displacement curve which occurs at an applied stress, σ_{0*}, corresponds to the debonding event. The debonding shear stress, τ_*, can be calculated using σ_{0*} with the following relation [39]:

$$\tau_* = \frac{n\sigma_{0*}\coth(nL/r)}{2} \tag{10.6}$$

where r is the fiber radius; n is a dimensionless constant; n is given by

$$n = \left[\frac{2E_m}{E_f\left(1+v_m\right)\ln(1/f)}\right]^{1/2} \tag{10.7}$$

in which E_m and E_f are the matrix and fiber modulus respectively; f is the fiber volume fraction; v_m is the matrix of Poisson's ratio.

10.3.2.2   Microbond Test

The microbond test is based on the fiber pull-out test adapted to small diameter fibers. In this technique, a small droplet of resin is applied to the fiber and the specimen is positioned between two knife-edge-shaped plates. A tensile force is then applied to fiber free end to pull the resin droplet against the knife edges so that the load is transferred to the fiber-matrix interface. The same equation as in the pull-out method is used to calculate the IFSS. This technique requires extremely small droplets and is applicable with very soft matrices [40].

10.3.2.3   Full-Fragmentation Test

In the full-fragmentation technique, a single fiber is embedded in a matrix and the matrix is strained in tension parallel to the fiber axis. As a result, the fibers break into a number of fragments with a wide variation in fragment lengths. The shortest fiber length that can be achieved upon the application of tensile load is referred to as critical fiber length, L_c. Shear strength, τ, is deduced from the aspect ratios of the fiber fragments considering the Weibull modulus of the fiber [30]. Fiber fragmentation test provides the best approach to the actual situation of the fiber in a composite among all the single-fiber tests. It also takes into account the heterogeneity of natural fibers [40].

10.4   Interfacial Modification of PP-Based Biocomposites and Bionanocomposites

A wide array of physical and chemical techniques can be used to modify either natural fiber surface or PP resin to increase fiber-matrix adhesion. Fiber modification techniques generally aim to change the chemical structure of fiber surface as well as the fiber surface energy whereas matrix modification involves mostly grafting of new functional groups to increase fiber/resin compatibility.

10.4.1   Physical Modification Methods

The majority of physical modification methods aim to modify the structural and surface properties of the fibers without using chemical agents. The chemical composition of the fibers generally remains intact after the physical treatment. The enhancement in fiber-matrix interface is obtained as a result of increased mechanical interlocking between the fibers and the polymer resin. Physical modification methods include corona, plasma, and alkali treatments (mercerization).

10.4.1.1   Corona Treatment

This method is based on "corona discharge," i.e., an electrical discharge appearing around the surface of charged conductors caused by ionization of the surrounding fluid. This ionization creates radicals on the surface of the fibers and promotes fiber-matrix bonding.

10.4.1.2   Plasma Treatment

In this treatment, plasma of different gases is used to bring about changes on the surface of the fibers. It is similar to corona treatment in that an ionized region including excited species, such as ions and radicals, is formed around the fiber surface. This technique requires a vacuum chamber and gas feed to maintain the desired pressure and composition of the gas mixture. Yuan *et al.* [41] investigated the effect of plasma treatment on the performance of wood/PP composites. Figure 10.5 shows the schematic diagram of the plasma treatment setup used. They used argon and air plasma treatments to modify the fiber surface. The tensile strength and modulus as well as storage modulus of the composite increased after plasma treatment. SEM observations suggested that the fiber/matrix adhesion improved as a result of plasma treatment.

Couto *et al.* [42] used oxygen plasma to enhance the compatibility between sisal fibers and PP. A plasma treatment applied to the matrix before compounding significantly improved the interaction between fiber and matrix by making PP more hydrophilic.

10.4.1.3   Alkali Treatment

One of the most common methods of fiber modification is alkali treatment, which is effective, low cost and easy to apply. In this method, the

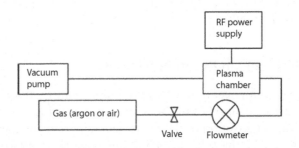

Figure 10.5 Schematic diagram of the plasma treatment setup [41].

fibers are treated with a dilute solution of sodium hydroxide (NaOH) with varying concentrations, temperatures and durations. Alkali treatment has been shown to remove the hemicellulose polymers as well as other impurities from the fiber surface, thus creating a rougher surface morphology. Consequently, the number of available sites for matrix penetration is increased, which results in an improved mechanical interlocking between fibers and matrix. Alkali treatment also causes fiber fibrillation by removing pectin and hemicellulose polymers that bind the individual fibers together to form fiber bundles. This leads to an increased fiber surface area for resin adherence. Gassan and Bledzki [43] investigated the alkali treatment of jute fibers and its effect on fiber structure and properties. Isometric NaOH treatment led to an increase in yarn tensile strength and modulus by 120% and 150% respectively. The change in the mechanical properties was attributed to changes in crystalline orientation, degree of polymerization and cellulose content. Yarn shrinkage plays an important role in that a high amount of shrinkage negatively affects the strength and modulus of the yarns.

Joseph *et al.* [44] investigated the effect of interface treatments, such as sodium hydroxide, maleic anhydride, urethane derivative of polypropylene glycol (PPG) and permanganate, on the mechanical properties of sisal fiber reinforced PP composites. They reported that all the treatments improved the tensile properties of the composites to varying extents. Figure 10.6 shows a possible reaction between the free isocyanate groups in urethane derivative of PPG and alkali treated sisal fibers. MAPP treated composites stand out for their excellent tensile properties when compared to other chemical treatments.

10.4.2 Chemical Modification Methods

Several chemical modification techniques have been developed to enhance the fiber-matrix adhesion in biocomposites. Some of these methods aim to change the chemical structure of the fibers and/or matrix in order to improve the compatibility between the two. Some involve the use of intermediary groups known as coupling agents, which act as a bridge between the fibers and matrix and link them together. Some of the most important interface modification methods are as follows:

- Esterification-based treatments
- Silane coupling agents
- Graft copolymerization
- Treatment with isocyanates
- Triazine coupling agents

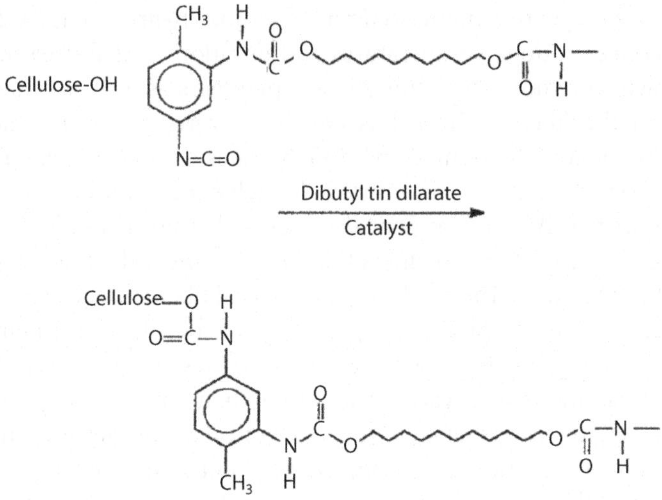

Figure 10.6 A possible reaction between the free isocyanate groups in urethane derivative of PPG and sisal fiber [44].

10.4.2.1   Esterification-Based Treatments

The product of the reaction between a carboxylic acid and an alcohol is called an ester. In esterification technique, hydroxyl groups, –OH, in cellulose macromolecules are reacted with the carboxyl groups, –COOH, which is the functional group found in carboxylic acids. As a result, the –OH groups, which give the fibers their hydrophilic character, are eliminated and the fibers gain a more hydrophobic character and hence become more compatible with most polymer resins. Esterification techniques include acetylation, benzylation, propionylation and treatment with stearates [40]. Acetylation is the most popular technique and it has been shown to be very effective as well as being low cost. Tserki *et al.* [45] investigated the effect of acetylation and propionylation surface treatments on flax, hemp and wood fibers. The spectroscopic analyses revealed that ester bonds formed on the fiber surface upon treatment. Furthermore, both treatments resulted in a removal of non-crystalline substances from the fiber surface, hence changing the surface topography. The moisture absorption of the fibers was reduced upon esterification. Figure 10.7 shows ester content versus reaction time plots of various treated fibers.

Rowell [6] made an attempt to improve the performance of natural fibers through acetylation. The hygroscopicity of the fibers greatly reduced after acetylation. Dimensional stability and resistance to biological attack

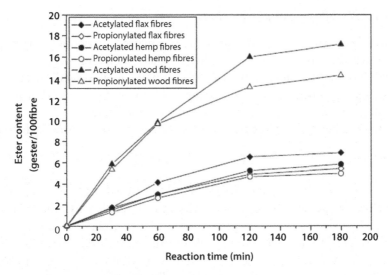

Figure 10.7 Ester content of chemically treated fibers [45].

also improved. The effect of acetylation on flax fiber and its PP composites was evaluated by Bledzki *et al.* [46]. Fiber surface morphology and moisture resistance were improved remarkably after the treatment. The tensile and flexural strength of composites increased with increasing degree of acetylation up to 18%. On the other hand, Charpy impact strength of the composites was found to decrease after acetylation. It was found that the cellulose content of the fibers increases gradually up to an acetylation degree of 12%, which is due to the extraction of lignin and other material. After that, degradation of cellulose molecules begins and the cellulose content decreases.

Ichazo *et al.* [47] investigated the properties of polyolefin blends with acetylated sisal fibers. The influence of acetylation on the mechanical, thermal and thermodegradational properties of sisal fiber reinforced PP, PP/ HDPE and PP/HDPE with functionalized and non-functionalized EPR composites was studied. Acetylation of the fiber improves the adhesion of fiber to polyolefin matrices. Tensile strength and modulus of acetylated sisal composites were increased in general. Acetylation favors the fiber-matrix interaction and fiber dispersion, which in turn improves the mechanical and thermal properties of the composites. Matsumura *et al.* [48] were the first to esterify the surfaces of cellulosic nanoparticles. They obtained high compatibility between cellulose I domains in a matrix of partially esterified cellulose, leading to a significant improvement in strength.

10.4.2.2 Silane Coupling Agents

As early as the 1940s, organosilanes were used as coupling agents to improve the strength and durability of glass fiber reinforced composite materials and they are still the largest group of coupling agents used in the composite industry today. A silane that contains at least one carbon-silicon bond (Si–C) structure is referred to as an organosilane.

The organosilane molecule can be represented by the following formula:

$$R-(CH_2)_n-Si(OR')_3 \tag{10.8}$$

where $n = 0$–3; R is a non-hydrolyzable functional organic group that is reactive toward various groups such as amino, epoxy, vinyl, methacrylate, sulfur; OR′ is a hydrolyzable group like an alkoxy group that can react with hydroxyl groups present in inorganic or organic substrates such as natural fibers. Organosilanes act as bridges between cellulose fibers and polymer matrices and can substantially improve adhesion between them.

The general mechanism of alkoxysilane reaction with fiber surface is shown in Figure 10.8 [49].

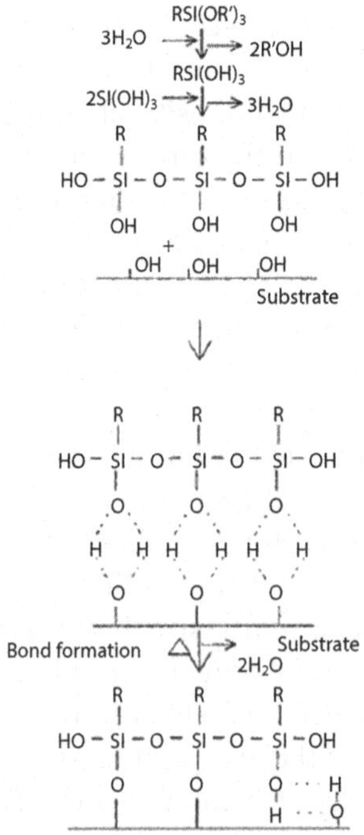

Figure 10.8 The general mechanism of alkoxysilane reaction with fiber surface [49].

Wang *et al.* [50] investigated the effect of various chemical treatments on the mechanical properties of jute fiber reinforced recycled PP composites. These treatments include alkali, MAPP and silane treatments. The best properties were achieved with combined alkali, MAPP and silane treatment. The tensile strength and impact toughness of the composites improved considerably upon chemical treatment. Asumani *et al.* [51] reported on the effects of alkali-silane treatment on the tensile and flexural properties of PP composites reinforced with kenaf fiber nonwoven fabrics. They used three forms of reinforcement, such as untreated, NaOH treated, and treated with NaOH solution, followed by three-aminopropyltriethoxysilane. Mechanical tests indicated that alkali treatment followed by three-aminopropyltriethoxysilane treatment significantly improved the tensile and flexural properties of the resulting PP composites. Figure 10.9 shows the specific mechanical properties of kenaf and glass fiber reinforced PP composites [51]. It is noteworthy that the specific tensile and flexural strength of the alkali-silane treated composites with 30% fiber fraction were only 4% and 11% lower than equivalent glass fiber composites.

Gassan *et al.* [52] conducted a detailed study on the surface characteristics of silane and corona treated jute fibers. They used dynamic contact angle (DCA), capillary rise, and resin adsorption techniques as well as inverse gas chromatography (IGC). It was shown that corona treatment increases the polarity of jute fibers, leading to an improvement in the wettability of the fibers by a polar resin. Table 10.5 lists the values of surface energies of the untreated and γ-glycidoxypropyltrimethoxy silane treated jute fibers by the capillary rise method and the results obtained by IGC. It

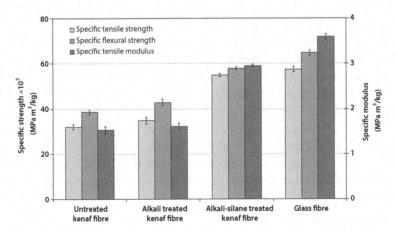

Figure 10.9 Specific mechanical properties of kenaf and glass fiber reinforced PP composites [51].

Table 10.5 Influence of silane treatment on the surface energy of jute fibers [52].

Treatment	IGC		Capillary rise (γ_s^p/γ_s)
	γ_s^d(mJ/m²)	I_{sp}(kJ/mol)	
Untreated	41	1.6	0.52
Silane treated (A-187)	37	5.1	0.61

can be shown that silane treatment of jute fiber leads to an increase in the polarity of the fiber surface. The results obtained by capillary rise methods indicated that the polarity of the silane-treated fibers is approximately 20% higher than that of the untreated fibers. The increase in specific interaction parameter (I_{sp}) as determined by IGC method was significantly higher when compared to the nondispersive part of the surface energy obtained by the capillary rise method.

Panaitescu [53] investigated the simultaneous effect of MAPP and 3-aminopropyl triethoxysilane treated hemp fibers on morphology, thermal and mechanical properties of high-flow PP modified with poly[styrene-b-(ethylene-co-butylene)-b-styrene] (SEBS). Thermal stability of hemp fibers improved after silane treatment. Better dispersion of fibers as well as enhanced static and dynamic mechanical properties of their composites were observed with treated fibers compared with untreated fibers and their composites. Gousse *et al.* [54] reported on surface silanization of cellulose microfibrils. Suspensions of cellulose microfibrils resulting from the homogenization of parenchymal cell walls were modified with isopropyl dimethylchlorosilane. The suspensions showed characteristic thickening and shear thinning, but no marked yield stress point. Panaitescu *et al.* [55] produced polymer nanocomposites from PP and two types of cellulose microfibrils. They used a silane coupling agent to enhance the compatibility between cellulose microfibrils and MAPP modified PP. Composite materials with enhanced properties were obtained. DSC measurements suggested important changes in melting and crystallization temperature of PP matrix in the presence of cellulose microfibrils and the MAPP coupling agent. The addition of MAPP significantly improved the strength and modulus of PP-cellulose composites. FTIR spectra and mechanical test results indicated the positive effect of silane treatment on the performance of the nanocomposites.

10.4.2.3 Graft Copolymerization

Graft copolymerization is among the most effective methods for interface modification. First, the cellulose material is treated with an aqueous solution of selected ions and is exposed to high energy radiation. This process results in the cleavage of cellulose macromolecules and formation of radical groups. Then the cellulose material is treated with a suitable polymer that is compatible with polymer matrix such as vinyl monomers, acrylonitrile, polystyrene, methyl methacrylate. The most important grafting method is the treatment of natural fibers with maleic anhydride grafted polypropylene (MAPP) copolymers. This process results in the formation of covalent bonds across fiber-matrix interface. The reaction takes place in two steps, as shown in Figure 10.10 [49]:

Gassan and Bledzki [56] investigated the influence of MAPP coupling agents on the mechanical properties of jute/PP composites. Flexural strength of the composites with MAPP was higher than that of untreated composites. The cyclic-dynamic test results showed that the coupling agent reduces the progress of damage. Dynamic strength of treated composites increased dramatically, i.e., 40%. SEM studies indicated that this increment is due largely to the improvement in fiber-matrix adhesion. Mohanty et al. [7] investigated the effect of MAPP coupling agent on the performance of jute/PP composites. They found that 30% fiber loading with 0.5% MAPP concentration give the best results. Flexural strength of treated composites increased nearly 72% with a reduction in moisture absorption. The DSC

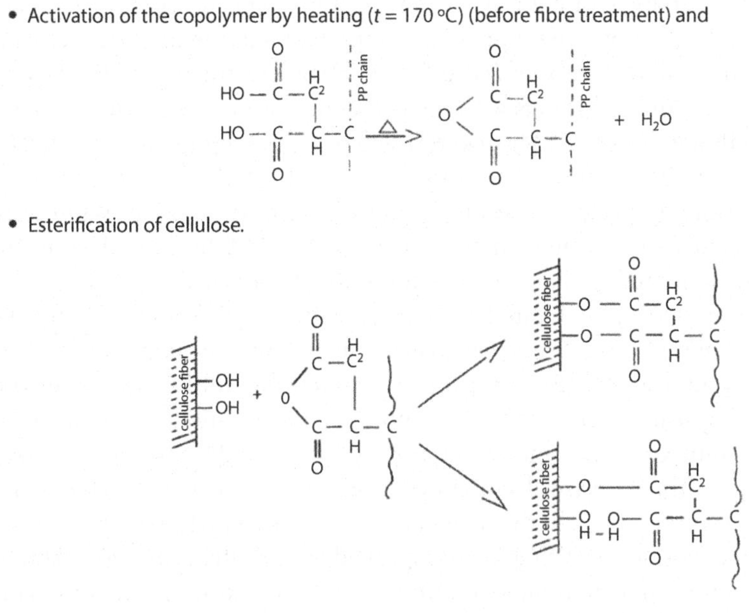

- Activation of the copolymer by heating ($t = 170$ °C) (before fibre treatment) and

- Esterification of cellulose.

Figure 10.10 The reaction between natural fibers and maleic anhydride grafted polypropylene (MAPP) copolymers [49].

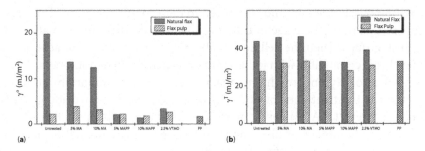

Figure 10.11 Polar component and the total surface energy of natural fibers before and after chemical treatments [57].

measurements showed that the melting point is raised after MAPP treatment, possibly due to the improvement in fiber-matrix interface. Biagiotti *et al.* [57] carried out a systematic study on the effect of MAPP and silane treatments on the properties of PP matrix composites reinforced with flax fiber and flax cellulose pulp. Several techniques were used for this purpose, including spectroscopic, mechanical and thermal tests, differential scanning calorimetry (DSC) and optical microscopy. Tensile and bending properties as well as the thermal stability of the composites significantly improved after the chemical treatments. The treatments increased the contact angle of natural fibers with water, thus decreasing its polarity. Figure 10.11 shows that the polar component and the total surface energy of natural fibers are reduced after chemical treatments. Bera *et al.* [58] obtained similar findings with jute/PP composites where MAPP and vinyl trimethoxy silane (VTMO) were used as surface modifiers after a NaOH treatment. They found that both MAPP and VTMO treatments improve the tensile properties of resultant jute/PP composites. Fiber pull-out tests indicated a better fiber/matrix interface for MAPP treated composites when compared to silane treated and untreated ones.

Khan *et al.* [59] treated jute fibers with 2-hydroxy ethylmethylacrylate (HEMA) and 2-ethyl-hexylacrylate (EHA) to improve the fiber/matrix adhesion in PP composites. The tensile and bending strength of the composites increased with monomer concentration and attained their maximum value at a monomer concentration of 3% and then decreased abruptly. The increase at low concentrations was attributed to the increase of crystallinity by orderly arrangement of short vinyl polymeric units on the fiber surface resulting in an improved fiber/matrix adhesion. At higher concentration of monomer-monomer reactions take over rather than fiber-monomer reactions. Then effect of chemical treatments on Alfa (*Stipa tenacissima*) fibers was investigated by Bessadok *et al.* [60]. Several chemical

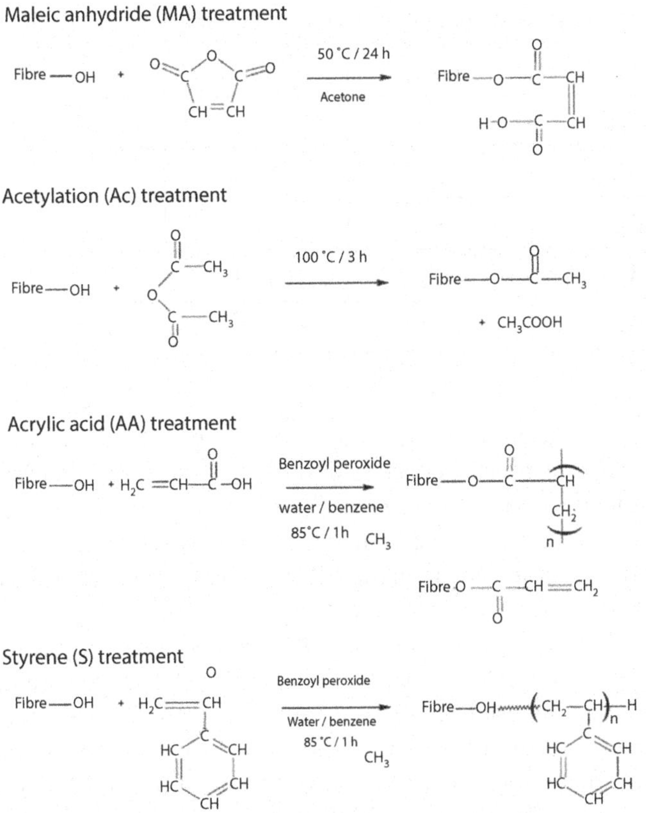

Maleic anhydride (MA) treatment

Acetylation (Ac) treatment

Acrylic acid (AA) treatment

Styrene (S) treatment

Figure 10.12 The chemical treatments used on Alfa fibers [60].

modification methods were performed, including acetylation (Ac), styrene (S), acrylic acid (AA), and maleic anhydride (MA) treatments. The details of the procedures and reactions are shown in Figure 10.12. The FTIR spectra of treated fibers indicated ester formation in the case of acetylation and acrylic acid treatments. The spectra of styrene-treated fibers showed signs of aromatic groups, showing the effectiveness of styrene treatment. Table 10.6 shows the surface energies of untreated and chemically treated Alfa fibers. Except for MA treatment, the total surface energy is reduced after treatments. The authors attributed this reduction to the polar component γ_p, which is reduced due to the decrease in the hydrophobic character. All the treatments reduced the water uptake of Alfa fibers.

Bledzki *et al.* [61] produced PP composites from abaca, flax and jute fibers. Tensile, flexural, Charpy impact and falling weight impact tests were carried out to determine the effect of MAPP coupling agent on composite

Table 10.6 Surface energies of untreated and chemically treated Alfa fibers [60].

	γ_d (mJ/m^2)	γ_p (mJ/m^2)	γ_T (mJ/m^2)	χ_p
Untreated	2.7	56.1	59	0.96
Treated MA	0.4	70.1	71	0.99
Treated AC	16.0	17.3	33	0.52
Treated AA	0.15	45.0	45	0.99
Treated S	25.3	16.3	42	0.39

performance. The tensile, flexural and falling weight impact properties were found to increase between 30 to 80% for different fiber loadings. No significant effect was observed for Charpy impact resistance with regard to MAPP addition. Authors explained this result by referring to the increased brittleness of composite after MAPP treatment as well as to local internal deformation in the composite material.

Bledzki *et al.* [62] investigated the effect of fiber physical, chemical and surface properties on barley husk and coconut shell reinforced PP composites. They used MAPP as a coupling agent. Tensile strength of the composites increased between 20–30% due to addition of coupling agent. This was attributed to the formation of ester linkages between PP and cellulose molecules through MAPP coupling agent. Elongation at break of the composites reduced because of the increased brittleness after the addition of MAPP. The notched Charpy impact strength of the composites was generally reduced with MAPP. This result is reasonable since a good fiber matrix interface generally has a negative effect on impact strength by hindering energy absorption mechanisms such as friction and fiber debonding.

Doan *et al.* [63] investigated the effect of matrix modification on the properties of jute/PP composites. They used two kinds of matrices, namely PP1 and PP2, and MAPP as a coupling agent. They formulated a modified rule of mixtures (ROM) theory which fits well to the experimental results. The addition of 2 wt% MAPP to PP matrices considerably increases the adhesion strength with jute fibers and hence the mechanical properties. The tensile modulus was less sensitive to variation in adhesion strength. Ljungberg *et al.* [64] developed a new nanocomposite material reinforced with cellulose whiskers in atactic PP. They investigated the surface and dispersion characteristics of MAPP treated and untreated nanocomposites. The mechanical properties above the glass transition temperature were dramatically improved over neat PP. MAPP treatment provided better dispersion of cellulose whiskers and also improved the mechanical properties.

Biswal et al. [65] investigated the mechanical, thermal and dynamic mechanical properties of banana fiber reinforced PP nanocomposites. The nanocomposites were produced by melt mixing using a twin-screw extruder followed by compression molding. MAPP was used to increase the compatibility between banana fibers, clay and PP matrix, and to enhance the exfoliation of organoclay and dispersion of fibers in PP matrix. They observed that 3 wt% of nanoclay and 5 wt% of MAPP in PP matrix increased the tensile and flexural strength by 41.3% and 45.6% respectively as compared with virgin PP. Incorporation of 30 wt% of banana fibers further increases the tensile and flexural strength by 27.1% and 15.8% respectively. SEM studies indicated a strong interaction between banana fibers, organoclay and PP matrix in the presence of MAPP coupling agent. Dynamic mechanical test results showed that the storage modulus and damping factor of the composites increased with the use of MAPP.

Spoljaric et al. [66] produced PP-microcrystalline cellulose (MCC) nanocomposites. They used a range of chemicals for interface modification, including MAPP, silicone oil, stearic acid and alkyltitanate. Infrared spectroscopy confirmed the effectiveness of surface treatments. The MCC content and MAPP increased the thermal stability and crystallization temperature (T_c) of PP but reduced the crystallinity due to cellulose II crystals. Tensile tests showed increased modulus with MCC content, MAPP, stearic acid and alkyltitanate. The MCC and MAPP reduced the creep deformation and increased the permanent strain. Storage modulus, loss modulus and glass transition temperature increased with MCC. The influence of organically modified nanoclay on the performance of pineapple leaf fiber (PALF) reinforced PP nanocomposites was investigated by Biswal et al. [67]. It was observed that the tensile, flexural and impact properties of PP increase with the increase in fiber loading from 10 to 30 wt%. Nanocomposites produced with 30 wt% PALF and 5 wt% MAPP showed optimum mechanical properties with an increase in tensile strength to 31% and flexural strength to 45% when compared to neat PP. Dynamic mechanical test results showed that the storage modulus and damping factor increase, confirming a strong interaction between fiber, nanoclay and MAPP. Wide angle X-ray diffraction studies indicated an increase in d-spacing in PP/PALF nanocomposite because of improved intercalated morphology.

$$O$$
$$R - N = C = O + H - O - Cell - R - HN - C - O - Cell$$

Figure 10.13 The reaction between cellulose and PMPPIC [49].

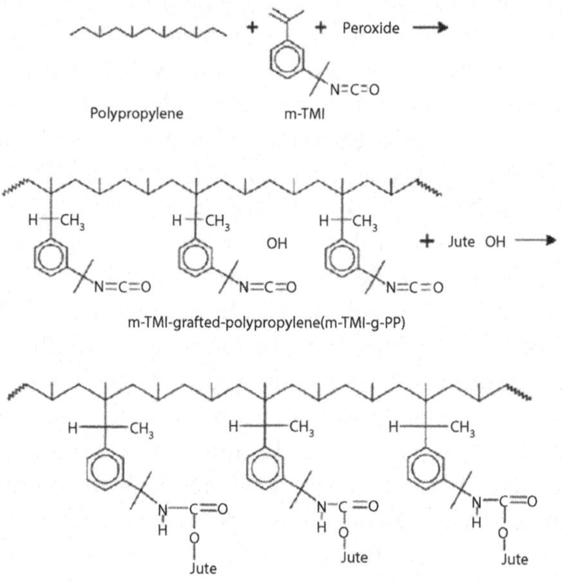

Figure 10.14 The coupling mechanism of m-TMI-g-PP with jute and PP [68].

10.4.2.4 Treatment with Isocyanates

The main use of isocyanates is in wood industry as binders. Polymethylene-polyphenyl isocyanates (PMPPIC) can make strong covalent bonds with –OH groups of cellulose through their –N=C=O functional groups (Figure 10.13) [49]:

The isocyanate treatment is very effective and can be used to modify both fibers and the polymer matrix. The reaction of isocyanates depends upon the catalysts and temperature. The main disadvantage of this method is the toxicity of the chemicals used. Aggarwal *et al.* [68] prepared jute/PP composites using m-isopropenyl-α-α-dimethylbenzyl-isocyanate (m-TMI)-grafted PP as the coupling agent. Figure 10.14 shows the coupling mechanism of m-TMI-g-PP with jute and PP. FTIR microscopy of coupling agent and composites suggested an effective coupling between jute fibers and PP matrix. The tensile and flexural strength of the composites was enhanced by 87% and 95% over virgin PP when isocyanate coupling agent was used. On the other hand, tensile and flexural strength of the composites showed no improvement with filler content without coupling agent. This strong result showed that the incorporation of jute fibers into the composites does not impart a positive effect on the strength values of the composite without the application of coupling agent. Tensile modulus increased with filler content whether the coupling agent is used or not, although the coupling agent had a positive effect on the modulus as well. The moisture absorption in composites without the coupling agent was two- to three-folds higher when compared to composites with coupling agent.

Stenstad *et al.* [69] treated microfibrillated cellulose with a range of chemicals using reactions both in aqueous and organic solvents. The modified MFC was characterized with Fourier transform infrared spectroscopy (FTIR), X-ray photoelectron spectroscopy (XPS) and transmission electron microscopy (TEM). Epoxy functionality was introduced onto the MFC surface by oxidation with cerium(IV) followed by glycidyl methacrylate grafting. Positive charge was introduced to the MFC surface via hexamethylenediisocyanate grafting followed by reaction with the amines. Succinic and maleic acid could be introduced onto the MFC surface as a monolayer by a reaction between the corresponding anhydrides and the surface hydroxyl groups of MFC. Figure 10.15 shows the reaction between hexamethylenediisocyanate and the surface hydroxyl groups of MFC. Isocyanate reacts with a hydroxyl group and a urethane linkage is formed. This group will react with a new isocyanate molecule in a second step, forming an allophanate group.

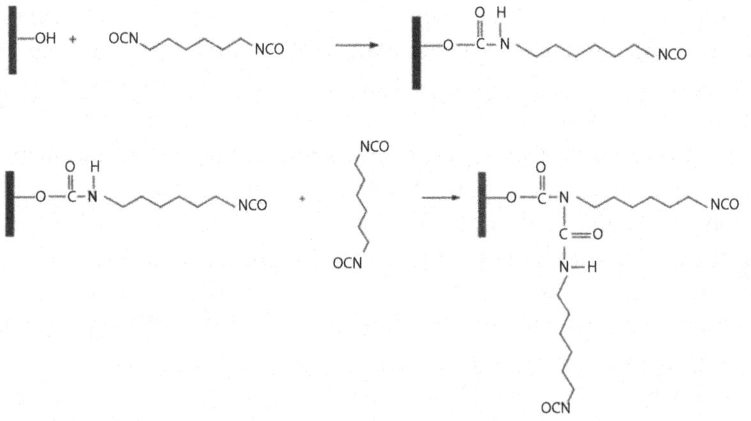

Figure 10.15 Reaction between hexamethylenediisocyanate and the surface hydroxyl groups of MFC [69].

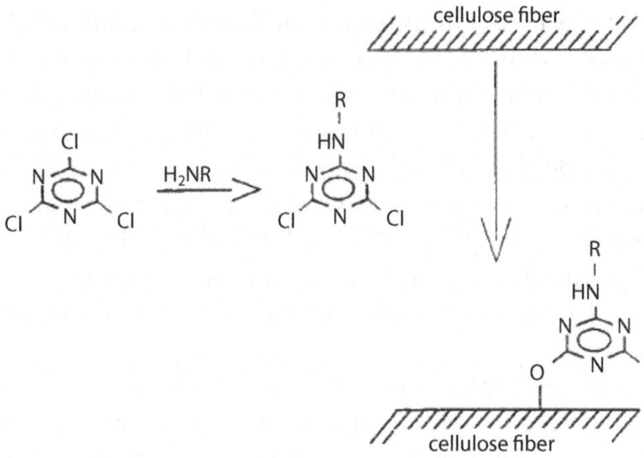

Figure 10.16 Reaction between cellulose fiber and triazine coupling agents [49].

10.4.2.5   Triazine Coupling Agents

Triazine coupling agent can form covalent bonds with cellulose fibers. Therefore the –OH groups on the cellulose fiber surface are covered and consequently the hydrophilic nature of the fibers is hindered (Figure 10.16) [49].

10.5   Conclusions and Future Trends

In recent years, natural fiber reinforced composites have attracted great attention for industrial usage thanks to their exceptional mechanical properties and environmentally friendly nature. Since their first industrial scale appearance in the 1990s, fiber-matrix interface modification has been a hot topic among researchers and still is, since it is the most problematic issue with these composites. Several fiber and matrix modification methods have been developed to increase the adhesion between natural fibers and polymer resins. Chemical and physical modification methods yielded good fiber-matrix adhesion and in turn improved mechanical properties in resulting composites. Among all thermoplastic matrices, polypropylene stands out for its low price and good applicability in natural fiber composites. Several studies have attempted to improve the performance of PP-based natural fiber composites by enhancing the fiber-matrix adhesion. These methods range from physical techniques, such as corona, plasma and alkali treatments, to chemical techniques such as esterification, silane treatment, graft copolymerization, isocyanate treatment and triazine coupling agent. It is important to note that this is by no means a full list of modification methods but only the most common ones whose effectiveness has been well-proven by numerous research studies. The list is constantly growing, as new modification methods are introduced every day. It is expected that more sophisticated techniques will come into play as nanotechnology is more widely used in this area. It is inevitable that, with the use of cellulose nanofibrils and nanowhiskers along with advanced nano-modification methods, stronger, more durable and environmentally friendly natural fiber composites will dominate the composite industry in the near future.

References

1. Joshi, S.V., Drzal, L.T., Mohanty, A.K., and Arora, S., Are natural fiber composites environmentally superior to glass fiber reinforced composites?. *Compos. Part A: Appl. Sci. Manuf.* 35, 371–376, 2004.
2. Wambua, P., Ivens, J., and Verpoest, I., Natural fibres: Can they replace glass in fibre reinforced plastics?. *Compos. Sci. Technol.* 63, 1259–1264, 2003.
3. Bledzki, A.K., Faruk, O., and Sperber, V.E., Cars from bio-fibres. *Macromol. Mater. Eng.* 291, 449–457, 2006.
4. Mehta, G., Mohanty, A.K., Thayer, K., and Misra, M., Novel biocomposites sheet molding compounds for low cost housing panel applications. *J. Polym. Environ.* 13, 169–175, 2005.
5. Xie, Y., Hill, C.A.S., Xiao, Z., Militz, H., and Mai, C., Silane coupling agents used for natural fiber/polymer composites: A review. *Compos. Part A: Appl. Sci. Manuf.* 41, 806–819, 2010.
6. Rowell, R.M., Acetylation of natural fibers to improve performance. *Mol. Cryst. Liq. Cryst.* 418, 153–164, 2004.
7. Mohanty, S., Nayak, S.K., Verma, S.K., and Tripathy, S.S., Effect of MAPP as a coupling agent on the performance of jute–PP composites. *J. Reinf. Plast. Compos.* 23, 625–637, 2004.
8. Onal, L., and Karaduman, Y., Mechanical characterization of carpet waste natural fiber-reinforced polymer composites. *J. Compos. Mater.* 43, 1751–1768, 2009.
9. Gassan, J., and Bledzki, A.K., Possibilities for improving the mechanical properties of jute/epoxy composites by alkali treatment of fibres. *Compos. Sci. Technol.* 59, 1303–1309, 1999.
10. Mohanty, A.K., Misra, M., Drzal, L.T., Selke, S.E., Harte, B.R., and Hinrichsen, G., Natural fibers, biopolymers and biocomposites: An introduction, in: *Natural Fibers, Biopolymers and Biocomposites*, Mohanty, A.K., Misra, M., and Drzal, L.T. (Eds.), p. 1, CRC Press: Boca Raton, 2005.
11. Ansell, M.P., and Mwaikambo, L.Y., The structure of cotton and other plant fibres, in: *Handbook of Textile Fibre Structure, Vol. 2: Natural, Regenerated, Inorganic and Specialist Fibres*, Eichhorn, S.J., Hearle, J.W.S., Jaffe, M., and Kikutani, T. (Eds.), Woodhead Publishing Limited: Cambridge, 2009.

12. Dinwoodie, J.M., *Timber: Its Nature and Behaviour*, E & FN Spon: London, 2000.

13. McLaughlin, E.C., and Tait, R.A., Fracture mechanism of plant fibres. *J. Mater. Sci.* 15, 89–95, 1980.

14. Aziz, S., and Ansell, M., Optimising the properties of green composites, in: *Green Composites: Polymer Composites and the Environment*, Baillie, C. (Ed.), Woodhead Publishing Ltd.: Cambridge, 2004.

15. Mwaikambo, L.Y., Review of the history, properties and application of plant fibres. *Afr. J. Sci. Technol.* 7, 120–133, 2006.

16. Satyanarayana, K.G., Arizaga, G.G.C., and Wypych, F.. Biodegradable composites based on lignocellulosic fibers—An overview. *Prog. Polym. Sci.* 34, 982–1021, 2009.

17. Bismarck, A., Mishra, S., and Lampke, T., Plant fibers as reinforcement for green composites, in: *Natural Fibers, Biopolymers, and Biocomposites*, Mohanty, A.K., Misra, M., and Drzal, L.T. (Eds.), pp. 37–108, Taylor & Francis Group: Boca Raton, FL, 2005.

18. Mohanty, A.K., Misra, M., and Hinrichsen, G., Biofibres, biodegradable polymers and biocomposites: An overview. *Macromol. Mater. Eng.* 276–277, 1–24, 2000.

19. Ciechanska, D., Wesolowska, E., and Wawro, D., An introduction to cellulosic fibres, in: *Handbook of Textile Fibre Structure, Vol. 2: Natural, Regenerated, Inorganic and Specialist Fibres*, Eichhorn, S.J., Hearle, J.W.S., Jaffe, M., and Kikutani, T. (Eds.), Woodhead Publishing Ltd.: Cambridge, 2009.

20. Nevell, T.P., and Zeronian, S.H. (Eds.), *Cellulose Chemistry and its Applications*, p. 15, Ellis Horwood Ltd.: Chichester, 1985.

21. Akin, D.E., Chemistry of plant fibres, in: *Industrial Applications of Natural Fibres: Structure, Properties and Technical Applications*, Müssig, J. (Ed.), John Wiley & Sons Ltd., 2010.

22. Focher, B., Physical characteristics of flax fibre, in: *The Biology and Processing of Flax*, Sharma, H.S.S., and van Sumere, C.F. (Eds.), pp. 11–32, M Publications: Belfast, 1992.

23. Sarkanen, K.V., and Ludwig, C.H., *Lignins: Occurrence, Formation, Structure and Reactions*, Wiley-Interscience: New York, 1971.

24. Thielemans, W., Can, E., Morye, S.S., Wool, R.P., Novel applications of lignin in composite materials. *J. Appl. Polym. Sci.* 83, 323–331, 2002.

25. Sakai, T., Sakamoto, T., Hallaert, J., and Vandamme, E.J., Pectin, pectinase, and protopectinase: Production, properties, and applications, in: *Advances in Applied Microbiology*, Neidleman, S., and Laskin, A.I. (Eds.), pp. 213–294, Academic Press: New York.

26. Beleski-Carneiro, E., Alquini, Y., and Reicher, F., Pectins from the fruit of myrtaceae, in: *Proceedings of the 3rd International Symposium on Natural Polymers and Composites*, pp. 162–170, Sao Pedro, Brazil, 2000.

27. Windham, W.R., Petersen, J.C., and Terrill, T.H., Tannins as anti-quality factors in forage, in: *Microbial and Plant Opportunities to Improve Lignocellulose*

Utilization by Ruminants, Akin, D.E., Ljungdahl, L.G., and Wilson, J.R. (Eds.), pp. 127–135, Elsevier: New York, 1990.

28. Morrison III, W.H., Holser, R., and Akin, D.E., Cuticular wax from flax processing waste with hexane and super critical carbon dioxide extractions. *Ind. Crops Prod.* 24, 119–122, 2006.

29. Atkins, P., Jones, L., and Laverman, L., *Chemical Principles: The Quest for Insight*, 6th ed., W.H. Freeman & Company, 2013.

30. Hull, D., and Clyne, T.W., *An Introduction to Composite Materials*, 2nd ed., Cambridge University Press: Cambridge, 1996.

31. Young, T., An essay on the cohesion of fluids. *Phil. Trans. R. Soc.* 95, 65–87, 1805.

32. Tran, L.Q.N., Fuentes, C.A., Verpoest, I., and Vuure, A.W.V., Interfacial compatibility and adhesion in natural fiber composites, in: *Natural Fiber Composites*, Campilho, R.D.S.G. (Ed.), pp. 127–155, CRC Press, 2016.

33. Tran, L.Q.N., Fuentes, C.A., Dupont-Gillain, C., Vuure, A.W.V., and Verpoest, I., Understanding the interfacial compatibility and adhesion of natural coir fibre thermoplastic composites. *Compos. Sci. Technol.* 80, 23–30, 2013.

34. Fuentes, C.A., Tran, L.Q.N., Hellemont, M.V., Janssens, V., Dupont-Gillain, C., Vuure, A.W.V., and Verpoest, I., Effect of physical adhesion on mechanical behaviour of bamboo fibre reinforced thermoplastic composites. *Colloids Surf. A* 418, 7–15, 2013.

35. Tran, L.Q.N., Yuan, X.W., Bhattacharyya, D., Rojas, C.F., Vuure, A.V., and Verpoest, I., The influences of fibre-matrix interfacial adhesion on composite properties in natural fibre composites, in: *11th Asia-Pacific Conference on Materials Processing*, Auckland, New Zealand, 2014.

36. Stamboulis, A., Baillie, C., and Schulz, E., Interfacial characterisation of flax fibre-thermoplastic polymer composites by the pull-out test. *Macromol. Mater. Eng.* 272, 117–120, 1999.

37. van de Velde, K., and Kiekens, P., Influence of fiber surface characteristics on the flax/polypropylene interface. *J. Thermoplast. Compos. Mater.* 14, 244–260, 2001.

38. Lawrence, P., Some theoretical considerations of fibre pullout from an elastic matrix. *J. Mater. Sci.* 7, 1–6, 1972.

39. Chua, P., and Piggott, M.R., The glass fibre-polymer interface: I–Theoretical considerations for single fibre pullout tests. *Compos. Sci. Technol.* 22, 33–42, 1985.

40. Zafeiropoulos, N.E., Engineering the fibre matrix interface in natural-fibre composites, in: *Properties and Performance of Natural-Fibre Composites*, Pickering, K.L. (Ed.), Woodhead Publishing Ltd., 2008.

41. Yuan, X., Jayaraman, K., and Bhattacharyya, D., Effects of plasma treatment in enhancing the performance of woodfibre-polypropylene composites. *Compos. Part A: Appl. Sci. Manuf.* 35, 1363–1374, 2004.

42. Couto, E., Tan, I.H., Demarquette, N., Caraschi, J.C., and Leao, A., Oxygen plasma treatment of sisal fibers and polypropylene: Effects on mechanical properties of composites. *Polym. Eng. Sci.* 42, 790–797, 2002.

43. Gassan, J., and Bledzki, A.K., Alkali treatment of jute fibers: Relationship between structure and mechanical properties. *J. Appl. Polym. Sci.* 71, 623–629, 1999.
44. Joseph, P.V., Joseph, K., and Thomas, S.E., Effect of processing variables on the mechanical properties of sisal-fiber-reinforced polypropylene composites. *Compos. Sci. Technol.* 59, 1625–1640, 1999.
45. Tserki, V., Zafeiropoulos, N.E., Simon, F., and Panayiotou, C., A study of the effect of acetylation and propionylation surface treatments on natural fibres. *Compos. Part A: Appl. Sci. Manuf.* 36, 1110–1118, 2005.
46. Bledzki, A.K., Mamun, A.A., Lucka-Gabor, M., and Gutowski, V.S., The effects of acetylation on properties of flax fibre and its polypropylene composites. *Express Polym. Lett.* 2, 413–422, 2008.
47. Ichazo, M.N., Albano, C., and Gonzalez, J., Behavior of polyolefin blends with acetylated sisal fibers. *Polym. Int.* 49, 1409–1416, 2000.
48. Matsumura, H., Sugiyama, J., and Glasser, W.G., Cellulosic nanocomposites. I. Thermally deformable cellulose hexanoates from heterogeneous reaction. *J. Appl. Polym. Sci.* 78, 2242–2253, 2000.
49. Bledzki, A.K., and Gassan, J., Composites reinforced with cellulose based fibres. *Prog. Polym. Sci.* 24, 221–274, 1999.
50. Wang, X., Cui, Y., Xu, Q., Xie, B., and Li, W., Effects of alkali and silane treatment on the mechanical properties of jute-fiber-reinforced recycled polypropylene composites. *J. Vinyl Addit. Technol.* 16, 183–188, 2010.
51. Asumani, O.M.L., Reid, R.G., and Paskaramoorthy, R., The effects of alkali-silane treatment on the tensile and flexural properties of short fibre nonwoven kenaf reinforced polypropylene composites. *Compos. Part A: Appl. Sci. Manuf.* 43, 1431–1440, 2012.
52. Gassan, J., Gutowski, V.S., and Bledzki, A.K., About the surface characteristics of natural fibres. *Macromol. Mater. Eng.* 283, 132–139, 2000.
53. Panaitescu, D.M., Vuluga, Z., Ghiurea, M., Iorga, M., Nicolae, C., and Gabor, R., Influence of compatibilizing system on morphology, thermal and mechanical properties of high flow polypropylene reinforced with short hemp fibers. *Compos. Part B: Eng.* 69, 286–295, 2015.
54. Gousse, C., Chanzy, H., Cerrada, M.L., and Fleury, E., Surface silylation of cellulose microfibrils: Preparation and rheological properties. *Polymer* 45, 1569–1575, 2004.
55. Panaitescu, D.M., Donescu, D., Bercu, C., Vuluga, D.M., Iorga, M., and Ghiurea, M., Polymer composites with cellulose microfibrils. *Polym. Eng. Sci.* 47, 1228–1234, 2007.
56. Gassan, J., and Bledzki, A.K., The influence of fiber-surface treatment on the mechanical properties of jute-polypropylene composites. *Compos. Part A: Appl. Sci. Manuf.* 28, 1001–1005, 1997.
57. Biagiotti, J., Puglia, D., Torre, L., Kenny, J.M., Arbelaiz, A., Cantero, G., Marieta, C., Ponte, R.L., and Mondragon, I., A systematic investigation on the influence of the chemical treatment of natural fibers on the properties of their polymer matrix composites. *Polym. Compos.* 25, 470–479, 2004.

58. Bera, M., Alagirusamy, R., and Das, A., A study on interfacial properties of jute-PP composites. *J. Reinf. Plast. Compos.* 29, 3155–3161, 2010.

59. Khan, M.A., Hinrichsen, G., and Drzal, L.T., Influence of novel coupling agents on mechanical properties of jute reinforced polypropylene composite. *J. Mater. Sci. Lett.* 20, 1711–1713, 2001.

60. Bessadok, A., Marais, S., Gouanve, F., Colasse, L., Zimmerlin, I., Roudesli, S., and Metayer, M., Effect of chemical treatments of Alfa (*Stipa tenacissima*) fibres on water-sorption properties. *Compos. Sci. Technol.* 67, 685–697, 2007.

61. Bledzki, A.K., Mamun, A.A., and Faruk, O., Abaca fibre reinforced PP composites and comparison with jute and flax fibre PP composites. *Express Polym. Lett.* 1, 755–762, 2007.

62. Bledzki, A.K., Mamun, A.A., and Volk, J., Barley husk and coconut shell reinforced polypropylene composites: The effect of fibre physical, chemical and surface properties. *Compos. Sci. Technol.* 70, 840–846, 2010.

63. Doan, T.T.L., Gao, S.L., and Mader, E., Jute/polypropylene composites I. Effect of matrix modification. *Compos. Sci. Technol.* 66, 952–963, 2006.

64. Ljungberg, N., Bonini, C., Bortolussi, F., Boisson, C., Heux, L., and Cavaille, J., New nanocomposite materials reinforced with cellulose whiskers in atactic polypropylene: Effect of surface and dispersion characteristics. *Biomacromolecules* 6, 2732–2739, 2005.

65. Biswal, M., Mohanty, S., and Nayak, S.K., Mechanical, thermal and dynamic-mechanical behavior of banana fiber reinforced polypropylene nanocomposites. *Polym. Compos.* 32, 1190–1201, 2011.

66. Spoljaric, S., Genovese, A., and Shanks, R.A., Polypropylene–microcrystalline cellulose composites with enhanced compatibility and properties. *Compos. Part A: Appl. Sci. Manuf.* 40, 791–799, 2009.

67. Biswal, M., Mohanty, S., and Nayak, S.K., Influence of organically modified nanoclay on the performance of pineapple leaf fiber-reinforced polypropylene nanocomposites. *J. Appl. Polym. Sci.* 114, 4091–4103, 2009.

68. Aggarwal, P.K., Raghu, N., Karmarkar, A., and Chuahan, S., Jute–polypropylene composites using m-TMI-grafted-polypropylene as a coupling agent. *Mater. Des.* 43, 112–117, 2013.

69. Stenstad, P., Andresen, M., Tanem, B.S., and Stenius, P., Chemical surface modifications of microfibrillated cellulose. *Cellulose* 15, 35–45, 2008.

Index

Also of Interest

Handbook of Engineering and Specialty Thermoplastics 4
Volume 4: Nylons
Edited by Sabu Thomas and Visakh P.M
Published 2012. ISBN 978-0-470-63925-2

Handbook of Engineering and Specialty Thermoplastics 3
Volume 3: Polyethers and Polyesters
Edited by Sabu Thomas and Visakh P.M.
Published 2012. ISBN 978-0-470-63926-9

MPM 220419
Printed in Singapore